Organic Fertilizers - New Advances and Applications

Edited by Khalid Rehman Hakeem

Published in London, United Kingdom

Organic Fertilizers - New Advances and Applications
http://dx.doi.org/10.5772/intechopen.1001521
Edited by Khalid Rehman Hakeem

Contributors
Tooba Abedi, Hadi Modaberi, Mukhtar Iderawumi Abdulraheem, Jiandong Hu, Shakeel Ahmed, Linze Li, Syed Muhammad Zaigham Abbas Naqvi, Tazbeen Tabara Nitu, Tasnim Binte Rayhan Promi, Syed Aflatun Kabir Hemel, Habtamu Tadele Belay, Birtukan Amare Kebede, Xiao-Lin Lu, Kai Ding, Xiao-Xia Dong, Gang Li, Jun Ma, Emrul Kayesh, Joydeb Gomasta, Nadira Bilkish, Khadiza Akter Koly, Sharmila Rani Mallick, Pratik Ramteke, Vijay Gabhane, Prakash Kadu, Vilas Kharche, Samrat Ghosh, Mohammad Atauzzaman, Quazi Hamidul Bari, Kajol Yadav, Lovely Bharti, Ashok Kumar Chaubey, Víctor Jesús Albores Flores, Julieta Grajales Conesa, Leopoldo Cruz López, José Alfonso López García, Eduardo Lozano Guzmán, Ebido Nancy Ekene, Ndubuaku Mabel Uchenna, Nelson Mauricio Espinel Pérez, Rahul Kumar, Renu Yadav, Rajender Kumar Gupta, Kiran Yodha, Sudhir Kumar Kataria, Pooja Kadyan, Pooja Sharma, Simran Kaur, Hamyana Yana, Kliwon Hidayat, Keppi Sukesi, Yayuk Yuliati, Barbara Čeh, Lucija Luskar, Julija Polanšek, Ana Karničnik Klančnik, Žan Trošt, Mami Irie, Tomomi Sugiyama, Ornela Maria Munoz Millet

First published in London, United Kingdom, 2023 by IntechOpen
IntechOpen is the global imprint of INTECHOPEN LIMITED, registered in England and Wales, registration number: 11086078, 5 Princes Gate Court, London, SW7 2QJ, United Kingdom
Printed in Croatia

British Library Cataloguing-in-Publication Data
A catalogue record for this book is available from the British Library

Additional hard and PDF copies can be obtained from orders@intechopen.com

Organic Fertilizers - New Advances and Applications
Edited by Khalid Rehman Hakeem
p. cm.

This title is part of the Agricultural Sciences Book Series, Volume 1
Topic: Agronomy and Horticulture
Series Editor: W. James Grichar
Topic Editor: Ibrahim Kahramanoglu
Associate Topic Editors: Murat Helvaci and Olga Panfilova

Print ISBN 978-1-83769-562-1
Online ISBN 978-1-83769-561-4
eBook (PDF) ISBN 978-1-83769-563-8
ISSN 3029-052X

IntechOpen Book Series

Agricultural Sciences

Volume 1

Aims and Scope of the Series

The importance of agriculture cannot be overstated. It helps sustain life, as it gives us the food we need to survive and provides opportunities for economic well-being. Agriculture helps people prosper around the world and combines the creativity, imagination, and skill involved in planting crops and raising animals with modern production methods and new technologies. This series includes two main topics: Agronomy and Horticulture, and Animal Farming. This series will help readers better understand the intricacies of production agriculture and provide the new knowledge that is required to be successful. The success of a farmer in modern agriculture requires knowledge of events happening locally as well as globally that impact input decisions and ultimately determine net profit.

Meet the Series Editor

W. James Grichar has been employed with Texas A&M AgriLife Research for over 45 years with an emphasis on research in agronomy, plant pathology, and weed science. He obtained his BS from Texas A&M in 1972 and his Masters of Plant Protection in 1975. He has published 195 journal articles, over 330 research reports and briefs, 11 book chapters, and over 300 abstracts of profession meetings. He also directs research in many crops including corn, grain sorghum, peanuts, and sesame. He has held various positions in different professional societies including the American Peanut Research and Education Society, Southern Weed Science Society, and Texas Plant Protection Conference in addition to being Associate Editor for Peanut Science and Weed Technology. Significant accomplishments have included spearheading efforts to determine the optimum planting time for soybean production along the upper Texas Gulf Coast. These efforts have shown growers that soybean yields can be improved by 10 to 20% by following a late March to early April plant date. He also has been instrumental in developing a herbicide program for peanut production in the south Texas growing region. Through the development and use of herbicides that are effective against major weed problems in the south Texas region, peanut yields have increased by 25 to 30%.

Meet the Volume Editor

Dr. Khalid Rehman Hakeem, Ph.D., FRSB, is a professor at King Abdulaziz University, Jeddah, Saudi Arabia. He obtained a Ph.D. in Botany with a specialization in Plant Ecophysiology and Molecular Biology from Jamia Hamdard, New Delhi, India in 2011, after which he worked as an assistant professor at the University of Kashmir, Srinagar for a short period. Later he joined Universiti Putra Malaysia, Selangor, Malaysia and worked there as a post-doctorate fellow in 2012 and fellow researcher (associate professor) from 2013 to 2016. He joined King Abdulaziz University in 2016 and was promoted to professor in 2019. Dr. Hakeem has more than 15 years of teaching and research experience in plant eco-physiology, biotechnology and molecular biology, medicinal plant research, plant–microbe–soil interactions, and environmental studies. He is the recipient of several national and international fellowships. He has recently been elected as a fellow of the Royal Society of Biology, London. Dr. Hakeem has been a visiting scientist at Fatih University, Istanbul, Turkey as well as at Jinan University, Guangzhou, China. He is currently involved with several international research projects with different government organizations. To date, Dr. Hakeem has authored and edited more than ninety books. He also has 185 research publications in peer-reviewed international journals and 80 book chapters to his credit.

Contents

Preface

Organic fertilizers have become increasingly popular due to their ability to improve soil health, enhance plant growth, and reduce environmental impact. *Organic Fertilizers – New Advances and Applications* explores the latest research and development in organic fertilizers, along with their use, production, and advantages for crop growth.

It is crucial to find sustainable and environmentally friendly methods to boost productivity without endangering the environment because agriculture is an essential component of the world economy and the well-being of the population. In this book, experts in the field of agricultural science examine the effects of various organic fertilizers on plant growth, nutrient uptake, and soil health. These organic fertilizers include compost, animal dung, and biofertilizers. The book also discusses recent advances in the manufacture of organic fertilizers, including bioreactor technology and microbial inoculants.

I am highly grateful to all our contributors for sharing their knowledge, research, and expertise in composing the chapters of this book. I also thank the team at IntechOpen, particularly Ivana Barac and Filip Lovricevic for their generous cooperation at every stage of the book's production.

I am hopeful that this book will be a valuable resource for agricultural scientists, researchers, farmers, and policymakers who are interested in the latest advances and applications of organic fertilizers.

Lastly, thanks are also due to our well-wishers, research students, and family members for their moral support, blessings, and inspiration. I dedicate this book to my beloved mother Hajrah Begum, who left so early from this world. May Allah rest her soul in peace and grant her Jannah-tul-Firdous (Aameen).

Khalid Hakeem
Faculty of Science,
Department of Biological Sciences,
King Abdulaziz University,
Jeddah, Saudi Arabia

Section 1

New Advances in Organic Fertilizers

Chapter 1

Investigation of Growth and Biomass Response of Five Tree Species under Irrigation with Compost Leachate

Tooba Abedi and Hadi Modaberi

Abstract

In this study, the growth and biomass of *Populus deltoides* Marshall., *Alnus glutinosa* (L.) Gaertn., *Populus euramericana* Guinier., *Salix alba* L. and *Taxodium distichum* (L.) Rich. were analyzed with compost leachate irrigation. Cuttings were collected at the beginning of the growing season and planted in pots with 40 cm depth in Safrabasteh Poplar Research Centre in Guilan Province, Iran. Three treatments were used consist of: tap pure compost leachate (P), water (control), and water to compost ratio of 1:1 (50% water + 50% compost leachate) treatments. Biomass and growth parameters including height, diameter, aboveground and underground biomass were calculated at the end of growing season. The results show that highest diameter growth was observed in *T. distichum* and *A. glutinosa* with compost leachate treatment and also showed the highest amount of height growth in tap water and 1:1 treatment. The highest absorption of elements in aboveground and root biomass was observed in *T. distichum*, *A. glutinosa* and *P. euramericana* with 1:1 treatment. According to results of this study, it is concluded that plants absorption of leachate elements can be used as an attractive method to reduce damages to the soil and ecosystem and in consequence increase the quality of life.

Keywords: leachate, nutrient, salinity, seedlings growth, solid waste

1. Introduction

Various pollutants such as ammonia, nitrogen, heavy metals, inorganic salts and chlorinated organic materials have been exposed to leachates [1]. The increase in solid waste in cities has encouraged resource managers to use plant/soil systems to treat landfill leachates before discharge [2]. Untreated leachate discharge, such as leachate components, hazardous contamination or water eutrophication, can be harmful to the environment [3]. Due to the high amount of contaminants, leachate treatment demonstrates an enormous cost in solid waste management to reach prescribed emission standards [4]. Traditional intensive forestry and waste management provide several goals such as bioenergy production, soil/water remediation and carbon sequestration [5].

One of the technology to landfill rededications is phytoremediation, and it can both stabilize soil and remediate landfill leachate by using plants abilities to accumulate toxic contaminants [6]. The phytoremediation principle is to match the suitable species to the contaminated sites regarding the soil and microclimate conditions [7]. Phytoremediation is accepted as an alternative solution to conventional engineering methods due to its several advantages such as cost-effective, environmentally friendly and less damaging to the soil and ecosystem [6].

Phytoremediation has long been employed for leachate treatment around the world [2, 4, 5, 8–10]. For example, Guidi Nissim et al. [11] report the results of a two-year project where poplar and willow grown in mesocosm were tested for their ability to withstand and remove specific pollutants from different (Low: 7% and High 15%) amounts of landfill leachate. Poplar showed, on average, significantly higher extraction rates for Cd, Cu, P and N than willow. Moreover, under high landfill leachate treatment, poplar also seemed more efficient than willow in decreasing the concentration of specific pollutants (BOD and COD) in output effluent. Lucero-Sobarzo et al. [4] performed a field trial on a real scale by landfill leachate used as a source of nutrients for the growth of maize by precipitation of struvite. Marginal higher maize yield was achieved in two sites (6.36% and 2.16%) compared to the commercial fertilizer. Struvite did not cause the presence of pathogens or heavy metals in the crops. The aim of Koda et al. [12] work was to find the relationship between the composition and leachate seepage points and determine the possibilities of their practical utilization for the assessment of the applied mineral sealing of landfill surfaces. The results indicate that the presence of leachates alters the plant species composition. The composition shows increasing representation of species tolerant to salinization. Shabir et al. [13] introduced *Acacia nilotica* as a phytoremediation potential species in cadmium-contaminated soil with saline and non-saline conditions in Pakistan. Askary et al. [14] also showed the potential effect of petroleum pollution of soil (0%, 1%, 2%, 3% and 4% V/W) on the proline, total protein, lead, cadmium and zinc contents in *Robinia pseudoacacia* L. leaves. Based upon these results, *R. pseudoacacia* L. can be used as bioaccumulation in petroleum pollution and was selected for further investigation of the phytoremediation of pb-contaminated soil. Alizadeh et al. [15] investigated the influence of soil amendment on cadmium accumulation responses in one-year-old *Populus alba* L. seedling. The results indicated that higher biomass productions in amended substrates compared to control led to an increase of total cadmium uptake two times more than that in the control substrate at 150 mg kg-1 cadmium supply. In some cases, there was no significant difference in cadmium accumulations among substrates. Sammons and Struve [5] investigated the effects of near-zero leachate irrigation on growth and water-use efficiency and nutrient uptake of container-grown baldcypress (*Taxodium distichum* (L.) Rich.) plants. Results show root dry mass ratios and fertilizer and irrigation interaction did not affect water-use efficiency. The higher fertilizer rate increased the whole plant N and K concentrations. Plant tissue mineral nutrient concentrations and water-use efficiency increased.

Zalesny et al. [9] mentioned *Populus* as an ideal species for phytoremediation because of their extensive root systems, fast growth and high-water usage rate. Several approaches have been developed to improve the tolerance and/or accumulation of Potentially Toxic Elements (PTEs) in the plant.

This study aims to compare the growth and biomass of different species with regard to different concentrations of compost leachate from green and municipal organic waste. The study objectives were to:

- Assessing the growth in diameter and height of species using leachate irrigation
- Determining the amount of aboveground and root biomass under compost leachate irrigation treatments

2. Materials and methods

The study was conducted in Poplar Research Centre of Safrabasteh in the eastern part of Gilan Province at Northern part of Iran (37° 19'N, 49° 57'E). In this research, five different species namely, *P. deltoides 69/55*, *Populus euramericana* I-sieres, *S. alba, A. glutinosa* and *T. distichum* were selected. The cuttings were collected from the nursery in the middle of March with the length of 20 cm from 1-year old saplings and planted in pots with 40 cm depth in sandy-loam soil.

The compost leachates were collected from the collection reservoir, which contains organic municipal waste, gardening and plant waste. The collection reservoir is located in the Compost Plant of Municipal Waste Management of Rasht (37° 10'N, 49° 34'E), Northern part of Iran. The leachate color was dark brown and had a putrid odour. The leachate was analyzed in the Laboratory of Guilan Department of Environment (Rasht, Iran) using approved Standard Methods for the Examination of Water and Wastewater (**Table 1**) [16].

Parameter	Unit	Amount
pH	—	5.22
EC	mS cm-1	1.26
N total	mgL-1	21.384
NO_2	mgL-1	0.08
NO3	mgL-1	21.3
SO4	mgL-1	7101
PO4-P	mgL-1	22.11
Na	mgL-1	310
K	mgL-1	250
Ca	mgL-1	152
Mg	mgL-1	1103
Pb	mgL-1	0.27
Ni	mgL-1	0.342
Cd	mgL-1	0.0047
Cr	mgL-1	Trace
COD	mgL-1	260,500
BOD	mgL-1	130,000
TSS	mgL-1	3060.6
Turbidity	mgL-1	12,500

Table 1.
Composition of pure compost leachate.

Three different treatments of irrigation were applied on each species, with five replicates for plant growth and three replicates for biomass and elements (the number of replicates for biomass and elements was limited due to the high costs of laboratory analysis). Three treatments consist of:

P (Pure compost leachate).

1:1 ratio (50% water +50% compost leachate).

Tap water (Control).

Water (control) from the study area was applied to all cuttings via hand irrigation for a settlement period of eight weeks. After the settlement, experiments were started in the middle of May with either leachate, water or 1:1 (50% water +50% compost leachate) treatment and lasted till December. The plants were irrigated with the respective water mixtures to the water holding capacity of the substrate in the pot (0.5 L per pot) in the first weeks of the experiment. With the growth of the plants, the amount of water added in a daily irrigation event was adjusted to the plant's demands. Pure leachate was the leachate without dilution. The tap water for treatment (C) and for the preparation of the water mixtures was used from the public drinking water supply.

The sapling growth (diameter and height) was monitored bimonthly and recorded. The diameter growth was measured from the sprout-out of the principal shoot, and the height growth was measured from ground level to the base of the apical bud on the terminal shoot of 125 seedlings.

All seedlings were harvested in December at the end of the growing season. The harvested saplings were divided into two portions as, aboveground (leaf + stem) and underground (root section). Root systems were washed carefully to remove soil particles, and then all the plant sections were dried at 70°C.

Root and groundmass fractions were calculated as the ratio between belowground dry mass, aboveground dry mass and total tree dry mass [10]. The amount of elements such as N, P, K and Ca at both aboveground and underground sections were measured with three replicates. Total N analyses with Kjeldahl method, P with Olsen and Sommers [17] for details on the Na2CO3 fusion method and K with flame photometric method. Soil experiments were performed according to the instructions for laboratory analysis of soil samples of the Soil and Water Research Institute.

The experiments were arranged in randomized complete design with five species and five replicates of each treatment for plant growth parameters and three replicates for biomass and elements. The data were analyzed using SAS and Analysis of Variance (ANOVA) to analyze the differences between treatments of each plant species and between the plant species for each treatment. Tukey HSD test was carried out for differences between means that were considered at different probability values of $P < 0.05$.

3. Results

The leachate characteristics are shown in **Table 1**.

A one-way ANOVA was conducted to compare the effect of irrigation treatments on plant height in pure compost leachate (P), 1:1 (50% leachate and 50% water) and tap water (Control) conditions. The results showed that there was a significant effect of irrigation treatments on plant height (P < 0.0001), and there was a significant effect on plant height (P < 0.000) (**Table 2**).

Tukey test results between five species illustrate in **Table 3**.

3.1 Seedlings growth

All pairwise mean comparisons were performed using the Tukey test between five species with a degree of significance of 0.05. Results showed that the highest average of seedlings height was 112.8 cm in *A. glutinosa* in tap water followed by 101 cm in *T. distichum* in 50% leachate +50% water treatment and the lowest average was 27.9 cm in *P.euramericana* (**Table 3**).

Diameter growth data were subjected to one-way ANOVA to test for differences among the five species. The ANOVA results showed that there was a significant difference between irrigation treatments (P < 0.0001) and species (P < 0.0001) on diameter growth (**Table 3**).

Tukey results explained the highest amount of diameter growth in tap water treatment was 2.05 cm (with no significant difference with 50% leachate +50% water treatment) for *A. glutinosa,* and the lowest amount was 0 in compost leachate treatment (P < 0.0001) for *P. deltoides* and *S. alba* in P treatment (**Table 3**).

3.2 Aboveground dry mass

ANOVA and Tukey procedures results showed a significant difference between species and treatment on aboveground dry mass (P < 0.0001) (**Table 4**). Comparing the mean aboveground dry mass between five plant species in treatments. *A. glutinosa* exhibited the maximum dry mass in all of the treatments (with the exception of 50% leachate +50% water treatment for *T. distichum* with 47.47 ± 11.85). *S. alba* exhibited the minimum dry mass with the exception of the 50% leachate +50% water treatment. 50% leachate +50% water treatment indicates the highest mean of N, P, K and Ca among the treatments (**Table 4**).

Trait	Source of variations		
	Treatment	**Species**	**Treatment ×Species**
Height (cm)	<0.0001	<0.0001	<0.0001
Diameter (cm)	<0.0001	<0.0001	<0.0001
Aboveground biomass (gr)	<0.0001	<0.0001	**0.0054**
Root biomass (gr)	0.0007	<0.0001	**0.0208**
Aboveground elements	<0.0001	<0.0001	<0.0001
Root elements	0.0011	<0.0001	<0.0001

Insignificant value is in bold.

Table 2.
Probability values from analysis of variance testing of the main effects of species, treatment and species ×treatment interaction on tree growth (height and diameter), biomass (aboveground and root biomass) and elements (aboveground and root) of five species irrigated with pure composite leachate, 1:1 (50% leachate and 50% water) and tap water.

Species	Treatment	Height (cm)	Diameter (cm)	Biomass component (gr)	
				Aboveground	Root
P. deltoides	P	0 ± 0 g	0 ± 0f	0 ± 0e	0 ± 0d
	1:1	52.33 ± 1.63def	1 ± 0.12cde	20.42 ± 4.14bcde	1.14 ± 0.87c
	C	67.33 ± 7.56bcde	0.77 ± 0.1e	15.86 ± 11.37cde	1.05 ± 0.68bc
A. glutinosa	P	91.50 ± 10.62abc	2.05 ± 0.52a	39.53 ± 12.14ab	4.20 ± 1.77bc
	1:1	90.60 ± 12.70abc	1.58 ± 0.13abc	44.36 ± 8.05a	7.78 ± 1.12a
	C	112.80 ± 16.97a	1.98 ± 0.1ab	47.12 ± 11.49a	5.5 ± 1.78abc
Populus euramericana	P	32.30 ± 2.65efg	0.68 ± 0.08e	1.42 ± 0e	0.44 ± 0ab
	1:1	27.90 ± 12.42 fg	0.64 ± 0.18e	0.77 ± 0e	0.45 ± 0ab
	C	59.20 ± 19.07cdef	0.84 ± 0.19de	29.5 ± 11.85abcd	1.53 ± 0.18bc
S. alba	P	0 ± 0 g	0 ± 0f	0 ± 0e	0 ± 0d
	1:1	73.80 ± 12.56bcd	0.8 ± 0.2de	6.70 ± 2.97e	1.44 ± 0.47bc
	C	79 ± 13.53abcd	0.68 ± 0.13e	13.39 ± 4.59de	0.78 ± 0.32bc
Taxodium distichum	P	97.20 ± 5.40ab	1.68 ± 0.22ab	31.20 ± 6.38abcd	3.91 ± 0.61bc
	1:1	101 ± 16.81ab	1.53 ± 0.45abc	47.47 ± 11.85a	5.17 ± 1.66abc
	C	87.20 ± 19.97abcd	1.40 ± 0.28bcd	35.90 ± 9.41abc	2.75 ± 0.31ab

P, pure compost leachate; 50% leachate +50% water; C, tap water (control).
Mean values of zero show P. deltoides and S. alba seedlings have died in leachate treatment.

Table 3.
Final mean height, diameter and biomass components of five plants in three different treatments.

Species	Treatment	Aboveground elements(mg/kg)			
		N	P	K	Ca
P. deltoides	P	0 ± 0d	0 ± 0b	0 ± 0 g	0 ± 0d
	1:1	5.16 ± 0.33a	0.36 ± 0.05a	5.85 ± 0.51a	5.95 ± 0.63abc
	C	3.69 ± abc	0.29 ± 0.01a	4.21 ± 0.36abc	6.1 ± 1abc
A. glutinosa	P	4.69 ± 1.14abc	0.27 ± 0.06a	2.4 ± 0.43def	5.33 ± 0.18bc
	1:1	5.38 ± 0.26a	0.25 ± 0a	1.13 ± 0.16 fg	4.62 ± 1.24c
	C	4.98 ± 0.31ab	0.25 ± 0.03a	1.46 ± 0.32efg	5.72 ± 0.58abc
Populus euramericana	P	4.77 ± 0ab	0.28 ± 0a	5.32 ± 0ab	6.52 ± 0abc
	1:1	3.82 ± 0abc	0.23 ± 0a	4.85 ± 0ab	7.52 ± 0ab
	C	3.08 ± 0.77bc	0.28 ± 0.06a	2.88 ± 0.67cde	7.1 ± 1.71abc
S. alba	P	0 ± 0d	0 ± 0b	0 ± 0 g	0 ± 0d
	1:1	5.36 ± 1.27a	0.29 ± 0.02a	3.71 ± 1bcd	8.35 ± 1.08a
	C	3.71 ± 0.7c	0.29 ± 0.02a	4.02 ± 0.95bcd	6.92 ± 1.62abc
Taxodium distichum	P	3.96 ± 0.2abc	0.29 ± 0.06a	5.14 ± 0.63ab	7.68 ± 1.27ab
	1:1	4.48 ± 0.59abc	0.34 ± 0.04a	5.34 ± 0.82ab	6.67 ± 0.31abc
	C	2.88 ± 0.74c	0.27 ± 0.06a	1.23 ± 0.65efg	6.97 ± 1.08abc

Different letters were significant at P < 0.05.
Zero number of mean show P. deltoides and S. alba seedlings have died in leachate treatment.

Table 4.
Mean of aboveground elements absorption of five species under irrigation treatments.

The amount of four elements, for example, N, P, K and Ca was analyzed in the aboveground section of seedlings after the growing period, and differences between treatments were tested by ANOVA followed by Tukey test using the SPSS with the effect of species, treatment and species × treatment on the elements. The results indicated that the effect of species ($P < 0.0001$), treatment (<0.0001), and species × treatment were significant on aboveground element absorption.

Tukey test determines that the mean score for the aboveground N element was significantly different between treatments, with the highest amount for 50% leachate +50% water treatment (M = 5.38). Comparing the mean N element between the five seedlings species in all treatments, *A. glutinosa* exhibited the greatest N (5.38) and followed by *s. alba* (5.36). *P. deltoides* and *s. alba* exhibited the lowest N (0).

The results of the aboveground mean P element showed a significant difference between treatments, with the highest amount for 50% leachate +50% water treatment (M = 0.36) and the lowest amount was achieved in compost leachate (M = 0) ($P < 0.0001$). Comparing the mean P element between the five plant species in all treatments, *P. deltoides* exhibited the greatest P (M = 0.36) and *T. distichum* (M = 0.34). *S. alba* exhibited the lowest it is (M = 0).

The mean score for the aboveground K element indicated a significant difference between treatments, with the highest amount for 50% leachate +50% water treatment (M = 5.85) and the lowest amount was achieved in compost leachate (M = 1.13) ($P < 0.0001$). Comparing the mean K element between the five plant species in all treatments, *P. deltoides* exhibited the greatest P (5.85), and *A. glutinosa* exhibited the lowest K (1.13).

Post hoc comparisons using Tukey test showed that the mean score for the aboveground Ca element was significantly different between treatments, with the highest amount for 50% leachate +50% treatment (M = 8.35) and the lowest amount was achieved in compost leachate (M = 4.62) ($P < 0.0001$). Comparing the mean Ca element between the five plant species in all treatments, *s.alba* exhibited the greatest Ca (8.35), and *A. glutinosa* exhibited the lowest Ca (4.62).

3.3 Root dry mass

Root dry mass results expressed there was a significant effect of irrigation treatment ($P = 0.0007$) and species ($p < 0.0001$). (**Table 5**).

Post hoc comparisons using Tukey test showed that the mean score for the root dry mass was significantly different between treatments with the highest amount of 50% leachate +50% water treatment (M = 7.78 gr) followed by water treatment with 5.5 gr ($P < 0.0001$), and the lowest amount was achieved 0 in compost leachate (**Table 5**). Comparing the mean root dry mass between the five plant species in all treatments, *A. glutinosa* and *P. euramericana* exhibited the greatest and lowest dry mass, respectively (**Figure 1**).

Mean of root elements absorption of five species under irrigation treatments demonstrate in **Table 5**. The amount of four elements, for example, K, N, P and Ca was analyzed in the root of plant species after the growing seasons, and the results analyzed with ANOVA with the effect of species, treatment, and species × treatment on the elements. ANOVA results showed that the effect of species ($P < 0.0001$) and treatment (<0.0001) was significant on root elements.

Species	Treatment	Root elements (mg/kg)			
		N	P	K	Ca
P. deltoides	P	0 ± 0 g	0 ± 0d	0 ± 0d	0 ± 0d
	1:1	1.62 ± 0.1cde	0.23 ± 0a	0.23 ± 0a	3.32 ± 0.29c
	C	1.15 ± 0.1ef	0.19 ± 0.02abc	0.19 ± 0.02abc	4.08 ± 0.73bc
A. glutinosa	P	2.24 ± 0.1abcd	0.12 ± 0.03c	1.13 ± 0.11b	4.26 ± 1.24bc
	1:1	2.53 ± 0.45ab	0.18 ± 0.05abc	0.28 ± 0.01c	5.98 ± 0.27a
	C	2.41 ± 0.17abc	0.16 ± 0.03abc	0.29 ± 0.01c	4.78 ± 0.78abc
Populus euramericana	P	1.58 ± 0cde	0.24 ± 0a	2.45 ± 0a	4.99 ± 0ab
	1:1	1.02 ± 0ef	0.22 ± 0ab	2.14 ± 0a	3.81 ± 0.71bc
	C	1.17 ± 0.19ef	0.24 ± 0.05a	2.35 ± 0a	3.91 ± 0.2bc
S. alba	P	0 ± 0 g	0 ± 0d	0 ± 0c	0 ± 0d
	1:1	1.82 ± 0.13bcde	0.24 ± 0.01a	2.36 ± 0.19a	3.81 ± 0.71bc
	C	1.43 ± 0.67def	0.29 ± 0.03a	2.55 ± 0.12a	3.92 ± 0.2bc
Taxodium distichum	P	2.93 ± 0.44a	0.22 ± 0.04ab	2.67 ± 0.49a	3.86 ± 0.82bc
	1:1	1.77 ± 0.52bcde	0.23 ± 0.03a	2.36 ± 0.22a	4.56 ± 0.91abc
	C	0.68 ± 0.16 fg	0.14 ± 0.02bc	0.24 ± 0.02c	5.17 ± 0.22ab

Different letters were significant at P < 0.05.
Zero number of mean show P. deltoides and S. alba seedlings have died in leachate treatment.

Table 5.
Mean of root elements absorption of five species under irrigation treatments.

Figure 1.
Seedlings growth after irrigation treatments.

The highest average of N and Ca in the root system was found in *A. glutinosa. P. euramericana* root system showed the highest mean of P and K.

4. Discussion

In our study, seedlings height was negatively affected by leachate irrigation, and plant species showed higher plant height in irrigation treatments of tap water and 50% leachate +50% water. Diameter growth showed a better response to compost leachate than tap water. The 50% leachate +50% water treatment in this study showed a positive effect on dry root mass for aboveground than to leachate irrigation (P).

4.1 Seedlings growth

Regarding the plant species, the maximum amount of seedlings growth and biomass was shown in *A. glutinosa* and *Taxodium distichum*, whereas the maximum diameter and height growth occurred in (P) and (C) treatments for *A. glutinosa* followed by *T. distichum*. These two species also showed the maximum average of biomass compared to other species.

Many researchers around the world indicated the positive and negative effects of leachate irrigation on plant growth and biomass. For example, Rosenkranz [18] and Guidi Nissim et al. [11] found that the *Salix* sp. growth in controlled water was much better than plants irrigated with leachate. Zalesny and Bauer [1] also indicated that *Populus* clone NM6 had a better response to water treatment than leachate. The result is the same in this study about *P. deltoides*, all seedlings of *P. deltoids* died in pure leachate treatment.

In contrast, Justin et al. [19] found that landfill leachate positively affected *Salix* and *Populus* growth with increased biomass production. Zalesny et al. [20] and Zalesny et al. [21] also found a positive effect of compost leachate irrigation on *Populus* growth. Zalesny and Bauer [1] concluded that *Salix* clones S287 and S566 showed better growth rate with leachate irrigation. Alizadeh et al. [15] indicated the high biomass growth of *Populus alba* in cadmium treatments.

Dimitriou et al. [19] investigated the growth rate of five *Salix* clones after irrigation with three different landfill leachate (1:2, 1:3, 1:4, 1 unit of leachate and 2, 3, and 4 units of tap water) and found that plants irrigated with tap water has higher growth rate compared to plants irrigated with landfill leachate. Their study showed a significant difference between control plants and leachate irrigation and found insignificant difference between the other leachate concentrations. They concluded that the degree of dilution had a minor importance on plant growth.

The same result occurred in our study where plant diameter had developed in tap water with no significant difference in dilution degree. Therefore, the dilution degree showed a minor influence on plant diameter growth. However, the dilution degree showed a significant difference between tap water and 50% leachate +50% water concentration at the height of seedlings. Aboveground biomass responds the same trend as the diameter to different dilutions. In contrast, root mass positively responded to 50% leachate +50% water concentration more than other treatments. Therefore, the dilution degree showed a positive effect on root biomass. Therefore, small and non-significant differences between tap water plant growth parameters and dilution degree growth parameters showed that dilution of compost leachate could not be considered as a conventional means of fertilizer for mentioned species except for root biomass [21]. The plant roots in this study may have contributed to the greater availability of elements concentrated in leachate irrigation treatment resulting in higher root dry mass in leachate treatment compared to controlled water. Under a leachate irrigation that leachate volume would be decreased, leachate electrical conductivity values and water-use efficiency would be increased, and at recommended fertilizer rates plant growth would be decreased.

Plant growth predictions in leachate treatments are difficult to make. Biomass production and plant growth rate are suitable indicators of imposed stress [19]. In our study, significant differences between controlled water and leachate treatments (P) of seedlings growth and biomass (except for root biomass) indicate stress on plants treated with compost leachate. Plant growth processes are strongly related to the salt effects; therefore, plant growth rate indicates a suitable way to understand salt stress [22]. The measurement of chemical components of leachate in this study showed the highest amount of salt concentration, leading to less growth rate than other treatments.

4.2 Above ground and root dry mass

The differences between the concentration of leachate treatments in this study and other [23–25] and their effects on plant growth and biomass can be taken into account that the concentration of wastewater can be used in irrigation that depend on wastewater and soil, and the nutrient demand of plants [20]. Dimitriou et al. [19] mentioned that in Sweden, the leachate treated to Salix irrigation, the plants have either died or suffered. Therefore, designing leachate irrigation treatments on plant vitality and growth must be considered.

In this study, there was a significant difference between plant growth and biomass for leachate and tap water (control) in all species with the exception of root mass biomass (**Table 4**). Therefore, the higher amount of toxic concentrations in leachate treatment prevents the development of the species. *P. deltoides* and *S. alba* seedlings have died in leachate treatment (P), which can be considered a visual sign of stress in leachate treatment, leading to damage and destruction of species. This can be attributed to an imbalance of nutrients, low pH and high salinity in leachate treatment [7, 10, 24].

The amount of four elements, for example, N, P, K and Ca was analyzed in aboveground and root of plant species in different treatments. The highest absorption of elements was carried out by *A. glutinosa*, *T. distichum* and *Populus euramericana* in 50% leachate +50% water treatment, and the lowest amount was exhibited in leachate treatment (P). The lowest amount of element absorption by plant sections in leachate treatment (P) is due to the fact that increasing the salinity concentration in component leads to decreased concentrations of K, P and Ca [4]. Zalesny et al. [20] mentioned that landfill leachate could be used as fertilizer as some of the essential elements available in the components. Zalesny and Bauer [1] conducted that the concentration of wastewater which could be used as fertilizer depending on soil, wastewater and nutrient demand of plants. Therefore, in this study, 50% leachate +50% water treatment can be considered a conventional fertilizer. Licht and Isebrands [26] mentioned that plants grow better when irrigated with leachate containing all the essential plant nutrients, petrochemical organic compounds, and salts below toxicity thresholds.

In conclusion, *A. glutinosa* and *T. distichum* species show higher diameter growth in leachate treatment compared with other species, making them suitable plants for contaminated areas. It can be concluded that hydrophilic species have shown a better growth response to different leachate treatments. Due to the higher content of nutrients in compost leachate, it is necessary to investigate the appropriate ratio of compost leachate and water in irrigation treatment. It is, therefore, important to consider different compost leachate concentrations on plant growth and biomass, which can be suggested in future research.

Author details

Tooba Abedi* and Hadi Modaberi
Academic Center for Education, Culture and Research, Environmental Research Institute, Rasht, Iran

*Address all correspondence to: t.abedi@acecr.ac.ir

References

[1] Zalesny RS, Wiese AH, Bauer EO, Riemenschneider DE. Ex situ growth and biomass of *Populus* bioenergy crops irrigated and fertilized with landfill leachate. Biomass and Bioenergy. 2009b;**33**(1):62-69. DOI: 10.1016/j.biombioe.2008.04.012

[2] Aronsson P, Dahlin T, Dimitriou I. Treatment of landfill leachate by irrigation of willow coppice - plant response and treatment efficiency. Environmental Pollution. 2010;**158**(3): 795-804. DOI: 10.1016/j.envpol.2009.10.003

[3] Navarro JM, Tornero OP, Morte A. Alleviation of salt stress in citrus seedlings inoculated with arbuscular mycorrhizal fungi depends on the rootstock salt tolerance. Journal of Plant Physiology. 2014;**60**:127-137. DOI: 10.1016/j.jplph.2013.06.006

[4] Lucero-Sobarzo D, Beltran-Villavicencio M, Gonzalez-Aragon A, Vazquez-Morillas A. Recycling of nutrients from landfill leachate: A case study. Heliyon. 2022;**8**:e09540

[5] Sammons J, Struve D. The effects of near-zero leachate irrigation on growth and water use efficiency and nutrient uptake of container grown Baldcypress (*Taxodium distichum* (L.) rich.) Plants. Journal of Environmental Horticulture. 2010;**28**(1):27-34. DOI: 10.24266/0738-2898-28.1.27

[6] Kim KR, Owens G. Potential for enhanced phytoremediation of landfills using biosolids - a review. Comprehensive Biotechnology, Second Edition. 2011;**6**(4):239-247. DOI: 10.1016/B978-0-08-088504-9.00373-1

[7] Abedi T, Moghaddami S, Lashkar BE. Growth of *Populus* and *Salix* species under compost leachate irrigation. Ecologia Balkanica. 2014;**6**(October):57-65

[8] Zaman B, Mangkoedihardjo S. Plant growth rate In evapotranspiration continuous system reactors as the 2nd treatment at anaerobic-evapotranspiration system with high strength ammonium in leachate influent. International Journal of Science and Engineering (IJSE). 2014;**6**(April):2-5. DOI: 10.12777/ijse.6.2.xx-xx

[9] Zalesny JA, Zalesny RS, Coyle DR, Hall RB, Bauer EO. Clonal variation in morphology of *populus* root systems following irrigation with landfill leachate or water during 2 years of establishment. Bioenergy Research. 2009a;**2**(3):134-143. DOI: 10.1007/s12155-009-9037-y

[10] Justin MZ, Pajk N, Zupanc V, Zupančič M. Phytoremediation of landfill leachate and compost wastewater by irrigation of *Populus* and *Salix*: Biomass and growth response. Waste Management. 2010;**30**(6):1032-1042. DOI: 10.1016/j.wasman.2010.02.013

[11] Guidi NW, Palm E, Pandolfi C, Mancuso S, Azzarello. Willow and poplar for the phyto-treatment of landfill leachate in Mediterranean climate. Journal of Environmental Management. 2021;**277**:111454

[12] Koda E, Winkler J, Wowkonowicz P, Cerny M, Kiersnowska A, Pasternak G, et al. Vegetation changes as indicators of landfill leachate seepage locations: Case study. Ecological Engineeroing. 2022;**174**:106448

[13] Shabir R, Abbas G, Saqib M, Shahid M, Shah GM, Akram M, et al. Cadmium tolerance and

phytoremediation potential of acacia (*Acacia nilotica* L.) under salinity stress. International Journal of Phytoremediation. 2018;**20**(7):739-746. DOI: 10.1080/15226514.2017.1413339

[14] Askary M, Noori M, Biegi F, Amini F. Evaluation of the phytoremediation of Robinia pseudoacacia L. in petroleum-contaminated soils with emphasis on the some heavy metals. Journal of Cell and Tissue. 2012;**2**(4):437-442

[15] Alizadeh SM, Zahedi AG, Savaghebi-Firoozabadi G, Etemad V, Shirvany A, Shirmardi M. Influence of soil amendment on cadmium accumulation responses in one-year old *Populus alba* L. seedling. Iranian Journal of Forest. 2012;**3**(4):355-366. DOI: 10.22059/JNE.2015.56933

[16] Eaton AD, Clesceri LS, Franson MAH, Rice EW, Greenberg AE. Standard Methods for the Examination of Water and Wastewater. American Public Health Association. 2005;21:256-269

[17] Olsen SR, Sommers LE. Phosphorus. In: Page AL, editors. Methods of Soil Analysis, Part 2. Chemical and Microbiological Properties. Second ed. Madison, Wisconsin, USA: American Society of Agronomy; 1982. pp 403-430

[18] Rosenkranz T. Phytoremediation of Landfill Leachate by Irrigation to Willow Short-Rotation Coppice. Master's Thesis. Swedish University of Agricultural Sciences; 2013

[19] Dimitriou I, Aronsson P. Nitrogen leaching from short-rotation willow coppice after intensive irrigation with wastewater. Biomass and Bioenergy. 2004;**26**(5):433-441. DOI: 10.1016/j.biombioe.2003.08.009

[20] Zalesny JA, Zalesny RS, Coyle DR, Hall RB. Growth and biomass of *Populus* irrigated with landfill leachate. Forest Ecology and Management. 2007;**248**(3):143-152. DOI: 10.1016/j.foreco.2007.04.045

[21] Zalesny RS, Bauer EO. Evaluation of *Populus* and *Salix* continuously irrigated with landfill leachate II. Soils and early tree development. International Journal of Phytoremediation. 2007;**9**(4):307-323. DOI: 10.1080/15226510701476594

[22] Larcher W. Physiological Plant Ecology. Ecophysiology and Stress Physiology of Functional Groups. Berlin, Heidelberg, New York: Springer; 1995

[23] Zhou C, Wang R, Zhang Y. Fertilizer efficiency and environmental risk of irrigating impatiens with composting leachate in decentralized solid waste management. Waste Management. 2010;**30**(6):1000-1005. DOI: 10.1016/j.wasman.2010.02.010

[24] Alaribe FO, Agamuthu P. Fertigation of Brassica rapa L. using treated landfill leachate as a nutrient recycling option. South African Journal of Science. 2016;**112**(3-4):1-8. DOI: 10.17159/sajs.2016/20150051

[25] Dimitriou I, Aronsson P, Weih M. Stress tolerance of five willow clones after irrigation with different amounts of landfill leachate. Bioresource Technology. 2006;**97**(1):150-157. DOI: 10.1016/j.biortech.2005.02.004

[26] Licht LA, Isebrands JG. Linking phytoremediated pollutant removal to biomass economic opportunities. Biomass and Bioenergy. 2005;**28**(2):203-218. DOI: 10.1016/j.biombioe.2004.08.015

Chapter 2

Advances in the Use of Organic and Organomineral Fertilizers in Sustainable Agricultural Production

Mukhtar Iderawumi Abdulraheem, Jiandong Hu, Shakeel Ahmed, Linze Li and Syed Muhammad Zaigham Abbas Naqvi

Abstract

To achieve high production in the current agricultural environment and increase soil quality for the plants, new technologies must be developed. It is becoming more and more important to adopt systems that can deliver higher production along with efficiency in the provision of nutrients in order to achieve sustainable agriculture. Despite improvements and novel techniques in modern agricultural practices, most of the global agriculture sectors still use conventional methods; as a result, it has sustainability and fertility issues. Since many yield-limiting problems are seriously hampering global agriculture, urgent action is needed to rehabilitate these soils; hence, the use of both organic and synthetic fertilizers was decreased by the usage of organomineral fertilizers (OMFs), which worked in concert with one another. When compared to inorganic fertilizers, the OMF improved soil physical qualities, as evidenced by a decrease in bulk density, high temperature, and preservation of soil moisture. The OMF increased soil organic matter, nutrient content, and cation availability compared to inorganic fertilizer. To increase the soil's nutrient content, pH, and crop nutrient uptake, organic wastes like poultry manure, oil palm bunch ash, cocoa pod ash, kola pod husk, and sawdust ash were successfully blended with inorganic fertilizers for efficient productivity.

Keywords: organic, organominerals, soil fertility, sustainability, agricultural production, fertilizers

1. Introduction

In Africa, where population growth is growing at a pace of 3% annually, this problem of the adequate food supply is especially critical. Additionally, more than 50% of Africans live in rural areas and are totally reliant on locally cultivated food crops that are taken from their immediate surroundings [1]. The traditional bush fallow period for preserving soil productivity under shifting cultivation is no longer practical due to the growing strain of population on the land, causing a shift toward marginal lands and an intensification of cultivation on productive fields [2]. The quest for agricultural production has increased as a result of the growing world population. This has caused a shift into marginal lands and an intensification of cultivation on fertile lands [2] and resulting in an increase in the reliance on added nutrients. In the years 2016–2021, it is predicted that the global demand for N, P, and K will rise by 1.4%, 2.2%, and 2.6% annually, respectively [3].

Utilizing organic waste to create organomineral fertilizers (OMFs) is one promising alternative. OMF is the result of physically combining organic and mineral sources of nutrients. The proportion of the primary macronutrients (N, P, and K) or their combination with other nutrients such (NP, NK, PK, or NPK) should be at least 10% in these fertilizers. OMFs may be supplemented with supplies of P, N, or K in order to comply with the law [4]. The Food Agricultural Organization recently promoted integrated plant nutrition (IPN), often known as integrated nutrient delivery or integrated nutrient management system. In order to supply crops with nutrients, a system of soil conservation farming and both organic and inorganic fertilizers is used. The idea is mostly supported in the tropics, especially in the wet tropics, where chemical or inorganic fertilizers have not revolutionized agricultural productivity as expected [1]. Chemical fertilizer (CF) has continually increased crop productivity over the world, notably in temperate agriculture, with an average yield gain of 50% being attributed to it (CF). In addition to driving up production costs, excessive use of CFs jeopardizes food safety and exacerbates problems like energy depletion, resource scarcity, and environmental damage. These issues, which have become more pressing in recent years, pose significant challenges. The significance of organic matter (OM) returning to the field and the development of organic fertilizers (OFs) on this basis in agricultural output have attracted new attention in the modern context of ecological agriculture's rapid development and environmental protection [5].

In comparison to inorganic fertilizer, OMFs, a new fertilizer that combines the benefits of organic and synthetic fertilizers, have an effect on nutrient release that coincides with the crop's growth phase [6]. When compared to CFs, OMFs can lessen the loss of some nutrients like potassium leaching, nitrogen volatilization, and phosphorus fixation. When compared to typical OFs, OMFs are abundant in the minerals needed for agricultural growth [7]. A successful soil fertility management method has been demonstrated to involve complementary fertilizers usage [8]. To replenish the soil and enhance plant fertilization, OF may be produced with inorganic minerals added. It disperses nutrients into the soil in a way that makes them simple for plants to absorb. It may also encourage soil microorganisms and boost their population, which will speed up the breakdown of organic compounds. Increased plant growth, healthier crops, and improved fruit production will result from this. It lowers the demand for CF, resulting in cheaper production costs and indirect income growth [9].

An important fertilizer technology is an ability to combine organic and mineral sources into a single formulation because it increases the effectiveness of the nutrient sources, offers better protection from the elements, and encourages the monitoring of the physical, chemical, and microbiological properties of the soil [10]. Applications of OMF are essential steps in the growth of organic agriculture in karst mountain areas. Yet it is still unclear how OMF impacts soil microbial diversity's organization and functioning, as well as how these elements connect to crop output and quality [2]. It is known that CFs have not had an essential influence on tropical agriculture, particularly in Nigeria, where it is impossible to produce crops sustainably using only CFs [11–13].

CF use efficiency and effectiveness issues are widely known. They include soil acidity caused by the repeated use of acid-forming substances, applications that are frequently made without considering the results of soil tests, an imbalance in the supply of nutrients, physical degradation of the soil, scarcity, high cost, and insufficient use as well as losses from volatilization, erosion, and leaching [13–15]. There are benefits and cons to both conventional and organic farming. Combining organic and mineral inputs is one interesting strategy that could be used [16]. This may be the ideal alternative to increase agronomic effectiveness and crop output while maintaining the long-term fertility and health of the soil, according to several research studies [17–20]. Theoretically, both resources are compatible. While interest in this concept has grown recently, a brand-new category of fertilizer product known as OMF has been developed.

Although modern biofertilizers like azolla and mycorrhiza as well as organic manures including sewage water, green manure, animal wastes, fallow fields, human excreta, and crop residues (e.g., in Ghana and China) has been essential to crop production for centuries, total dependence on OFs is hindered by issues like insufficient availability, heavy metal pollution, unsuitable or low quality, high C:N ratio, bulkiness, and gradual nutrients release [21, 22].

However, the only source utilized in small-scale cropping is organic wastes, which led to the accumulation of OM and nutrients, biological buildup, and significant advancements in the soil's physical characteristics, as shown by a fall in bulk density and a rise in porosity, a decrease in runoff and erosion due to an increase in infiltration, the stabilization of the soil structure, and an increase in soil pH, etc. High levels of macro- and micronutrients can be found in crop and animal waste. OMFs are frequently noted for their many advantageous effects on agrosystems, including their capacity to increase plant physiological features and soil physico-chemical and biological functions [23–29].

Combining organic and inorganic fertilizers is essential for maintaining soil fertility. A long-term increase in soil productivity and quality will result from efforts in this direction. It is believed that these elements will increase crop growth, yield, and quality while reducing the price of production and fertilizing and making sure that plants receive adequate nutrients [30, 31]. Utilizing agricultural, human, and municipal wastes, among other things, IPN will also help with environmental cleanliness. It is a program to turn waste into wealth. As compared to CFs, the physical, chemical, and biological properties of the soil will be improved overall using IPN [23]. The IPN improves crop performance by integrating the benefits of both organic and inorganic fertilizers.

Studies on soil fertility have recently concentrated on IPN in developing and tropical nations. Since the 1960s, India has been setting the bar higher in this area. IPN research is still very new in Nigeria. State governments in Nigeria started producing organic and OMFs after realizing the value of OM in sustaining soil production and the need to enhance its quality and nutrient release. Utilizing agricultural and municipal trash at their facilities is another goal to promote environmental sanitation. Nutrient cycling and conservation in farm systems are ensured by using agricultural wastes to create organic and OMFs. Additionally, nutrients that have already been taken away by towns can at least partially be recovered. The organic and OMF companies already functioning in Nigeria include those of Oyo and Ondo States, which, respectively, are named Pacesetter and Sunshine fertilizer companies.

2. Current fertilizer trends

In order to address these nutrient deficits, it has been recommended for crop production in the tropics to utilize CFs intensively at the recommended rate (NPK

15:15:15) [18]. Currently, employing CFs as soil amendment has grown expensive and is frequently associated with adverse environmental repercussions (causing residual effects on the soil). Given these issues, several alternatives to CFs include crop rotation, organic manure, agricultural waste, legumes, and green manure [21]. The bulk of the world's agriculture industry continues to use conventional ways despite advancements and contemporary approaches in current agricultural practices; as a result, it suffers sustainability and fertility problems. Additionally, the usage of CFs improperly to improve agricultural yields and the expansion of various cropping methods are both contributing to the continued reduction in soil fertility [32]. The application of abundant refuse in our community is an outcome of the capability of biological substances to boost output, enhance soil richness, and stimulate crop proliferation [33]. According to [34], the use of cow dung in combination with NPK manure was more potent in enriching soil chemical properties, nutrient uptake, growth, and yield of crops than using either biological (cow dung) or artificial (NPK) manure alone

Since many yield-limiting problems are seriously hampering global agriculture, urgent action is needed to rehabilitate these soils (a rise in disease outbreaks, an invasion of pests and insects, and unpredictable weather). Therefore, it is debatably believed that agricultural operations constitute a significant source of CO_2 that damages the climate and contributes to global warming. The amount of residues thrown away rose due to the huge agricultural expansion in the twenty-first century. Farmers started recycling or reusing materials in greater quantities as a result, discovering the synergistic benefits of organic and mineral fertilizers [16]. However, it is currently not economically feasible to spread significant amounts of OFs across a wide area. Mineral components are added to the biofertilizer to enrich it, which results in higher yields and lower costs. Farmers and businesses in the sector must deal with genuine challenges such as logistics, infrastructure for manufacturing, and the availability of raw materials suitable for enriching mineral sources. The method is hampered by constraints related to composition knowledge and residue treatment [16].

Natural organic elements are used to create OFs without being significantly altered. Depending on the source of the OM, they have varying levels of nitrogen, phosphorous, potassium, and other nutrients. In addition to providing the plants with nutrients, these fertilizers improve the physical, chemical, and biological properties of the soil, assisting in the growth of the plants. Organic fertilizers come in a wide range of varieties. The primary sources of OFs are farmyard manure, which includes animal excrement and litter, domestic organic wastes, wastes from the food and agriculture industries, wastes from wood processing and harvesting, compost, mosses, vermicomposts, treatment sludge, and waste from the production of biogas [35, 36].

CFs can quickly and considerably boost agricultural yield and soil fertility. CFs impair soil microbial populations and biological activities when they are applied in excess over an extended period of time, which lowers soil quality, increases crop growth's reliance on fertilizer nutrients, and exacerbates agricultural surface source pollution [37, 38]. More than 200 million tons of fertilizer (N, P, and K) will be used annually around the world by the year 2020 [39]. This is concerning since fertilizers have detrimental effects on nearby ecosystems and soils. Additionally, the use of insecticides to stop and manage plant diseases is expanding at a riskier rate. A new breakthrough is the creation of smart fertilizers based on nanotechnology, with a focus on controlled-release and/or carrier/delivery systems to synchronize nutrient availability with plant demands and minimize environmental losses (**Figure 1**) [41].

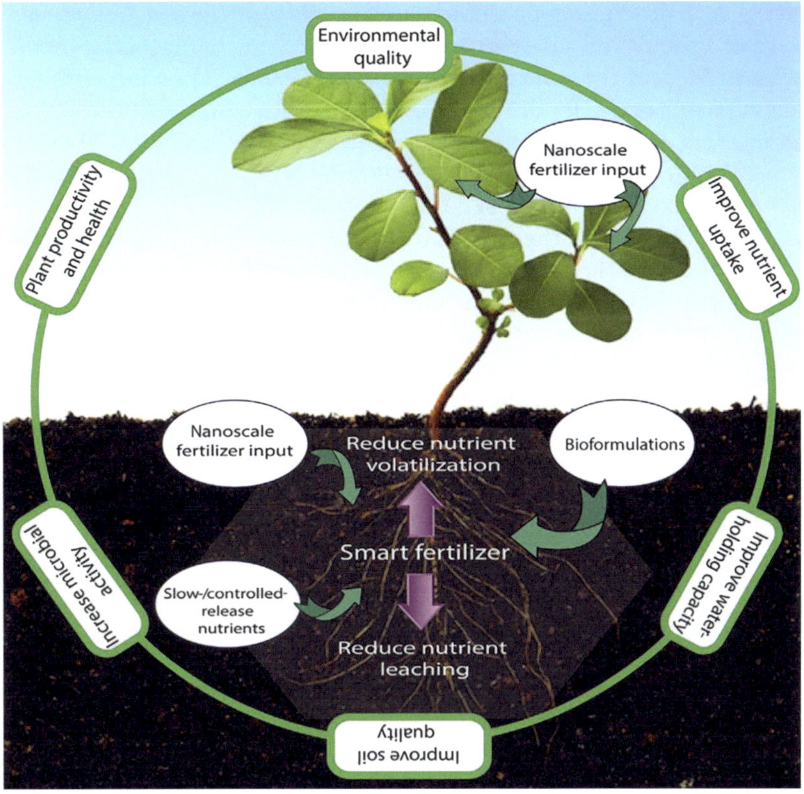

Figure 1.
Effects of smart fertilizers on the soil-plant system are shown schematically [40].

When CFs are irresponsibly used to boost production, it seriously jeopardizes the long-term soil fertility, the soil environment, and each of its constituent parts. However, the soil microbial community performs a significant and crucial function in promoting soil health and plant growth [4]. While CFs, especially N and P fertilizers, have a significant negative impact on the microbial community, they also have a profound negative impact on plant growth. Sustainable agricultural methods can preserve soil fertility and productivity while reducing the depletion of natural resources. Moreover, they safeguard against soil degradation, facilitate the development of beneficial soil particles, and can loosen the ground when applied as a natural fertilizer [42, 43].

Given that it is predicted that there will be 9.3 billion people on the planet by the year 2050, there will likely be a 60% rise in the need for food during that time [44]. The production of agriculture and food is one of the primary factors in the depletion of natural resources. The era of affordable feedstock has ended. Because demand grows more quickly than available production capacity, resource scarcity drives up input and production costs and tightens the market. Food security may be achieved through the efficient and sustainable utilization of natural resources [45].

In the current situation, the decline in fertility brought on by intensive agricultural methods is of enormous significance. Increased productivity might be achieved by raising fertility in intensive farms or on non-arable land, which would increase food output without further taxing the environment. The secret to balancing crop production on a commercial basis is fertile soil. Only a small percentage of agricultural soils contain enough nutrients to meet crop productivity requirements. Most of them start to rely increasingly heavily on fertilizer applications that are made often. By the end of 2018, it is anticipated that the overall fertilizer use (N + P_2O_5 + K_2O), which was expected to be 183,200,000 t in 2013, will reach 200,500,000 t [3]. **Figure 2** shows the projected global demand for all nutrients in fertilizers from 2016 to 2021.

China produced 33% of the world's fertilizer in 2021, according to FAOSTAT [40], making it the world's top producer. The Russian Federation (9%), India (9%), and the United States (which contributed 10% each to global fertilizer production) were other nations with large contributions. Prices for fertilizer are anticipated to be high [45]. Commercial fertilizer manufacturing uses a lot of energy. Around 74% of the energy used to produce fertilizer is provided by natural gas [46]. The key element in the manufacture of nitrogen fertilizer is natural gas. The cost of ammonia will rise when natural gas prices rise [47]. On the other hand, fertilizer prices are substantially impacted by transportation expenses. In contrast to the year 2010, the fertilizer price index will rise by 15% in 2014 (**Figure 3**) [3].

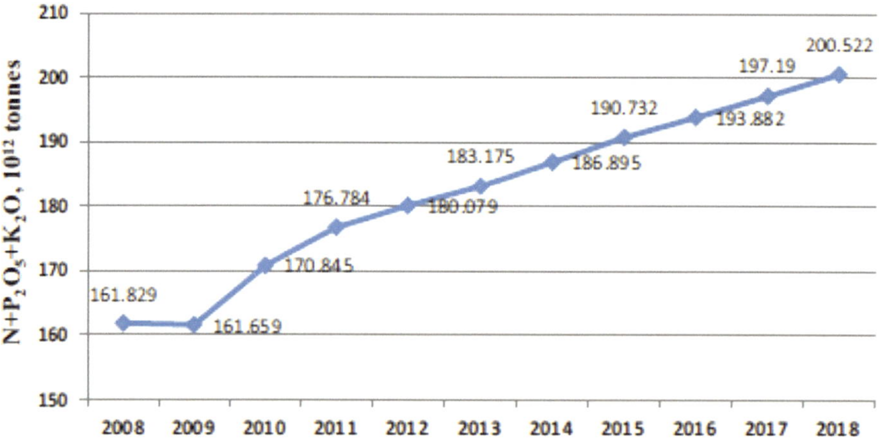

Figure 2.
Global nutrients consumption (N + P_2O_5 + K_2O).

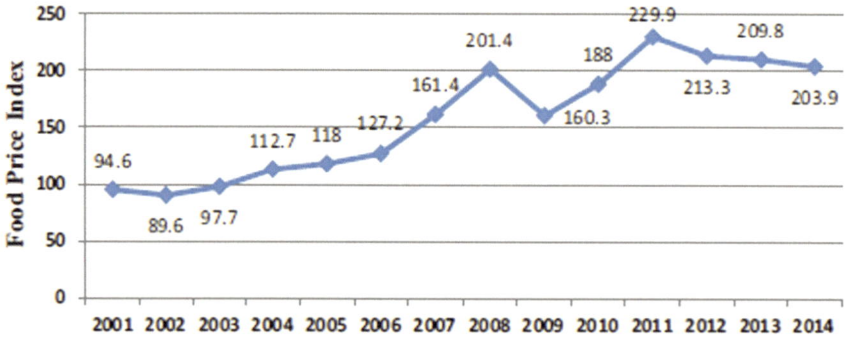

Figure 3.
Annual food price index.

3. Case studies with processed organic and organomineral fertilizers

Natural infertility, a lack of soil OM, and the rapid rate at which fertility diminishes during intensive farming make Sub-Saharan African soils unsuitable for cultivation [48]. These issues have been solved by using different fertilizers [49, 50]. It has been discovered that using inorganic fertilizer to boost yield is efficient in the near term but requires regular application over time.

Sugarcane output is increased when organic and mineral fertilizer are applied together [51] and larger residual advantages are encouraged, which have an impact on how each fertilizer is used differently. In this respect, OMF displays a comparatively lower reactive chemical potential than mineral fertilizer. Its solubilization happens gradually throughout the course of the culture's development although its agronomic effectiveness might be higher than that of sources of soluble minerals [16].

Subsistence farmers, who dominate Nigeria's agricultural sector, cannot afford inorganic fertilizers because they are not only unsightly but also expensive [52]. Using minerals that naturally contain fertilizer ingredients is a very effective way to reduce emissions from processing chemical companies. During mineral weathering, organic materials from natural sources such as animal manure, agricultural biomass, and a source of microbial culture and a catalyst for the release of nutrients into the soil can be other organic materials [5]. OFs frequently completely or partially replace CFs to safeguard soil biodiversity and maintain the ecological balance of the soil [53, 54]. Intensifying the use of OFs has the potential to increase soil biological activity, plant nutrients, and soil carbon storage, all of which are essential for decreasing global warming and promoting sustainable agricultural output [55, 56].

Due to scarcity and high prices, peasant farmers have recently drastically decreased their use of mineral fertilizers [57, 58]. Also, their use has been discouraged due to the negative side effects on the soil, such as acidity and aluminum toxicity [59, 60]. Numerous research employing the OMFs Pacesetter and Sunshine have been conducted on various crops [11, 13, 59, 61–63].

Moreover, excessive use of CFs interferes with the rhizosphere's regular microbial population functioning (nutrient cycling, OM creation, and soil nutrient improvement). Large tracts of fertile soils, however, were either degraded or converted into non-agricultural activities as a result of the rapid development of industry, agriculture, and population growth [64], and numerous areas of recently reclaimed soil—the majority of which are poor soils—have been modified for plant cultivation. Many regenerated soils, like soil properties, have adequate nutrient quantities (K, Ca, and Mg), yet as these nutrients are distinct chemical phases and are not utilized by plants, they often exist as plant nutrients. Furthermore, the problem of heavy metal pollution in agricultural soil is made worse by continuing to utilize CFs [5].

The Department of Agronomy University of Ibadan also developed a poultry-based urea-fortified OMF [63, 65]. The yield, nutrient content, and nutritional quality of maize and vegetables were all observed to increase with the use of OMFs [66]. As part of a field study at the University of Ibadan, the agronomic and financial evaluation of five OMFs that combined poultry manure (PM), sorted trash, phosphate rock, and urea was evaluated. When using the formulations, maize yield improved to roughly 3.17 t/ha as opposed to 3.46 t/ha when using 300 kg/ha of NPK fertilizer.

In contrast to synthetic fertilizers, the chemical properties of OMFs are not predetermined or fixed; they vary based on the method of production [67]. For example, urea contains 46% nitrogen, while fertilizers such as monoammonium phosphate

(MAP) and diammonium phosphate (DAP) have unique chemical compositions of 11% nitrogen and 52% P_2O_5 and 18% nitrogen and 46% P_2O_5, respectively. Each OMF product may differ since different nutrient ratios are utilized during production [35, 68].

Using organic manures in addition to mineral fertilizers has been shown to be a successful method for controlling soil fertility in many countries across the world [69]. When organic manures and inorganic fertilizer were applied combined, improvements to the soil's organic content, structure, water-holding capacity, nutrient availability, maintenance of soil nutritional status, cation exchange capacity (CEC), and soil organisms were seen [70]. This procedure enhances nutrient density, enabling the utilization of reduced dosages and yielding exceptional, immensely durable, and upgraded merchandise [71, 72]. The soil's contents of organic carbon, nitrogen, phosphorous, and potassium increased when both organic and inorganic manure were applied together. Examining the benefits of all the various integrated nutrition management techniques is crucial rather than concentrating simply on the short-term benefits. As a result, the combined use of organic and inorganic fertilizer is a sustainable strategy for nutrient management that increases the effectiveness of CF while minimizing nutrient losses (**Table 1**) [73].

OMF is a low-input method that can help tropical soils' poor nutritional conditions for sustained crop production. It combines both sources' positive qualities to increase yield. The emphasis is increasingly moving away from the straightforward use of either organic or inorganic fertilizers toward combinations used in many places of the world. As a result of the recent flooding that wreaked havoc in some areas of the nation and the rapidly diminishing productivity of our soils, Nigeria is currently confronting its worst challenges with regard to food insecurity. In order for our many farmers to meet their yield expectations and get the greatest benefits from their introduction, it is crucial to encourage and introduce the combined use of organic and inorganic fertilizers (OMFs). Recent studies on OMFs have revealed improved yield results compared to their single-use (**Table 2**) [26, 62, 74].

In conclusion, BOMFs are fertilizers made from little processed materials (chemical agents such as urea, phosphates, etc.). When compared to traditional OFs or bio-fertilizers, BOMFs are mostly made up of bio-organic components like animal excreta (cow dung, chicken feces, etc.) and mineral sources. **Table 3** lists the elements and key distinctions between organic and conventional CFs.

Recent and current research suggest that OM, OF, dust, mineral powder (MP), and pasture legume cultivation as green manure can all improve soil quality [75–77]. This is in response to the numerous obstacles.

Treatment	**Soil moisture (%)**	**Soil temperature (°C)**	**Soil bulk density (mg/m^3)**
Control	7.1	33.0	1.05
SOF	8.3	31.6	0.92
SOMF	7.4	31.8	0.98
NPK	7.1	32.0	0.04
LSD (0.05)	0.1	0.25	0.004

SOF = sunshine organic fertilizer; SOMF = sunshine organomineral fertilizer; and NPK = nitrogen, phosphorus, and potassium.

Table 1.
Soil physical properties as influenced by sunshine organic and OMFs.

Treatment	pH	N	Avail. P	K	Ca (cmol/kg)	Mg (cmol/kg)
Control	6.7	0.04	2.9	0.22	3.22	2.87
SOF	7.1	0.07	4.5	0.32	3.68	4.53
SOMF	7.0	0.09	6.4	0.39	3.70	6.37
NPK	6.8	0.05	4.2	0.23	2.93	4.16
LSD (0.05)	0.004	NS	NS	NS	NS	NS

SOF = sunshine organic fertilizer; SOMF = sunshine organomineral fertilizer; and NPK = nitrogen, phosphorus, and potassium.

Table 2.
Soil chemical characteristics as affected by sunshine organic and OMFs [15].

Different distinguishing components	Organomineral fertilizer	Conventional fertilizer
Mineral source (P, K, Ca, B, Mg, etc.)	Natural	Chemical
Mineral (dust or powder)	Available	Not available
Microbial agents	Available	Not available
Growth-promoting microorganisms for plants	Available	Not available
Plant-defense microorganisms	Available	Not available
Organic carbon	Natural (agri. waste)	Not available
Nitrogen source	Natural (animal waste)	Chemical (urea)
synthetic additives	Not available	Chemical additives (P, K, etc.)
Environmentally harmful conditions	Not available	available
Potential to sequester carbon dioxide	Available	Not available
Positively impacts soil properties	Available	Not available

Table 3.
Variations between standard synthetic fertilizers and OMFs in terms of composition and functionality [5].

Two soil types that were gathered from Ikorodu and Ojoo in Lagos State were used in a greenhouse experiment by [78] along with amaranthus. The treatments included ground kola pod husk (GPH), pacesetter organic fertilizer (POF), and NPK fertilizer (NPK) mixed at 50:50 or 75:25. OF and inorganic fertilizers are applied together for the greatest yield and longest-lasting results; organic and OMFs greatly boosted amaranthus growth as compared to control. Additionally, combined applications significantly increased the nutrient content and nutritional quality (proximate study) [55, 68]. According to research by [65], when applied at 3t/ha to slightly acidic soil that is poor in OM, N, and P, sunlight organic (SOF) and organomineral fertilizers (SOM) significantly enhanced soil moisture, decreased temperature and bulk density, and elevated pH compared to NPK fertilizer and control. The percentages of N and Mg in garden eggplant leaves also dramatically rose. The application of OF and OMF increased the yield and nutritional value of garden eggplants while also enhancing the soil's fertility and physical qualities [79].

The effects of pacesetter OMF on soil and maize were examined in two locations in Ilesa. The soil was treated with 200 kg/ha of NPK fertilizer (NPK) and 0, 2.5, 5.0, 7.5, and 10.0 t/ha of OMF, respectively. OMF and NPK increased plant height, leaf area, grain production, cob and ear weight, soil, and plant N, P, and K, with 10.0 t/ha OMF producing the highest output. The soil's pH was at its ideal level, and the 7.5 and 10 t/ha rates provided adequate amounts of N, P, and K in the soil and on the leaves. The 2.5, 5.0, 7.5, and 10.0 t/ha OMF and NPK, respectively, increased the grain yield by 20%, 29%, 35%, 94%, and 46% [58, 60].

4. Studies with raw organic fertilizers

4.1 Sawdust ash

In order to produce an adequate and high-quality output, plants in organic agriculture need to absorb plant nutrients at their optimal level or be provided to the soil. However, in organic farming, the soil must be amended with organic forms of both macro and micronutrients. Important macro elements like potassium, nitrogen, and phosphorus must be available in their micro and macro forms. For optimum plant development and high production, plant nutrients alone are insufficient.

In addition to research using synthetic or produced OMF, studies using raw organic waste and CFs were also carried out. [11, 59] studied the integrated application of sawdust ash (SDA) with urea and compared different combinations of their reduced levels with urea or SDA alone. Tomato was the test crop. The co-application of urea, SDA (4.5 t/ha), and SDA increased the soil's OM levels of N, P, K, Ca, and Mg. The SDA offered the highest soil pH, OM, accessible P, exchangeable K, Ca, and Mg. Using urea, soil pH was decreased. The simultaneous application of urea and SDA, as well as the reduced amounts of each, improved leaf N, P, K, Ca, and Mg. The highest leaf N was delivered by urea, while SDA produced the highest amounts of leaf P, Ca, and Mg. When compared to controls, SDA, urea, and their mixtures all significantly raised growth metrics and the number of fruits, while combinations of 3.0 t/ha SDA + 120 kg/ha U and 1.5 t/ha SDA + 80 kg/ha U significantly increased fruit weight. It is ascertained that SDA is an effective source base element and can be integrated with urea at reduced levels to maximize yield, soil, and plant N, P, K, Ca, and Mg content (**Table 4**) [18].

Results show that adding urea and SDA alone enhanced soil pH although the increases were not statistically significant. In 2013, its growing OM and impact were tremendous. In 2012 and 2013, urea alone, along with 1.5 t/ha SDA + 180 kg/ha U, 3.0 t/ha SDA + 150, 1.5 t/ha SDA + 180 kg/ha U, 3.0 t/ha SDA + 120 kg/ha U, 4.5 t/ha SDA + 60 kg/ha U, and 6.0 t/ha SDA, significantly increased N. Urea significantly increased P in 2013 as well [59].

4.2 Poultry manure

Because animal dung, regardless of form, has a favorable effect on soil and crops, its usage cannot be overstated. A major goal of modern agriculture is to increase soil fertility and improve environmental quality. There is a global trend toward creating agricultural production systems that use inputs more effectively and produce less waste in an effort to create a more environmentally friendly, ecologically sound, and economically viable agricultural system. An innovative idea for managing animal waste includes mixing animal feces with fertilizers that are mineral-based to produce fertilizers that are organomineral [80]. In Akure, Nigeria, researchers compared the effects of 300 kg/ha of NPK 15-15-15 fertilizer, 7 t/ha of PM, six combinations of lower amounts of N-P-K 15-15-15 and PM, and control (no fertilizer) over the course

Treatment	Soil pH	OM (%)	N (%)	P (%)	Ca (cmol/kg)	Mg (cmol/kg)
Control	5.7a	0.46b	0.11c	11.0d	0.42c	0.47b
240 kg/ha-1U	5.7a	0.64ab	0.15b	10.6d	2.10b	1.03b
1.5 t/ha SDA + 180 kg/ha U	6.1a	1.23a	0.24a	18.9a	3.47ab	1.27ab
3.0 t/ha SDA + 120 k/ha U	5.9a	0.79ab	0.17b	15.6b	4.93ab	1.20ab
4.5 t/ha SDA + 60 kg/ha U	5.8a	0.89ab	0.12c	20.5a	4.80ab	2.00ab
6.0 t/ha SDA	5.9a	0.67b	0.11c	13.5c	5.47a	3.63a

SDA = sawdust ash; U = urea.
According to the findings of the Duncan's Multiple Range Test, the means within each column that are separated by the same letter are not significantly different (P = 0.05).

Table 4.
Effects of urea (U) and SDA on soil nutrients contents at 12 weeks after application [59].

Experiment	pH	OM (%)	N (%)	P (mg/kg)	K (cmol/kg)	Ca (cmol/kg)	Mg (cmol/kg)
Control	5.85	2.91	0.09	9.12	0.21	2.55	2.53
5 t/ha CDM	6.13	6.72	0.24	12.61	0.39	4.40	3.09
10.0 t/ha CDM	6.11	6.48	0.20	12.66	0.33	4.50	3.17
200 kg/ha NPK	5.94	4.74	0.23	11.69	0.31	3.27	2.63

CDM = cow dung manure; NPK = nitrogen phosphorus and potassium.

Table 5.
Effects of NPK Fertilizer and Cow Dung on Soil Nutrient Contents [18].

of two years [30, 72]. The soil's chemical characteristics, the yield of dry matter from maize, the yield of grain, plant height, leaf area, and nutrient uptake were all considerably improved by the PM, lower levels of PM and NPK, and NPK alone. The highest values were given by 3 t/ha PM + 260 kg/ha NPK fertilizer with regard to dry matter yield and nutrient uptake, resulting in the highest grain yields.

Cow dung applications of 5 and 10 t/acre considerably improved soil OM compared to other approaches (**Table 5**). The increase in soil OM brought on by the application of cow dung may be attributable to organic waste's capacity to enrich soil OM [18].

4.3 Oil palm bunch ash

Synthetic fertilizers frequently release nutrients inequitably and insufficiently into the soil, which tends to exacerbate soil acidity. These deficiencies can be compensated for by the synergistic effect of using organic ingredients such as oil palm bunch ash along with synthetic fertilizer. The application of both organic and inorganic nutrient sources together is anticipated to have positive effects, as opposed to complete reliance on any one source, which will result in lower costs for CFs, better-balanced plant nutrition, and soil acidity control [23]. Integrated application of NPK and oil palm bunch ash (OPBA) together. The study looked into the effects of applying OBA and NPK fertilizer together (NPK). Six treatments: NPK, OBA 4 t/ha, and control

Experiment	pH	OM (%)	N (%)	P (mg/kg)	K (cmol/kg)	Ca (cmol/kg)	Mg (cmol/kg)
Control	5.66b	1.43f	0.14f	9.70j	0.14h	1.06g	0.88e
100% U	5.60b	1.60f	0.68a	20.00i	0.17gh	3.37e	1.20cd
75% U + 25% OPBA	5.85b	2.20d	0.64ab	21.60h	0.23fgh	4.80d	1.30c
50% U + 50%OPBA	5.93b	2.73ab	0.58abc	22.00h	0.26f	5.25cd	2.08b
25% U + 75% OPBA	5.98ab	2.25d	0.50c	24.90g	0.30f	4.90d	2.05b
100% OPBA	6.06a	2.41bcd	0.47cd	36.70d	0.64d	3.90e	2.63a
100% NPK	5.68b	1.90f	0.59abc	47.40a	0.75c	2.60fe	1.00cd
75% NPK + 25% OPBA	5.83b	2.40cd	0.53bc	44.20b	0.85b	6.20a	1.25c
50% NPK + 50%OPBA	5.96b	2.79a	0.50c	41.60c	0.96a	6.00ab	2.07b
25% NPK + 75% OPBA	5.99ab	2.78a	0.48cd	35.80cd	0.92ab	5.50bc	2.03b
100% OPBA	6.07a	2.40bcd	0.48cd	36.70d	0.64d	3.90e	2.62a
75% OPBA	6.03ab	2.37bcd	0.37e	30.10e	0.62d	3.45e	2.50a
50% OPBA	6.00ab	2.30cd	0.30e	26.30f	0.60d	3.20f	2.59a
25% OPBA	5.93b	2.10e	0.28e	23.70g	0.41e	3.11f	2.23ab

According to Duncan's Multiple Range Test, means in the same columns that are not separated by the same letters are substantially different at the 5% level of significance.
OPBA = oil palm bunch ash; U = urea; NPK = nitrogen phosphorus and potassium.

Table 6.
Effect of OPBA and its combined use with urea and NPK fertilizer on soil nutrient composition [22, 47, 65].

(15-15-15) At NIFOR and Ekiadolor, 300 kg/ha of maize received applications of 25% NPK + 75% OBA, 50% NPK + 50% OBA, and 75% NPK + 25% OBA. Other treatments boosted SOM, N, P, K, Ca, Mg, and pH plant nutrients, growth, and yield in comparison to controls (**Table 6**) [65, 81].

The use of OPBA alone or in combination with urea or NPK generally led to higher soil K, Ca, and Mg concentrations, which in turn raised pH. Using OPBA had a liming impact and decreased soil acidity in this manner. Usman et al. [82] states that OPBA is alkaline and has relatively high quantities of K, Ca, and Mg but low levels of OM, N, and P.

4.4 Kola pod husk

Eight fertilization methods were applied to amaranthus, including control, pacesetter grade B organic fertilizer (PGB) applied at a rate of 3 t/ha (100%), 300 kg/ha NPK fertilizer (NPK), PGB + NPK applied at a rate of 75:25, PGB + NPK applied at a rate of 50:50, kola pod husk (KPH) applied at a rate of 3 t/ha (100%), and KPH + N (50:50). Study was done on the residual impact on the second and third crops. A close analysis was conducted. Crude protein (CP) and EE (ether extract) were considerably increased both immediately and subsequently by the PGB, KPH alone, or in combination with lowered NPK. Reduced crude fiber from organic materials alone or in combination with NPK (CF). NPK produced the least CP, ash, CF, and EE in

Plant height (cm)	No. of leaves (cm^2)	Leaf area (t/ha)	Stover yield (t/ha)	Grain yield (t/ha)	Root dry (%)	Matter	Increase in grain
Control	72.60e	8.00c	14d	3.23c	2.84c	0.67b	—
2.5 t/ha OG	89.70e	9.33bc	20c	3.59c	3.00b	0.93a	5.63
5 t/ha OG	107.90d	9.23c	19c	3.97bc	3.11b	0.97a	9.51
10 t/ha OG	149.40c	12.0b	32b	4.99b	4.25a	0.99a	49.65
2.5 t/ha OMF	129.40c	12.20b	44a	5.34a	4.55a	1.10a	60.21
5 t/ha OMF	169.20b	14.59a	31b	5.36a	4.78a	1.00a	68.31
10 t/ha OMF	164.10b	12.4b	30b	4.63b	3.94a	0.97a	38.72
300 kg/ha NPK	194.00a	12.3b	24c	4.23b	3.44ab	0.93a	12.13

Means with the same letter in the same column are not significantly different at 5% using Duncan Multiple Range Test.

Table 7.
Effect of OG, OMF, and NPK fertilizers on growth and yield of maize [83].

comparison to organic materials. In contrast to NPK, which failed to sustain appropriate CP and EE in the second crop, OFs and OMF did so in both the first and second crops [58, 60]. All amounts of OG, OMF, and NPK fertilizers significantly increased (P0.05) the height, number of leaves, leaf area, stover, dry matter, and grain yields of maize when compared to the control (**Table 3**).

The highest plant height was achieved with the application of 300 kg/ha NPK fertilizer, followed by 5 t/ha OMF for the highest number of leaves, stover yield, and grain production; 2.5 t/ha OMF for the highest leaf area; and the lowest results with the control experiment (**Table 7**) [56, 83]. [56] To produce African eggplant, cocoa pod husk that has been treated with urea is used (Solanum macrocarpon). Aleshinloye grade A and Sunshine grade A fertilizers were employed by [81] to improve the fluted pumpkin's growth, yield, and nutritional makeup. In Ilorin, North-central Nigeria, little to no research has been done on how these fertilizers affect the soil and amaranth plants.

The yield data at the Ikorodu and Ojo sites in Lagos State showed that other treatments significantly increased plant height, number of leaves, and girth in comparison to the control. In comparison to KPHT, PGB, and NPK alone, soil treated with KPH + NPK (50:50) primarily boosted growth parameters, while PGB + NPK (50:50) also had a better residual effect on yield parameters in Ikorodu. The combined treatments had the highest yields and were recommended [84].

5. Conclusions

From an economic and environmental perspective, using organic waste as fertilizer is a tempting option. Recycling the nutrients found in the farmer-accessible organic materials can replace the application of expensive conventional fertilizers while reducing the likelihood of environmental damage due to improper waste disposal from agricultural activities. Furthermore, it is debatably believed that agricultural operations constitute a significant source of CO_2 that damages the climate and contributes to global warming. As a result, this chapter reviews recent modifications to Nigeria's usage of organic and OMFs. So it may be concluded that a variety of organic wastes could be utilized either independently of or in combination with mineral fertilizers. More lasting effects, balanced nutrition, and an increase in the physicochemical qualities of the soil are all guaranteed by the inclusion of organic manures in OMFs. Utilizing the two resources together has a synergistic impact and lowers spending on rare and pricey mineral fertilizers. It is a sustainable strategy for assuring high agricultural output and soil productivity.

Acknowledgements

Special appreciation to the Henan Agricultural University, Zhengzhou, China. The authors are also thankful to the National Natural Science Foundation of China (No. 32071890) and supported by Henan Center for Outstanding Overseas Scientists (No. GZS2021007).

Conflict of interest

The authors declare no conflict of interest.

Author details

Mukhtar Iderawumi Abdulraheem[1,2,3], Jiandong Hu[1,2,3]*, Shakeel Ahmed[1,2,3], Linze Li[1,2,3] and Syed Muhammad Zaigham Abbas Naqvi[1,2,3]

1 Department of Electrical Engineering, Henan Agricultural University, Zhengzhou, China

2 Henan International Joint Laboratory of Laser Technology in Agriculture Science, Zhengzhou, China

3 State Key Laboratory of Wheat and Maize Crop Science, Zhengzhou, China

*Address all correspondence to: jdhu@henau.edu.cn

References

[1] Iderawumi AM, Kamal TO. Green manure for agricultural sustainability and improvement of soil fertility. Farming and Management. 2022;**7**(1):1-8

[2] Li Y et al. Organomineral fertilizer application enhances Perilla frutescens nutritional quality and rhizosphere microbial community stability in karst mountain soils. Frontiers in Microbiology. 2022;**13**:1-19

[3] FAO. World fertilizer trends and outlook to 2018. Rome: FAO; 2015

[4] Martins DC et al. Organomineral phosphorus fertilization in the production of corn, soybean and bean cultivated in succession. 2017

[5] Syed S et al. Bio-organic mineral fertilizer for sustainable agriculture: current trends and future perspectives. Minerals. 2021;**11**(12):1336

[6] de Souza Magalhães CA et al. Eficiência de fertilizantes organominerais fosfatados em mudas de eucalipto. Scientia Agraria. 2017;**18**(4):80-85

[7] Aguilar AS et al. Influence of organomineral fertilization in the development of the potato crop cv. Cupid. Bioscience Journal. 2019;**35**(1):199-210

[8] Law-Ogbomo KE, Remison SU, Jombo EO. Effects of organic and inorganic fertilizer on the productivity of Amaranthus cruentus in an ultisol environment. International Journal of Plant Physiology and Biochemistry. 2011;**3**(14):247-252

[9] Worthington V. Nutritional quality of organic versus conventional fruits, vegetables, and grains. The Journal of Alternative & Complementary Medicine. 2001;**7**(2):161-173

[10] Magela MLM et al. Efficacy of organomineral fertilizers derived from biosolid or filter cake on early maize development. Australian Journal of Crop Science. 2019;**13**(5):662-670

[11] Abdulraheem MI, Ojeniyi SO, Charles EF. Integrated application of urea and sawdust ash: Effect on soil chemical properties, plant nutrients and sorghum performance. International Organization of Scientific Research-Journal of Agriculture and Veterinary science (IOSR-JAVS). 2012;**1**(4):38-41

[12] Abdulraheem MI et al. Comparative performance of different varieties of maize under organic, inorganic and combined (organic and inorganic) fertilization. Agricultural Studies International Technology and Science Publications. 2018;**2**(2):20-29

[13] Ojeniyi SO, Oyatoye A, Abdulraheem MI. use of ash for soil fertility improvement: effect on Cowpea. Nigeria Journal of Soil Science. 2017;**27**:216-221

[14] Olutumise AI et al. Determinants of health management practices' utilization and its effect on poultry farmers' income in Ondo State, Nigeria. Sustainability. 2023;**15**(3):2298

[15] Iderawumi AM et al. Innovative techniques of operating school farm. Farming and Management. 2021;**6**(1):21-28

[16] Moraes ER de, Mageste JG, Lana RMQ, Silva RV da, Camargo R de. Sugarcane: Organo-Mineral Fertilizers and Biostimulants. InTech; 2018. DOI: 10.5772/intechopen.71493

[17] Laird DA et al. Impact of biochar amendments on the quality of a typical

Midwestern agricultural soil. Geoderma. 2010;**158**(3-4):443-449

[18] Abdulraheem MI, Omogoye AM, Charles EF. Influence of NPK fertilizer and cow dung manure application on soil chemical properties, growth and yield of sweet potato (Ipomoea batatas). European Modern Studies Journal. 2018;**2**(5):1-7

[19] Ichami SM et al. Fertilizer response and nitrogen use efficiency in African smallholder maize farms. Nutrient cycling in agroecosystems. 2019;**113**:1-19

[20] Liu L et al. Combined application of organic and inorganic nitrogen fertilizers affects soil prokaryotic communities compositions. Agronomy. 2020;**10**(1):132

[21] Ayeni LS. Effect of combined cocoa pod ash and NPK fertilizer on soil properties, nutrient uptake and yield of maize (Zea mays). Journal of American Science. 2010;**6**(3):79-84

[22] Iderawumi AM. Growth and yield responses of okra (Abelmoscus esculentum L.) as influenced by sawdust ash and ammonium nitrate. Sumerianz Journal of Agriculture and Veterinary. 2018;**1**(1):8-13

[23] Ayeni LS, Adetunji MT. Integrated application of poultry manure and mineral fertilizer on soil chemical properties, nutrient uptake, yield and growth components of maize. Nature and Science. 2010;**8**(1):60-67

[24] Carvalho RP et al. Organomineral fertilization on the chemical characteristics of Quartzarenic Neosol cultivated with olive tree. Scientia Horticulturae. 2014;**176**:120-126

[25] Pawlett M, Deeks LK, Sakrabani R. Nutrient potential of biosolids and urea derived organo-mineral fertilisers in a field scale experiment using ryegrass (Lolium perenne L.). Field Crops Research. 2015;**175**:56-63

[26] Bouhia Y et al. Conversion of waste into organo-mineral fertilizers: current technological trends and prospects. Reviews in Environmental Science and Bio/Technology. 2022;**21**(2):425-446

[27] Zainab MA et al. Effects of organic and inorganic fertilizers on the growth of NH-Ae 47-4 variety of okra. Journal of Applied Sciences and Environmental Management. 2016;**20**(1):201-206

[28] Silva SM et al. Organo-mineral fertilization effects on biomass and essential oil of lavender (Lavandula dentata L.). Industrial Crops and Products. 2017;**103**:133-140

[29] Farrar MB et al. Short-term effects of organo-mineral enriched biochar fertiliser on ginger yield and nutrient cycling. Journal of Soils and Sediments. 2019;**19**:668-682

[30] Adeniyan ON, Ojeniyi SO. Comparative effectiveness of different levels of poultry manure with NPK fertilizer on residual soil fertility, nutrient uptake and yield of maize. Moor Journal of Agricultural Research. 2003;**4**(2):191-197

[31] Abdulraheem MI, Fudzagbo J, Abdulazeez RA, Oyetoro BA. Rapid advancement on integrated application of sawdust ash manure and urea fertilizer on soil chemical properties and sesame (*Sesamum Indicum L.*) performance. International Scientific Survey Journal. 2020;**3**(2):1-8

[32] Lin W et al. The effects of chemical and organic fertilizer usage on rhizosphere soil in tea orchards. PloS One. 2019;**14**(5):e0217018

[33] Wahab A et al. Interactions of metal-based engineered nanoparticles

with plants: An overview of the state of current knowledge, research progress, and prospects. Journal of Plant Growth Regulation. 2023;**42**(4):1-21

[34] Oladapo A, Afolami CA. Assessment of organic fertilizer usage by vegetable farmers in Ondo State South West, Nigeria. African Journal of Agricultural Research. 2021;**17**(3):371-377

[35] Abdi G et al. Cyanobacteria for marine-based biomolecules. In: Biomanufacturing for Sustainable Production of Biomolecules. Springer; 2023. pp. 189-209. DOI: 10.1007/978-981-19-7911-8_10

[36] Fudzagbo J, Abdulraheem MI. Vermicompost technology: impact on the environment and food security. Agriculture and Environment. 2020;**1**(1):87-93

[37] Gomiero T, Paoletti MG, Pimentel D. Energy and environmental issues in organic and conventional agriculture. Critical Reviews in Plant Sciences. 2008;**27**(4):239-254

[38] Jinal NH, Amaresan N. Evaluation of biocontrol Bacillus species on plant growth promotion and systemic-induced resistant potential against bacterial and fungal wilt-causing pathogens. Archives of Microbiology. 2020;**202**(7):1785-1794

[39] Han Y et al. Global and regional estimation of net anthropogenic nitrogen inputs (NANI). Geoderma. 2020;**361**:114066

[40] Calabi-Floody M et al. Smart fertilizers as a strategy for sustainable agriculture. Advances in Agronomy. 2018;**147**:119-157

[41] Bley H et al. Nutrient release, plant nutrition, and potassium leaching from polymer-coated fertilizer. Revista Brasileira de Ciência do Solo. 2017;**41**:1-11

[42] Su R et al. Minimalizing non-point source pollution using a cooperative ion-selective electrode system for estimating nitrate nitrogen in soil. Frontiers in Plant Science. 2022;**12**:810214

[43] Himics M et al. Does the current trade liberalization agenda contribute to greenhouse gas emission mitigation in agriculture? Food Policy. 2018;**76**:120-129

[44] Lee R. The outlook for population growth. Science. 2011;**333**(6042):569-573

[45] Visser W. Sustainable Frontiers: Unlocking Change through Business, Leadership and Innovation. 1st ed. London: Routledge; 2015. DOI: 10.4324/9781351284080

[46] Blanco M. Supply of and access to key nutrients NPK for fertilizers for feeding the world in 2050. Madrit: UPM; 2011

[47] Omogoye AM, Abdulraheem MI. Soil properties and performance of Sorghum (Sorghum bicolor L Moench) in response to sole and combined applications of NPK, cocoa pod husk and cocoa pod ash manures. In: Proceedings of the 40th Annual Conference of Soil Science Society of Nigeria, Calabar. pp. 341-346

[48] Shiyam JO, Binang WB. Effect of poultry manure and plant population on productivity of fluted pumpkin (Telfaiaria occidentalis Hook F.) in Calabar, Nigeria. Journal of Organic Systems. 2013;**8**(2):29-35

[49] Ojetayo AE et al. Effect of fertilizer types on nutritional quality of two cabbage varieties before and after storage. Journal of Applied Biosciences. 2011;**48**(5):3322-3330

[50] Senjobi BA, Ande OT, Akindolie MS. Performance of Abelmoschus Esculentus (L Moech) as influenced by different

organic manure amendment in Yelwa enclave of Ogun State Nigeria. Envirotropica: International Environ Tropical Journal. 2012;**8**:60-72

[51] Ramos LA et al. Effect of organo-mineral fertilizer and poultry litter waste on sugarcane yield and some plant and soil chemical properties. African Journal of Agricultural Research. 2017;**12**(1):20-27

[52] Oyedeji S et al. Effect of NPK and poultry manure on growth, yield, and proximate composition of three Amaranths. Journal of Botany. 2014;**6**:1-6

[53] Megali L, Glauser G, Rasmann S. Fertilization with beneficial microorganisms decreases tomato defenses against insect pests. Agronomy for Sustainable Development. 2014;**34**:649-656

[54] Pant LP, Adhikari B, Bhattarai KK. Adaptive transition for transformations to sustainability in developing countries. Current Opinion in Environmental Sustainability. 2015;**14**:206-212

[55] Gattinger A et al. Enhanced top soil carbon stocks under organic farming. Proceedings of the National Academy of Sciences. 2012;**109**(44):18226-18231

[56] Seufert V, Ramankutty N, Foley JA. Comparing the yields of organic and conventional agriculture. Nature. 2012;**485**(7397):229-232

[57] Akinrinde EA, Okeleye KA. Short- and long-term effects of sparingly soluble phosphates on crop production in two contrasting Nigerian Alfisols. West African. Journal of Applied Ecology. 2005;**8**(1)

[58] Atere CT, Olayinka A. Effect of organo-mineral fertilizer on soil chemical properties, growth and yield of soybean. African Journal of Agricultural Research. 2012;**7**(37):5208-5216

[59] Abdulraheem MI, Ojeniyi SO. Combined application of urea and sawdust ash in okra production effects on yield and nutrients availability. Nigerian Journal of Soil Science. 2015;**25**:146-154

[60] Naqvi SMZA et al. Green synthesis of silver nanoparticles and anti-oxidant activity in plants under semiarid condition - a review. Pakistan Journal of Weed Science Research. 2022;**28**(3)

[61] Ojeniyi SO et al. Field study of effect of organomineral fertilizer on maize growth, yield soil and plant nutrient composition in Ilesa, Southwest Nigeria. Nigerian Journal of Soil Science. 2009;**19**(1):11-16

[62] Ojeniyi S et al. Effect of organic, organomineral and NPK fertilizer on nutritional quality of Amaranthus in Lagos, Nigeria. Nigerian Journal of Soil Science. 2009;**19**(2):129-134

[63] Olowokere FA. Respond of pepper and tomato intercrop to different rates and methods of application of poultry based organomineral fertilizer. In: Proc. 29th Annual Conf. of Soil Sci. Soc. of Nigeria. 6-10 Dec 2004. Abeokuta: UNAAB; 2004. pp. 186-191

[64] Osman KT. Soil Degradation, Conservation and Remediation. Vol. 820. Switzerland: Springer; 2014

[65] Abdulraheem MI, Lawal SA. Combined application of ammonium nitrate and goat manure: effects on soil nutrients availability, okra performance and sustainable food security. Open Access Research Journal of Life Sciences. 2021;**1**:021-028

[66] Fagbola O, Ogunbe PW. Growth and yield response of some maize cultivars to organomineral fertilizer application

in simulated degraged soil under greenhouse conditions. Nigerian Journal of Soil Science. 2007;**17**:87-93

[67] Crusciol CAC et al. Organomineral fertilizer as source of P and K for sugarcane. Scientific Reports. 2020;**10**(1):5398

[68] Shaji H, Chandran V, Mathew L. Organic fertilizers as a route to controlled release of nutrients. In: Controlled Release Fertilizers for Sustainable Agriculture. US: Elsevier; 2021. pp. 231-245

[69] Okhumata S, Uzezi E, Idowu R. Effects of integrated application of inorganic and organic fertilizer on properties of soil planted with rice. World Journal of Advanced Research and Reviews. 2021;**11**(2):117-123

[70] Sharma S et al. Effect of organic manures on growth, yield, leaf nutrient uptake and soil properties of kiwifruit (Actinidia deliciosa Chev.) cv. Allison. Plants. 2022;**11**(23):3354

[71] Ahmed S et al. Molecular communication network and its applications in crop sciences. Planta. 2022;**255**(6):128

[72] Jakhar AM et al. Nano-fertilizers: a sustainable technology for improving crop nutrition and food security. NanoImpact. 2022:100411

[73] Udo EJ et al. Manual of Soil, Plant and Water Analysis. Lagos: Sibon Books, Publishers Ltd., Nigeria; 2009. p. 183

[74] Matveeva VA, Smirnov YD, Suchkov DV. Industrial processing of phosphogypsum into organomineral fertilizer. Environmental Geochemistry and Health. 2022;**44**(5):1605-1618

[75] Egodawatta WCP, Sangakkara UR, Stamp P. Impact of green manure and mineral fertilizer inputs on soil organic matter and crop productivity in a sloping landscape of Sri Lanka. Field Crops Research. 2012;**129**:21-27

[76] Du NXT. Effects of green manures during fallow on moisture and nutrients of soil and winter wheat yield on the Loss Plateau of China. Emirates Journal of Food and Agriculture. 2017:978-987

[77] Flores-Felix JD et al. Future perspective in organic farming fertilization: Management and product. In: Organic Farming. Elsevier; 2019. pp. 269-315

[78] Makinde EA, Ayeni LS, Ojeniyi SO. Morphological characteristics of amaranthus cruentus L. as influenced by kola pod husk, organomineral and NPK fertilizers in Southwestern Nigeria. New York Science Journal. 2010;**3**(5):130-134

[79] Olowoake AA et al. The Effect of farmyard manure and urea on grain yield and agronomic characteristics of maize (Zea mays). Ghana Journal of Agricultural Science. 2022;**57**(1):83-96

[80] Buam I, Hussain N. Chapter 5. Generation and Management of Agricultural Waste-Worldwide View. Chief Editor. 2021. p. 63. DOI: 10.13140/RG.2.27269.47847

[81] Ojeniyi SO, Awanlemhen BE, Adejoro SA. Soil plant nutrients and maize performance as influenced by oil palm bunch ash plus NPK fertilizer. Journal of American Science. 2010;**6**(12):456-460

[82] Usman M, Madu VU, Alkali G. The combined use of organic and inorganic fertilizers for improving maize crop productivity in Nigeria. International Journal of Scientific and Research Publications. 2015;**5**(10):1-7

[83] Ayeni LS, Adeleye EO, Adejumo JO. Comparative effect of organic, organomineral and mineral fertilizers on soil properties, nutrient uptake, growth and yield of maize (Zea mays). International Research Journal of Agricultural Science and Soil Science. 2012;**2**(11):493-497

[84] Oyekunle OJ, Abosede OT. Growth, yield and nutritional compositions of fluted pumpkin (Telfairia occidentalis Hook. f.) as affected by fertilizer types in Ogbomoso, south west Nigeria. Journal of Applied Biosciences. 2012;**56**:4080-4088

Chapter 3

Organic Agriculture: Global Challenges and Environmental Impacts

Tazbeen Tabara Nitu, Tasnim Binte Rayhan Promi and Syed Aflatun Kabir Hemel

Abstract

Agriculture has been intensified for years to provide food and nutrition security for the growing world population in emerging nations. Conventional methods, especially the widespread and ineffective use of N fertilizer, increase the cost of agricultural production and contribute to environmental degradation, including the production of greenhouse gases, ammonium volatilization, groundwater pollution, etc. In long term, intensive agricultural practices cause depletion of soil productivity by limiting its functions such as biomass production, carbon sequestration, etc. which may threaten our sustenance. In this crisis scenario, for sustainable intensification, organic agriculture has been proposed as a one-stop solution with enormous benefits. Many researchers have proved that organic fertilizer application in agriculture improves soil health by enhancing biogeochemical properties. Moreover, organic agriculture has been claimed as climate-smart agriculture. Contrarily, it is clear that organic fertilizer (as compost, manure, etc.) may cause heavy metal pollution and that organic particles may pollute the atmosphere. There is controversy on how OA affects biodiversity. Although just 1.5% of all land is organically grown (excluding organic non-agricultural land), the world has recently seen a surge in interest in OA. This chapter will concentrate on the present incarnation of organic agriculture worldwide, including advancements, benefits, and environmental consequences.

Keywords: organic agriculture, environment, GHG emission, global aspects, scientific debates

1. Introduction

In the twenty-first century, one of the biggest challenges that have been addressed globally is "food security". The Food and Agricultural Organization has reported that food production will have to increase by 70% for a 2.1 billion additional global population by 2050 [1]. Between 1970 and 2010, the world's grain production double to 2.5 billion tons, with an increase in the average yield of 1600 to 3030 kg per hectare [2]. However, this agricultural intensification is practiced worldwide mainly through the increase in global fertilizer usage from 32 to 106 Mt yr^{-1} [3]. In addition, Synthetic N fertilizer widely used for high-yield production contributes 38.8% of N_2O emission, a GHG with 265 times more global warming potential than CO_2 over a 100 years period [4]. Traditional farming methods for increasing agricultural productivity increase environmental problems such biodiversity loss, rapid soil erosion and deterioration, eutrophication, groundwater pollution, adverse effects of pesticides on wildlife and humans, etc. It is already abundantly evident that current "non-organic" agricultural methods, which mainly rely on intensive synthetic fertilizer input and pesticide application, cannot meet the task of providing enough food while maintaining a sustainable ecology [5]. Hence, ecological methods to sustainable intensification are the all-in-one solution for the production of food in the future. Nowadays, "organic agriculture" is recognized as a method of ecological intensification on a global scale [6].

Contrary to traditional farming, organic agriculture (OA) produces food in quantities insufficient for mass consumption, but it is environmentally benign and produces as nutritious or more nutrient-dense foods with less (or no) pesticide residues. Generally, when a natural ecosystem converts to farming practices, the topsoil losses about 20–40% of SOC following cultivation in the first few years [7]. OA helps to sequestrate SOC in soil and improves soil health by enhancing soil biological activities, nutrient dynamics, and availability [8]. It has been scientifically demonstrated that organic soils have 44% more long-term carbon storage than conventionally managed soils by evaluating over a thousand soil samples from farms maintained organically and conventionally in 48 states of the USA [9]. According to a recent research, organic farming reduces soil erosion by delivering less mean sediment than conventional farming [10]. OA helps to mitigate GHG emissions by reducing nitrous oxide by 50% and methane emissions by 70% [11]. OA may be viewed as a climate-smart agricultural strategy for reducing the consequences of climate change and enhancing soil health and quality, including nutrient and water retention, as well as boosting agricultural output.

Organic farming is a fast-growing sector. Since people are becoming health-conscious day by day, the demand for organic food is skyrocketing in the local market. The fact is, OA currently occupies only 1.5% of the global total agricultural land [12]. There is a concern that it is difficult to predict the impacts of widespread OA adoption due to mounting evidence that organic fertilizers can include significant levels of trace metals including Cu, Ni, Cr, Zn, As, Pb, and Cd [13]. Application of organic manure above 6.25 t ha^{-1} per season may contribute to greater CH_4 and N_2O emissions in the rice ecosystem [14]. Yet, the use of animal manures, the use of natural pesticides and fertilizers, the management of postharvest residues, irrigation, and tillage activities may all have a detrimental influence on the environment [15]. Further research is undergoing globally to understand the environmental impacts of long-term adaptation of OA on a large scale. This review will discuss the present scenario of OA worldwide and the current reports in the literature regarding its environmental impacts under the long-term adaptation in different management systems.

2. Organic agriculture (OA)

Considering ecological principles as the base for crop production, the concept of organic agriculture is analogous worldwide. The General Assembly of IFOAM defines "organic Agriculture as a production system that sustains the health of soils, ecosystems, and people. It relies on ecological processes, biodiversity and cycles adapted to local conditions, rather than the use of inputs with adverse effects. OA combines tradition, innovation, and science to benefit the shared environment and promote fair relationships and good quality of life for all involved". Although there are many meanings and interpretations of OA, common techniques to maintain soil fertility and produce high-quality products in OA include: (1) implementing suitable rotation programs; (2) adding composts; (3) using physical, mechanical, and biological mechanisms to control diseases, pests, and weeds; and (4) implementing organic practices in the feed and livestock production [16]. Cover crops and green manures, however, are crucial to the success of OA because they perform a variety of vital tasks such as fixing nitrogen (N), contributing organic matter (OM), and offering home for beneficial species [17]. Efficiency, environmental impact, economic feasibility, and social well-being are the four main sustainability criteria that Reganold and Wachter [18] have used to analyze the success of organic farming. They came to the conclusion that, although having lower yields, OA provides more environmental services and societal benefits. In order to create sustainable farming systems, which probably would include a combination of organic and non-organic methods, organic farming techniques must be taken into account.

2.1 Global scenario of OA

According to a study on "The World of Organic Agriculture – Statistics and Emerging Trends 2021", approximately 72.3 million hectares were under organic agricultural management worldwide in 2019. This is over 1.1 million hectares or 1.6% more than the previous year. 2019 saw a more than six-fold rise in the proportion of agricultural land that was organic compared to 1999. Half of the world's organic farming acreage is in Oceania. Around 23% of the world's organic agricultural land is in Europe, a region that has seen steady expansion in this area over the years, followed by South America with 12% (**Figure 1**). Organic farming is not growing as much in certain continents, including Asia and Africa, due to larger populations and more food demand, as organic farmers do not receive as much produce as conventional farmers do. In 2019 the agricultural land has significantly decreased in Asia (−7.1%) due to the drop in organic farmland in China. And Oceania (−0.3%) respectively.

At 35.69 million hectares, Australia has the most organic land worldwide. Argentina comes in second place with 3.67 million hectares, while Spain comes in at more over two million hectares. Yet, grazing land makes up 97% of Australia's agriculture, according to estimates. Almost 80% of all agricultural area used for organic production is found in the top 10 nations (**Figure 2**). In terms of percentage, 1.5% of all organic land in the

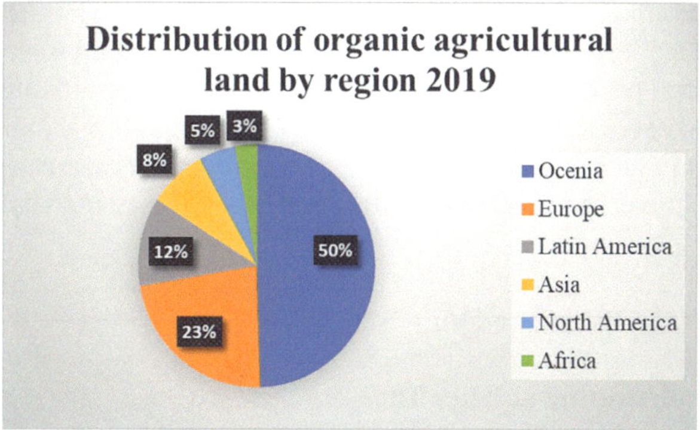

Figure 1.
2019 global organic agricultural land distribution by region (Source: FiBL survey 2021).

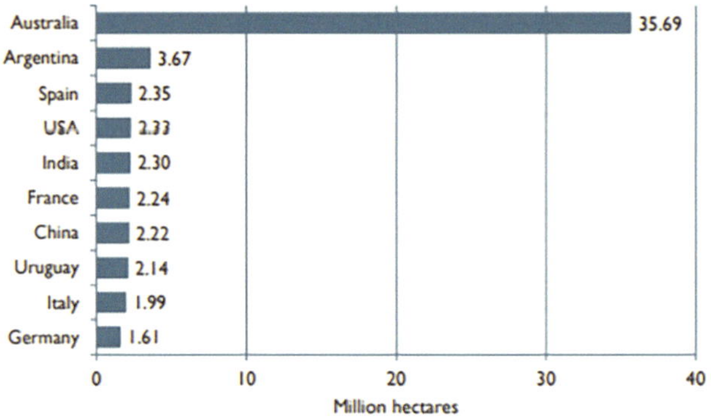

Figure 2.
The 10 countries with the largest area of organic agricultural land in 2019 (Source: FiBL survey 2021).

globe is organic. Liechtenstein has the greatest percentage of its agricultural land managed organically, at 41%. More than 10% of the organic share is held by the majority of European nations. Organic land expansion has accelerated in numerous nations, including India (18.6%) and Ukraine (54.6%) A total of 35 Mha of non-agricultural land—including land used for forestry, aquaculture, wild collecting, and grazing regions on non-agricultural land—is organic [12]. This data posits that there may be a good probability of increasing the popularity of organic farming among farmers in areas with lower population densities, strain on the land, overall food demand and shortfall.

Further, there are more than 3 million organic producers worldwide. Asia, Africa, and Europe together account for more than 91% of global producers. The country with the greatest number of organic producers is India (1.36 million), followed by Uganda and Ethiopia. 1–2% of all food sales globally are organic. According to the nation, future growth is anticipated to range from 10 to 50% yearly [19]. Almost 106 billion euros worth of organic food was sold at retail in total. The United States is the nation with the biggest market for organic food.

When compared to conventional agriculture, the adoption of OA is economically competitive. For instance, OA is more lucrative (22–35%) than conventional agriculture due to the higher cost of organic goods compared to their non-organic equivalents (20–24%). Though with lower organic yields of 10–18%, breakeven premiums for organic earnings must still be at least 5–7% to meet conventional profits. Reduced environmental costs (negative externalities) and improved ecosystem services brought about by the use of appropriate farming methods may also support OA financially [20].

In conclusion, while being modest globally, agricultural land is expected to grow. Organic agricultural practices have several advantages for sustainability and can help ensure global food security [20]. Moreover, it is becoming increasingly evident that OA can be crucial in tackling issues including land and soil erosion, climate variability, hunger relief, poverty reduction, health, and biodiversity management [21].

3. Environmental aspects of OA

3.1 Organic agriculture and soil health

One of the keystones of organic agriculture is to ameliorate and maintain soil health. Soil organic carbon (SOC) strongly affects soil health and functionality. From an agronomic and ecological perspective, proper SOC stock management is crucial for crop productivity in organic farming [22]. The main practices of OA including returning plant residues and manure application, and/or integrating cover crops into the system reduce SOC losses from topsoil, and either maintain SOC or increase SOC stocks [23]. For example, a meta-analysis contrasting conventional and organic farming methods under Mediterranean croplands revealed that the rate of SOC sequestration in topsoil (average soil depth 19.4 cm) rose by 0.97 Mg C ha^{-1} yr^{-1} under OA in comparison to those under conventional management [24]. According to a recent study, decreased tillage in organic farming raised cumulative SOC stocks by 1.7%, or 1.5 Mg ha^{-1} (0–50 cm), and 3.6%, or 4.0 Mg ha^{-1} (0–100 cm), compared to traditional tillage, with estimated mean C sequestration levels of 0.09 and 0.27 Mg ha^{-1} yr^{-1}, respectively [25]. An optimized conventional system with some "organic" tactics, like cover crops, crop rotation, and mulches, but without using mineral fertilizers, demonstrated higher sequestration in one research [26]. Yet, in a different study, only the biodynamic system with a high livestock density showed a noticeable advantage [27]. Using basic default values for carbon sequestration in forests, **Table 1** compares the sequestration rates in organic and conventional agriculture systems and demonstrates the efficiency of the organic system in the USA. These results show that in the studied locations, organic systems consistently retain more carbon than conventional systems. Yet, compared to organic farming, conventional farming paired with forests sequesters more carbon. Also, the relative production in organic farming is satisfactory, but more acreage is required for the profound respect.

Higher SOM concentrations in the topsoil layer on arable land under organic farming can positively influence soil stability, according to a number of studies [9, 29]. According to the findings of William et al. [30], organic farming reduces bulk density by 3% more than the conventional farming system. Additionally, it enhances water infiltration capacity and increases aggregate stability by 50% under green manure application than conventional farming. Seitz et al. [10] demonstrated that organic agricultural practices decline soil erosion by decreasing mean sediment delivery by 30% more than modern agricultural practices. It is revealed that conservation tillage may decrease soil erosion and improve soil structure [31]. In contrast, organic farming with minimum tillage may increase soil compaction in long term and decreases the earthworm population in the soil [32].

The incorporation of manure or compost in organic agriculture helps to avoid exceeding soluble nutrients release such as nitrogen and phosphorus, at the same time, contributing an essential source of carbon for the growth and activity of soil organisms. Organic N fertilizers significantly increase the potential for nitrification, nitrite oxidation, and denitrification [33]. Moreover, with particular management practices, seasonal variation affects soil N mineralization and potentially synchronizes soil available N supply with demand for cash crops, productivity, nutrient (N and P) loading, risk of losses, and nutrient use efficiency from organic manufacturing systems can vary [34]. The best strategy to increase microbial biomass and its activity is by amending the soil with organic materials [35, 36]. Organic fertilizer increases the microbial exoenzymatic activities involved in the mineralization of C, N, and P [37]. Certain microbial communities engaged in symbiotic nitrogen fixation and microbial stimulation of nutrient absorption at root surfaces are made more effective by OA [38].

OA ensures soil sustainability by improving soil bio-physiochemical activities. Though, caution is required for soil health-related issues in organic agriculture, as the on-farm management efficiency differs. Nevertheless, for eco-friendly agricultural production organic agriculture in the standard way can be considered a large potent system.

Reference		C sequestration rate (kg ha^{-1} yr^{-1})		Relative productivity of organic system		Potential carbon sequestration by afforestation (in kg ha^{-1} yr^{-1})	Total sequestration C (kg ha^{-1} yr^{-1})	
		Organic	Conventional	Relative yields	Additional area demand ha^{-1}		Conventional + forest	Advantage/disadvantage of organic system
Switzerland[1]	Min	−123	−207	84%	1.2	1700	98	−221
	Max	42	−207	83%	1.2	1700	117	−75
USA Pennsylvania[2]	Min	857	217	92%	1.1	3300	464	393
	Max	1218	217	97%	1.0	3300	309	909

[1] [27].
[2] [28].

Table 1.
A comparison of the sequestration potential of a "organic" and a "conventional + forest" system is carried out based on the measured sequestration rates of four comparable field studies in Europe and the USA.

3.2 Greenhouse gas fluxes under OA

It is assumed that agriculture is a major contributor to GHG emissions and is vulnerable to climate change. OA is suggested as a one-stop solution to mitigate emissions. Some past studies have produced controversial results regarding the potential of OA for mitigation [39]. For example, The amount of greenhouse gases (GHGs) emitted from agricultural output as a whole and the intensity of GHGs emitted per hectare of agricultural land are both rising in the USA, according to McGee [40]. According to Williams et al. [41] lower yields and higher rates of nitrate leaching counteract the reduced input consumption in most organic cropping systems in England, which create equivalent or higher GHG emissions per ton of crop than conventional processes.

By using meta-analysis, structural factors impacting GHG emissions for traditional and open architecture systems have been studied [42]. As compared to traditional farming, OA had fewer GHG emissions in nearly two-thirds of the 195 observations. In terms of GHG emissions for farms growing field crops, dairy products, and mixed crops, OA outperformed conventional farming. On farms that raised animals, grew vegetables, or produced fruit, OA was less likely to be more effective in terms of GHG emissions. Yet, the unit or foundation of measurement had a significant impact on how well OA reduced GHG emissions. Due to yield variations, output-based (ratio/Mg) metrics greatly lessened the superiority of GHG emissions impacts for OA compared to area-based (ratio/ha) measures. The fact that most studies were from Europe and did not take into account nutritional spillover effects in conventional-organic conversions was one of the drawbacks of this meta-analysis [42].

Using a life-cycle analysis, Smith et al. [39] evaluated the effects on net GHG emissions of a 100% switch to organic food production in England and Wales. He came to the conclusion that organic farming reduces direct GHG emissions, but when increased overseas land usage to make up for local supply shortages is taken into account, net emissions are higher (**Figure 3**).

3.2.1 CO_2 emission

Under irrigated conventional and irrigated OA management of common beans in the Mediterranean climate, Kontopoulou et al. [44] measured CO_2 emissions. In comparison to conventional management, OA had higher cumulative CO_2 emissions over the 84-day cropping period (2.5 and 2.8 Mg CO_2 ha^{-1} for high and low salinity irrigation water, respectively) (2.1 and 2.3 Mg CO_2 ha^{-1} for high and low-salinity irrigation water, respectively). Compost and other organic manures can improve soil C stocks [45] but may also result in higher CO_2 emissions [46]. Also, the use of compost may have improved the following factors: (1) soil structure and pore space continuity; (2) root penetration and gas and water movement; (3) root exudation and, therefore, a microbial activity that may have increased microbial respiration in the rhizosphere [44]. The global warming potential (GWP) per 1 m^2 of land in organic farming was determined to be 0.12 kg CO_2 eq., which was three times lower than in the conventional system, according to an environmental impact evaluation of organic and conventional leek production methods in Belgium (0.36 kg CO_2 eq.) [47]. The GWP in organic and conventional herbaceous farming systems was also studied in Spain. The organic system greatly reduced GHG emissions (between 35.9 and 64.7% and 16.3 and 41.9%, respectively) [24].

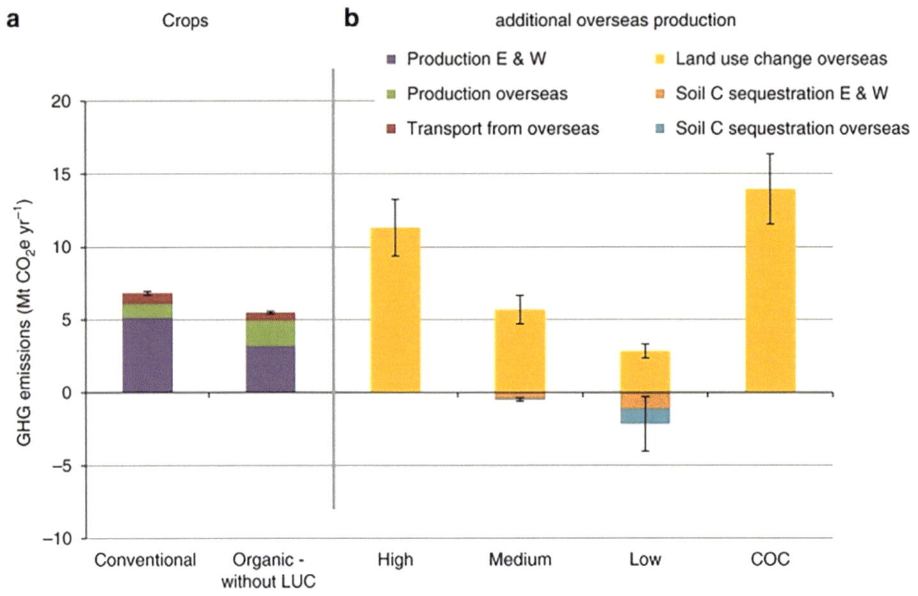

Figure 3.
Total greenhouse gas emissions (GHG) from farming in conventional and organic systems in England and Wales (E & W). (a) For food crops for human consumption both from home and overseas production. (b) Additional net emissions due to soil C sequestration (CS) and overseas land use changes (LUC) to compensate for shortfalls in home production: high = all LUC by conversion from grassland, no CS; medium = 50% of LUC by conversion from grassland, moderate CS; low = 25% of LUC by conversion from grassland, high CS; COC = carbon opportunity cost of Searchinger et al. [43] (Methods).

3.2.2 CH_4 emission

Based on the activity of certain CH_4^- and ammonium oxidizing bacteria and site-specific circumstances, well-aerated soils have the ability to behave as CH_4 sinks. A meta-study was carried out to compare area-scaled CH_4 emissions from organic and non-organic farming, however no appreciable differences between the two cultivation methods were discovered [2]. It has been shown that soils managed organically assimilate more CH_4 than soils managed conventionally (−0.61 versus −0.54 kg CH_4 ha^{-1} yr^{-1}; mitigating 20.2 and 18.0 kg CO_2 eq. ha^{-1}, respectively) [5]. The cause is that consistent treatment of piled cow dung boosts methanogenic archaea biomass and enzymatic activity [23]. Contrarily, rice paddies are a significant source of CH_4 emissions under both types of management, contributing 6023 kg CO_2 eq. ha^{-1} yr^{-1} under organic management and 4857 CO_2 eq. ha^{-1} yr^{-1} under conventional management. In the rice environment, emissions of CH_4 are caused by anaerobic microbial degradation of OM and organic fertilizer [48].

3.2.3 N_2O emission

The anticipation of decreased soil N_2O emissions is supported by the usually lower N input level for soils under OA compared to those under conventional management approaches [48]. In addition, Skinner et al. [3] observed that organic systems reduced N_2O emissions per hectare by 40.2% when compared to non-organic systems. It is expected that a significant portion of the ensuing N_2O emissions may begin to be effective beyond the vegetative period under investigation because of the delayed release of mineral N from organic sources [3]. As opposed to unamended and conventionally managed plots (0.64 kg N_2O ha^{-1}), manure-amended organic plots had greater cumulative N_2O emissions over the winter in plots that were grown with soybeans (1.63 kg N_2O ha^{-1}) [49]. Yet, under OA, denitrification efficiency improves most likely as a result of the following factors: (1) higher C inputs from grassland and fertilizer; (2) higher SOC and N contents; (3) bigger, more active microbial communities; and (4) variations in how the denitrifier communities work [50].

3.3 Organic fertilizer and heavy metal contamination

While if other activities (such air deposition or mining) might potentially result in soil buildup, organic fertilizer, such as manure and compost, can be the main source releasing heavy metals into the ecosystem [51]. There is a lot of data to support the claim that organic fertilizers (such manure, compost, and sludge) may contain significant amounts of trace metals including Cr, Ni, Cu, Zn, As, Cd, and Pb as well as have a high bio transfer potential [13]. Various trace metals have lower thresholds for non-agricultural usage and higher threshold limits for conventional and organic agriculture. For instance, 212 samples of Chinese organic fertilizers including cattle dung had Cr, As, Cd, and Pb contents that were 4.2%, 13.7%, 2.4%, and 1.4% higher than the recommended range, respectively [52]. According to some authors, organic amendments increase the mobility of Cd in soil, and copper follows a similar pattern in the soil horizon [53].

3.4 Water quality

Water is essential for both human and ecological wellbeing, as well as the long-term ecological and socioeconomic resilience of our food and agricultural systems. Water use and pollution are mostly the responsibility of the agriculture sector. Water usage and outflow in both plant and animal farms are major contributors to water pollution. The pesticides, fertilizers, and feed put to the ponds cause a multitude of contaminants to be carried in the wastewater [54]. Synthetic fertilizers cause numerous contaminants to combine with freshwater, but because they are not allowed in organic agriculture, the water quality is finally improved.

Organic farming employs a variety of methods to prevent soil erosion, water runoff, and nutrient leakage. Organic farming practices maintain nitrogen in crop plants used in rotation and stop nitrate leaching [55]. N leaching per area appears to be lessened in organic agriculture on average [56, 57]. When organic matter is added to the soil, soil organisms develop and reproduce more quickly and hold onto soil nitrogen in a more stable form [58]. Lower N inputs are often linked to lower N losses from organic systems [59]. In addition to that, higher levels of organic matter in organically maintained soils can increase their ability to store nitrogen [60, 61]. Non-leguminous plants are frequently used in organic farming as cover crops. It was observed that in cover-cropped areas with high levels of biological activity, the soil's capacity to retain nitrogen against leaching was also higher [62].

Effective remarks on phosphorus (P) leaching from organic versus conventional systems cannot be made due to the dearth of research [57, 63]. Because many organic inputs have poor N:P ratios, organic farmers frequently overfertilize for P while attempting to meet crop N requirements [64]. P surpluses in agricultural areas do not necessarily translate into P shortages since many soils depend on erosion rates and the absorption of P in organic inputs as well as having strong P buffering capacities (or P deficit in crop P restriction) [64].

Critics contend that some organic pesticides are more dangerous than synthetic pesticides [65], despite the common belief that organic management reduces pesticide burdens [18]. Despite having lower toxicity quotients, several organic pesticides, like sulfur and rotenone, can cause toxicity if they are used more often [66]. In contrast to that, organic farming generally uses integrated pest management [67] or less toxic pesticides [68]. It's probable that organic agriculture has less pesticide leaching than conventional agriculture does.

Utilizing methods that recycle and store nutrients within the farming system, organic farmers may prevent water pollution. When these procedures are carried out as a component of an integrated, systems-based strategy, they are both most efficient and long-lasting. In places where pollution is a real problem, organic agriculture is highly anticipated as a positive solution [69].

3.5 Air quality

By discharging nutrients, pathogens, heavy metals (including Cu), particulate matter, and toxic gases into the air, OA may contribute to air pollution [15]. For instance, the common practice of OA farming with traditional tillage may degrade air quality by dispersing fine dust and debris in the atmosphere. However, compared to conventionally managed farms, organically managed farms may have lower airborne particle concentrations due to usually less soil erosion [35]. Moreover, the processing and surface application of animal manures, as well as emissions from feedlots, have all been linked to air pollutants such particulate matter, oxides of N, C, and S, as well as NH_3, CH_4, and H_2S, volatile organic compounds, and pathogens [15]. Moreover, OA may degrade the air quality as a result of N losses from organic compost or green manures due to the volatilization of excess N that does not meet crop needs [35].

3.6 Impacts of OA on biodiversity

Comparing organic agriculture practices to traditional systems, several species and organism groupings may be found in greater numbers. According to certain studies, organic farming practices often boost biodiversity [70]. Most research comparing biodiversity in organic and conventional farming showed that organic farming has less of an impact on the ecosystem [56]. Organic farming raises agricultural landscape diversity, for instance by attracting carabid beetles [71, 72] vascular vegetation [73], or birds [74]. By utilizing toxic herbicides and pesticides that build up in ground and surface waterways, conventional agriculture contributes to the loss of biodiversity. This contaminates fisheries, pollinator habitats, and wildlife's natural habitats. Yet, organic farming preserves the health of ecosystems, soils, and people. Given that organic food is grown in accordance with nature, organic farmers are guardians of biodiversity on all scales, from seeds and worms to birds and bees. In research, it was found that predator abundances and predator-prey ratios were 20 times greater in organic fields than in conventional fields, the abundance of cereal aphids was five times lower in organic fields. This suggests that organic fields have a much better potential for biological pest management [75].

Organic agriculture ensures greater variety and abundance of floral species in the crop, crop perimeter, and uncultivated areas [76, 77]. In contrast with conventional farms, Shepherd et al. [78] discovered six times more species in the crop on organic agriculture. Hole et al. [79] found that rare agricultural species are more common on organic farms. Higher bee diversity [80], higher butterfly abundance [81], microbiological activity and biomass in soil [82] was found in organic farming compared to traditional one. Forfeiting the use of herbicides, using very little and fresher organic fertilizer, varying crop rotations with a higher proportion of clover grass, conservation tillage, and a more diverse farming structure are typical organic farming practices that most noticeably boost biodiversity. Nevertheless, organic farming seems to perform superior to traditional farming and offers significant environmental benefits like reducing water consumption, reducing carbon and ecological footprints, and stopping the use of dangerous chemicals and their spread across the environment and up the food system.

3.7 Controversial scientific debate on OA

In the scientific community, the effects of organic farming on the environment have been hotly debated for many years. Regarding how much organic farming can do to help with resource issues and if its promotion is a sensible course of action for addressing current socio-ecological issues, there are still divergent opinions. Two major objections have been made in the last 20 years or so that suggest organic agriculture is not superior to conventional agriculture in terms of its effects on the environment. As discussed further in detail below and illustrated schematically in **Figure 4**.

The yield gap between organic and conventional systems is therefore an important topic in these discussions, and it is mostly examined in light of a few important meta-studies [84–86]. Moreover, it has been stated that the idea of yield stability, which refers to the temporal unpredictability and dependability of output, is crucial when contrasting organic and conventional agriculture in terms of food security [87]. In general, it is becoming more and more apparent that the sustainability evaluation of various agricultural systems must take into account a wide range of complicated monetary and environmental interrelationships in addition to output [88, 89].

On the other hand, tradeoff assessments predominate in the debates of OA's environmental benefits. For instance, the common logic that local biodiversity benefits of OA are negated or even turn into disadvantages owing to larger land needs when extended dominates when it comes to the consequences on biodiversity. More comprehensive OA may result in other areas of land use intensification, which would have a net negative impact on the environment, such as increased greenhouse gas (GHG) emissions from LU change or biodiversity loss through habitat conversion [43, 90]. The assumption is that large-scale conversion would result in an increase in arable land to fulfill the unaltered (or rising) demand for agricultural goods because of yield gaps, which is why OA is condemned for increased nutrient leaching [91]. Several researchers also point out that a paucity of data, especially on water conservation, prevents drawing firm general conclusions. This is true even if studies have found decreased eutrophication potential in OA [92] and more efficient nutrient use on a specific region [93] owing to system boundaries [94]. Furthermore, it is claimed that evaluating the consequences of widespread OA adoption on biodiversity and GHG emissions is difficult since it is unclear how yield levels relate to land that is used for production or to convert natural habitat [18, 95, 96]. Besides which, there are arguments suggesting that comparison studies done to date may not account for OA's as-yet-unmeasured and perhaps beneficial benefits [97], such as the advantages of OA's broad continuous regions for biodiversity [98]. Hence, some writers contend that, in terms of integrated policy actions that concurrently address improvements in various environmental parameters, extending OA may be the most cost-effective method [99].

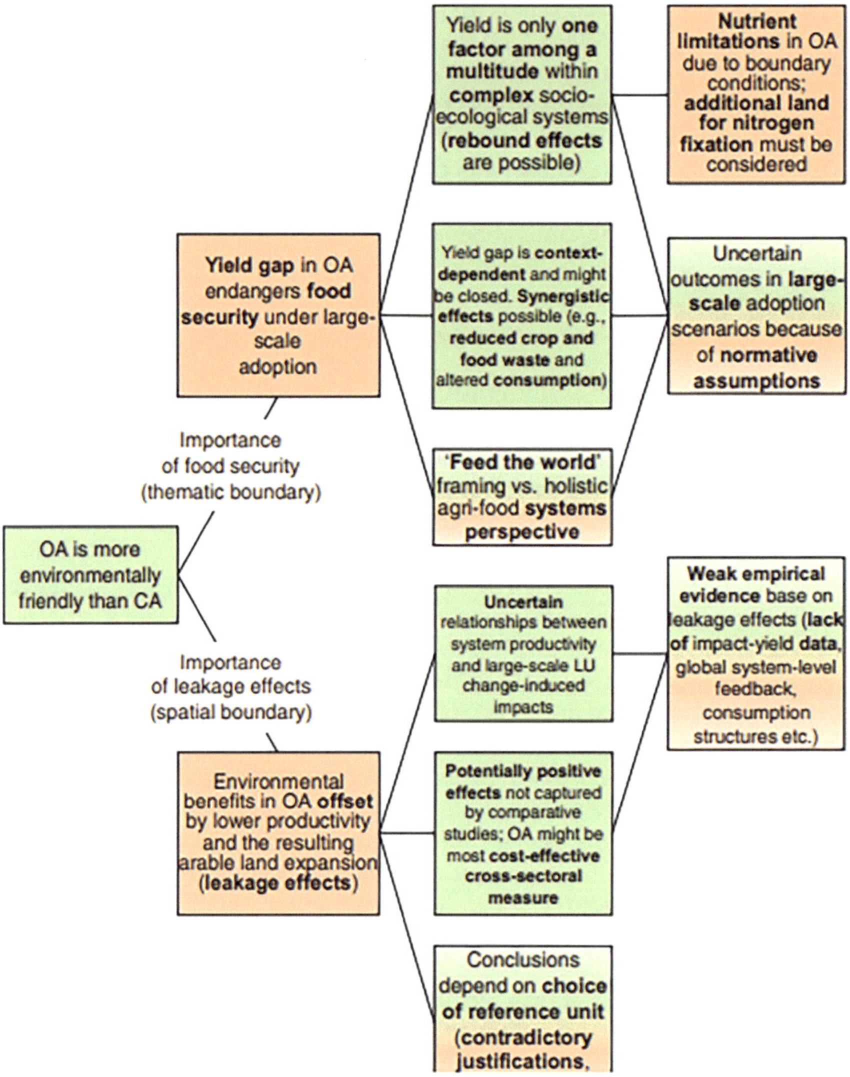

Figure 4.
There are two main discussion lines that can be traced back to two significant objections against the environmental advantages of organic farming (OA) [83].

4. Conclusion

Presently global agriculture is at a new threshold. It has become the leading source of environmental pollution in many countries. Nonetheless, OA systems are thought to be less harmful than typical conventional agriculture methods. Nevertheless, there is conflicting scientific data supporting these environmental benefits, and there have long been debates over OA. There is some evidence that the soils under OA reduce GHG emissions, although direct data are few and subject to bias due to geography. It enhances SOC stock and improves soil biogeochemical activities. In contrary, it is evident that organic fertilizer (such as compost, manure etc) may lead heavy metal contamination and air can be polluted by organic particulates. The effects of OA on biodiversity are

DOI: http://dx.doi.org/10.5772/intechopen.1001515

debatable. Though global organic agricultural land occupies only 1.5% organic share (except, organic nonagricultural land), switching to OA has gained new global momentum now a days. Since consumer demand for organic food continues to rise, more agricultural land will be used for organic farming in the future. Nevertheless, in order to more thoroughly evaluate the environmental effects of OA, long-term field studies in significant global agricultural regions using LCA are required.

Author details

Tazbeen Tabara Nitu[1]*, Tasnim Binte Rayhan Promi[1] and Syed Aflatun Kabir Hemel[2]

1 Department of Soil Science, Bangladesh Agricultural University, Mymensingh, Bangladesh

2 Bangladesh Jute Research Institute, Dhaka, Bangladesh

*Address all correspondence to: tazbeen127@gmail.com

References

[1] FAO. How to Feed the World in 2050. Food and Agriculture Organization; 2009. Available from: www.fao.org/3/a-ak542e/ak542e13

[2] Smith P et al. Chapter 11 - Agriculture, forestry and other land use (AFOLU). In: Climate Change 2014: Mitigation of Climate Change, IPCC Working Group III Contribution to AR5. Cambridge, United Kingdom and New York, NY, USA: Cambridge University Press; 2014

[3] Skinner C, Gattinger A, Krauss M, et al. The impact of long-term organic farming on soil-derived greenhouse gas emissions. Scientific Reports. 2019;**9**:1702. DOI: 10.1038/s41598-018-38207-w

[4] IPCC. Climate Change (2014) synthesis report. In: Pachauri RK, Meyer LA, editors. Core Writing Team. Geneva, Switzerland: Contribution of Working Groups I, II and III to the Fifth Assessment Report of the Intergovernmental Panel on Climate Change. IPCC; 2014. p. 151

[5] Lorenz K, Lal R. In: Donald LS, editor. Chapter three - Environmental Impact of Organic Agriculture In Advances in Agronomy. Vol. 139. Ohio, Columbia, USA: Academic Press; 2016. pp. 99-152

[6] Halberg N, Panneerselvam P, Treyer S. Eco-functional intensification and food security: Synergy or compromise? Sustainable Agriculture Research. 2015;**4**:126-139

[7] Davidson EA, Ackerman IL. Changes in soil carbon inventories following cultivation of previously untilled soils. Biogeochemistry. 1993;**20**:161-193

[8] Han WY, Wang DH, Fu SW, Ahmed S. Tea from organic production has higher functional quality characteristics compared with tea from conventional management systems in China. Biological Agriculture & Horticulture. 2018;**34**(2):120-131

[9] Ghabbour EA, Davies G, Misiewicz T, Alami RA, Askounis EM, Cuozzo NP, et al. National comparison of the total and sequestered organic matter contents of conventional and organic farm soils. Advances in Agronomy. 2017;**146**:1-35

[10] Seitz S, Goebes P, Puerta VL, et al. Conservation tillage and organic farming reduce soil erosion. Agronomy for Sustainable Development. 2019;**39**:4. DOI: 10.1007/s13593-018-0545-z

[11] IFOAM. Organic Farming, Climate Change Mitigation and Beyond: Reducing the Environmental Impacts of EU Agriculture. Brussels, Belgium: IFOAM; 2016. p. 10. Available from: https://www.organicseurope.bio/content/uploads/2020/06/ifoameu_advocacy_climate_change_report_2016

[12] IFOAM. The World of Organic Agriculture 2021: Statistics and Emerging Trends. Brussels, Belgium: FiBL and IFOAM; 2021. Available from: https://www.fibl.org/fileadmin/documents/shop/1150-organic-world-2021.pdf

[13] Lopes C, Herva M, Franco-Uría A, Roca E. Inventory of heavy metal content in organic waste applied as fertilizer in agriculture: Evaluating the risk of transfer into the food chain. Environmental Science and Pollution Research. 2011;**18**:918-939

[14] Sampanpanish P. Effect of organic fertilizer on CO2, CH4 and N2O emissions in a Paddy field. Modern Applied Science. 2012;**6**:239

[15] Udeigwe TK, Teboh JM, Eze PN, Stietiya MH, Kumar V, Hendrix J, et al. Implications of leading crop production practices on environmental quality and human health. Journal of Environmental Management. 2015;**151**:267-279

[16] Shi-ming M, Sauerborn J. Review of history and recent development of organic farming worldwide. Agricultural Sciences in China. 2006;**5**:169-178

[17] Abbott LK, Manning DAC. Soil health and related ecosystem services in organic agriculture. Sustainable Agriculture Research. 2014;**4**:116-125

[18] Reganold JP, Wachter JM. Organic agriculture in the twenty-first century. Nature Plants. 2016;**2**:15221. DOI: 10.1038/nplants.2015.221

[19] Mahanta D, Bisht JK, Kant L. Chapter 1 - Concept and global scenario of organic farming. Advances in Organic Farming: Agronomic Soil Management Practices. 2021:1-16

[20] Crowder DW, Reganold JP. Financial competitiveness of organic agriculture on a global scale. Proceedings of the National Academy of Sciences of the United States of America. 2015;**112**(24):7611-7616. DOI: 10.1073/pnas.1423674112

[21] Willer H, Lernoud J. The World of Organic Agriculture 2015: Summary. In: Willer H, Lernoud J, editors. The World of Organic Agriculture. Statistics and Emerging Trends 2015. Switzerland; Bonn, Germany: FiBL-IFOAM Report, Frick; 2015. pp. 24-30

[22] Brock C, Fließbach A, Oberholzer HR, Schulz F, Wiesinger K, Reinicke F, et al. Relation between soil organic matter and yield levels of nonlegume crops in organic and conventional farming systems. Journal of Soil Science and Plant Nutrition. 2011;**174**:568-575

[23] Gattinger A, Muller A, Haeni M, Skinner C, Fliessbach A, Buchmann N, et al. Enhanced top soil carbon stocks under organic farming. Proceedings of the National Academy of Sciences of the United States of America. 2012;**109**:18226-18231

[24] Aguilera E, Lassaletta L, Gattinger A, Gimeno BS. Managing soil carbon for climate change mitigation and adaptation in Mediterranean cropping systems: A meta- analysis. Agriculture, Ecosystems & Environment. 2013;**168**:25-36

[25] Krauss M et al. Reduced tillage in organic farming affects soil organic carbon stocks in temperate Europe. Soil & Tillage Research. 2022;**216**:105262. DOI: 10.1016/j.still.2021.105262

[26] Wells AT, Chan KY, Cornish PS. Comparison of conventional and alternative vegetable farming systems on the properties of a Yellow Earth in New South Wales. Agriculture, Ecosystems & Environment. 2000;**80**:47-60

[27] Fliessbach A, Oberholzer HR, Gunst L, Mäder P. Soil organic matter and biological soil quality indicators after 21 years of organic and conventional farming. Agriculture, Ecosystems & Environment. 2007;**118**:273-284

[28] Hepperly P, Douds J, Seidel R. The Rodale farming system trial 1981-2005: Long-term analysis of organic and conventional maize and soybean cropping systems. In: Raupp J, Pekrun C, Oltmanns M, Köpke U, editors. Long-term Field Experiments in Organic Farming. Bonn: International Society of Organic Agricultural Research (ISOFAR); 2006. pp. 15-32

[29] Six J, Elliott ET, Paustian K. Soil macroaggregate turnover and micro aggregate formation: A mechanism for C sequestration under no-tillage

agriculture. soil. Biology and Biochemistry. 2000;**32**:2099-2103. DOI: 10.1016/S0038-0717(00)00179-6

[30] Williams DM, Canqui HB, Francis CA, Galusha TD. Organic farming and soil physical properties: An assessment after 40 Years. Agronomy Journal. 2017;**109**(2):600-609. DOI: 10.2134/agronj2016.06.0372

[31] Zhang G, Chan K, Oates A, Heenan D, Huang G. Relationship between soil structure and runoff/soil loss after 24 years of conservation tillage. Soil & Tillage Research. 2007;**92**:122-128. DOI: 10.1016/j.still.2006.01.006

[32] Peigné J, Vian J-F, Payet V, Saby NPA. Soil fertility after 10 years of conservation tillage in organic farming. Soil & Tillage Research. 2018;**175**:194-204. DOI: 10.1016/j.still.2017.09.008

[33] Ouyang Y, Reeve JR, Norton JM. Soil enzyme activities and abundance of microbial functional genes involved in nitrogen transformations in an organic farming system. Biology and Fertility of Soils. 2018;**54**:437-450. DOI: 10.1007/s00374-018-1272-y

[34] Lynch DH. Nutrient cycling and soil health in organic cropping systems - Importance of management strategies and soil resilience Sustainable Agriculture. Research. 2015;**4**:80-88

[35] Gomiero T, Pimentel D, Paoletti MG. Environmental impact of different agricultural management practices: Conventional vs. organic agriculture. Critical Reviews in Plant Sciences. 2011;**30**:95-124

[36] Ros M, Hernandez MT, García C. Soil microbial activity after restoration of a semiarid soil by organic amendments soil. Biology and Biochemistry. 2003;**35**:463-469. DOI: 10.1016/S0038-0717(02)00298-5

[37] Francioli D, Schulz E, Lentendu G, Wubet T, Buscot F, Reitz T. Mineral vs. Organic amendments: Microbial community structure, activity and abundance of agriculturally relevant microbes are driven by long-term fertilization strategies. Frontier Microbiology. 2016;**7**:1446. DOI: 10.3389/fmicb.2016.01446

[38] Verbruggen E, Roling WFM, Gamper HA, Kowalchuk GA, Verhoef HA, van der Heijden MGA. Positive effects of organic farming on below-ground mutualists: Large-scale comparison of mycorrhizal fungal communities in agricultural soils. New Phytologist. 2010;**186**:968-979. DOI: 10.1111/j.1469-8137.2010.03230.x

[39] Smith LG, Kirk GJD, Jones PJ, et al. The greenhouse gas impacts of converting food production in England and Wales to organic methods. Nature Communications. 2019;**10**:4641. DOI: 10.1038/s41467-019-12622-7

[40] McGee JA. Does certified organic farming reduce greenhouse gas emissions from agricultural production? Agriculture and Human Values. 2015;**32**:255-263

[41] Williams AG, Audsley E, Sandars DL. Determining the Environmental Burdens and Resource Use in the Production of Agricultural and Horticultural Commodities. London: Defra; 2006

[42] Lee KS, Choe YC, Park SH. Measuring the environmental effects of organic farming: A meta-analysis of structural variables in empirical research. Journal of Environmental Management. 2015;**162**:263-274

[43] Searchinger TD, Wirsenius S, Beringer T, Dumas P. Assessing the

efficiency of changes in land use for mitigating climate change. Nature. 2018;**564**:249-253

[44] Kontopoulou CK, Bilalis D, Pappa VA, Rees RM, Savvas D. Effects of organic farming practices and salinity on yield and greenhouse gas emissions from a common bean crop. Scientia Horticulturae. 2015;**183**:48-57

[45] Paustian K, Lehmann J, Ogle S, Reay D, Robertson GP, Smith P. Climate smart soils. Nature. 2016;**532**(7597):49-57

[46] Ray RL, Singh VP, Singh SK, Acharya BS, He Y. What is the impact of COVID-19 pandemic on global carbon emissions? Science of the Total Environment. 2022;**816**:151503

[47] De Backer E, Aertsens J, Vergucht S, Steurbaut W. Assessing the ecological soundness of organic and conventional agriculture by means of life cycle assessment (LCA) - A case study of leek production. British Food Journal. 2009;**111**(10):1028-1061. DOI: 10.1108/00070700910992916

[48] Muller A, Aubert C. The potential of organic agriculture to mitigate the influence of agriculture on global warming - A review. In: Bellon S, Penvern S, editors. Organic Farming, Prototype for Sustainable Agricultures. Dordrecht, The Netherlands: Springer; 2014. pp. 239-259

[49] Phillips RL. Organic agriculture and nitrous oxide emissions at sub-zero soil temperatures. Journal of Environmental Quality. 2007;**36**:23-30

[50] Kramer SB, Reganold JP, Glover JD, Bohannan BJM, Mooney HA. Reduced nitrate leaching and enhanced denitrifier activity and efficiency in organically fertilized soils. Proceedings of the National Academy of Sciences of the United States of America. 2006;**103**:4522-4527

[51] Zawadzki J, Fabijnczyk P. Geostatistical evaluation of lead and zinc concentration in soils of an old mining area with complex land management. International Journal of Environmental Science and Technology. 2012;**10**:729-742

[52] Yang X, Li Q, Tang Z, Zhang W, Yu G, Shen Q, et al. Heavy metal concentrations and arsenic speciation in animal manure composts in China. Waste Management. 2017;**64**:333-339

[53] Schwab P, Zhu D, Bank M.K. Heavy metal leaching from mine tailing as affected by organic amendments. Bioresource Technology. 2017;**98**(15):2935-2941. DOI: 10.1016/j.biortech.2006.10.0121

[54] Anh PT. Water pollution by intensive brackish shrimp farming in south-east Vietnam: Causes and options for control. Agricultural Water Management. 2010;**97**(6):872-882

[55] Stolze M, Piorr A, Haring A, Dabbert S. The environmental impacts of organic farming in Europe. In: Organic Farming in Europe: Economics and Policy. Vol. 6. Stuttgart: University of Hohenheim; 2000. p. 127

[56] Tuomisto HL, Hodge ID, Riordan P, Macdonald DW. Does organic farming reduce environmental impacts?-A meta-analysis of European research. Journal of Environmental Management. 2012;**112**:309-320

[57] Mondelaers K, Aertsens J, Huylenbroeck GV. A meta-analysis of the differences in environmental impacts between organic and conventional farming. British Food Journal. 2009;**111**:1098-1119

[58] Drinkwater LE, Wagoner P, Sarrantonio M. Legume-based cropping systems have reduced carbon and nitrogen losses. Nature. 1998;**396**:262-265

[59] Kirchmann H, Bergström L. Do organic farming practices reduce nitrate leaching? Communications in Soil Science and Plant Analysis. 2001;**32**(7-8):997-1028

[60] Stockdale EA, Shepherd MA, Fortune S, Cuttle SP. Soil fertility in organic farming systems—Fundamentally different? Soil Use and Management. 2002;**18**:301-308

[61] Berry PM, Sylvester-Bradley R, Philipps L, Hatch DJ, Cuttle SP, Rayns FW, et al. Is the productivity of organic farms restricted by the supply of available nitrogen? Soil Use and Management. 2002;**18**:248-255

[62] Wander MM, Traina SJ, Stinner BR, Peters SE. Organic and conventional management effects on biologically active soil organic matter pools. Soil Science Society of America Journal. 1994;**58**:1130-1139

[63] Watson CA, Bengtsson H, Ebbesvik M, Løes A-K, Myrbeck A, Salomon E, et al. A review of farm-scale nutrient budgets for organic farms as a tool for management of soil fertility. Soil Use and Management. 2002;**18**:264-273

[64] Nelson NO, Janke RR. Phosphorus sources and management in organic production systems. Hort Technology. 2007;**17**:442-454

[65] Trewavas. Urban myths of organic farming. Nature. 2001;**410**:409-410

[66] Edwards-Jones G, Howells O. The origin and hazard of inputs to crop protection in organic farming systems: Are they sustainable? Agricultural Systems. 2001;**67**:31-47

[67] Zehnder G, Gurr GM, Kühne S, Wade MR, Wratten SD, Wyss E. Arthropod pest management in organic crops. Annual Review of Entomology. 2007;**52**:57-80

[68] Reganold JP, Glover JD, Andrews PK, Hinman HR. Sustainability of three apple production systems. Nature. 2001;**410**:926-930

[69] Ahmed Q, Yousuf F, Sarfraz M, Bakar NKA, Balkhour MA, Ashraf MA. Seasonal elemental variations of Fe, Mn, Cu and Zn and conservational management of Rastrelliger kanagurta fish from Karachi fish harbour, Pakistan. Journal of Food, Agriculture and Environment. 2014;**12**(3&4):405-414

[70] Paoletti MG, Pimentel D, Stinner BR, Stinner D. Agroecosystem biodiversity: Matching production and conservation biology. Agriculture, Ecosystems and Environment. 1992;**40**:3-23

[71] Dritschilo W, Wanner D. Ground beetle abundance in organic and conventional corn fields. Environmental Entomology. 1980;**9**:629-631

[72] Pfinner L, Niggli U. Effects of bio-dynamic, organic and conventional farming on ground beetles (Col. Carabidae) and other epigaeic arthropods in winter wheat. Biological Agriculture and Horticulture. 1996;**12**:353-364

[73] Hyvönen T, Ketoja E, Salonen J, Jalli H, Tiainen J. Weed species diversity and community composition in organic and conventional cropping of spring cereals. Agriculture, Ecosystems and Environment. 2003;**97**:131-149

[74] Freemark KE, Kirk DA. Birds on organic and conventional farms in

Ontario: Partitioning effects of habitat and practices on species composition and abundance. Biological Conservation. 2001;**101**:337-350

[75] Krauss J, Gallenberger I, Steffan-Dewenter I. Decreased functional diversity and biological pest control in conventional compared to organic crop fields. PLoS One. 2011:6

[76] Roschewitz I, Gabriel D, Tscharntke T, Thies C. The effects of landscape complexity on arable weed species diversity in organic and conventional farming. Journal of Applied Ecology. 2005;**42**:873-882

[77] Gabriel D, Roschewitz I, Tscharntke T, Thies C. Beta diversity at different spatial scales: Plant communities in organic and conventional agriculture. Ecological Applications. 2006;**16**(5):2011-2021

[78] Shepherd M, Pearce B, Cormack B, Philipps L, Cuttle S, Bhogal A, et al. An Assessment of the Environmental Impacts of Organic Farming. A Review for Defra-Funded Project OF0405. Mansfield: ADAS Consulting Ltd.; 2003

[79] Hole DG, Perkins AJ, Wilson JD, Alexander IH, Grice PV, Evans AD. Does organic farming benefit biodiversity? Biological Conservation. 2005;**122**:113-120

[80] Holzschuh A, Steffan-Dewenter I, Kleijn D, Tscharntke T. Diversity of flower visiting bees in cereal fields: Effects of farming system, landscape composition and regional context. Journal of Applied Ecology. 2007;**44**:41-49

[81] Rundlöf M, Smith HG. The effect of organic farming on butterfly diversity depends on landscape context. Journal of Applied Ecology. 2006;**43**:1121-1127

[82] Mader P, Fließbach A, Dubois D, Gunst L, Fried P, Niggli U. Soil fertility and biodiversity on organic farming. Science. 2002;**296**:1694

[83] Debuschewitz E, Sanders J. Environmental impacts of organic agriculture and the controversial scientific debates. Organic Agriculture. 2022;**12**:1-15. DOI: 10.1007/s13165-021-00381-z

[84] Ponisio LC, M'Gonigle LK, Mace KC, Palomino J, de Valpine P, Kremen C. Diversification practices reduce organic to conventional yield gap. Proceedings. Biological sciences. 2015;**282**:20141396. DOI: 10.1098/rspb.2014.1396

[85] de Ponti T, Rijk B, van Ittersum MK. The crop yield gap between organic and conventional agriculture. Agricultural Systems. 2012;**108**:1-21. DOI: 10.1016/j.agsy.2011.12.004

[86] Seufert V, Ramankutty N, Foley JA. Comparing the yields of organic and conventional agriculture. Nature. 2012;**485**:229-232. DOI: 10.1038/nature11069

[87] Knapp S, van der Heijden MGA. A global meta-analysis of yield stability in organic and conservation agriculture. Nature Communications. 2018;**9**:3632. DOI: 10.1038/s41467-018-05956-1

[88] Ponisio L, Ehrlich P. Diversification, yield and a new agricultural revolution: Problems and prospects. Sustainability. 2016;**8**:1118. DOI: 10.3390/su8111118

[89] Seufert V, Ramankutty N. Many shades of gray—The context-dependent performance of organic agriculture. Science Advances. 2017;**3**:e1602638. DOI: 10.1126/sciadv.1602638

[90] Kirchmann H. Why organic farming is not the way forward.

Outlook on Agriculture. 2019;**48**:22-27. DOI: 10.1177/0030727019831702

[91] Bergström L, Kirchmann H. Are the claimed benefits of organic agriculture justified? Nature Plants. 2016;**2**:16099. DOI: 10.1038/nplants.2016.99

[92] Schader C, Stolze M, Gattinger A. Environmental performance of organic farming. In: Boye JI, Arcand Y, editors. Green Technologies in Food Production and Processing. New York: Springer; 2012. pp. 183-210

[93] Mäder P, Fließbach A, Dubois D, Gunst L, Fried P, Niggli U. The ins and outs of organic farming [Response]. Science. 2002;**298**:1889-1890. DOI: 10.1126/science.298.5600.1889b

[94] Kusche D, Hoppe J, Hupe A, Heß J. Wasserschutz. In: Sanders J, Heß J, editors. Leistungen des ökologischen Landbaus für Umwelt und Gesellschaft. Braunschweig: Johann Heinrich von Thünen-Institut; 2019. pp. 59-91

[95] Ponisio LC, Kremen C. System-level approach needed to evaluate the transition to more sustainable agriculture. Proceedings. Biological sciences. 2016;**283**:20152913. DOI: 10.1098/rspb.2015.2913

[96] van der Werf HMG, Knudsen MT, Cederberg C. Towards better representation of organic agriculture in life cycle assessment. Nature Sustainability. 2020;**3**:419-425. DOI: 10.1038/s41893-020-0489-6

[97] Clark M, Tilman D. Comparative analysis of environmental impacts of agricultural production systems, agricultural input efficiency, and food choice. Environmental Research Letters. 2017;**12**:64016. DOI: 10.1088/1748-9326/aa6cd5

[98] Meng F, Qiao Y, Wu W, Smith P, Scott S. Environmental impacts and production performances of organic agriculture in China: A monetary valuation. Journal of Environmental Management. 2017;**188**:49-57. DOI: 10.1016/j.jenvman.2016.11.080

[99] Jespersen LM, Baggesen DL, Fog E, Halsnæs K, Hermansen JE, Andreasen L, et al. Contribution of organic farming to public goods in Denmark. Organic Agriculture. 2017;**7**:243-266. DOI: 10.1007/s13165-017-0193-7

Chapter 4

Review of the Relationship between Soil Health, Climate Change Mitigation, and Crop Production When Soils Are Enhanced with Organic Sources of Nutrients

Habtamu Tadele Belay and Birtukan Amare Kebede

Abstract

When soils are not properly maintained, the agriculture sector contributes significantly to global warming by raising greenhouse gases like CO_2 and N_2O. This review describes the relationship between organic fertilizers in improving soil health, crop production and mitigating climate change. Organic fertilizers are produced by the quick artificial decomposition from biological wastes. The synergy of using both poultry manure and nitrogen has proven to enhance the production of crops. The utilization of 4 t t ha^{-1} of poultry manure resulted in the most significant development and production of maize. Likewise, utilization of bio-slurry in both liquid and composted forms, either alone at a rate of 20 t ha^{-1} or in combination with the complete dose of chemical fertilizer at a rate of 10 t ha^{-1}, results in varying increases in crop yield of maize, soybean, wheat, sunflower, cotton, ground nut, cabbage, and potato compared to the control group. By utilizing organic sources of nutrients, the emissions of N_2O can be diminished through the enhancement of nitrogen utilization effectiveness. Organic source of nutrients possesses numerous characteristics that not only enhance crop yield but also serve as options for safeguarding the environment by enhancing soil organic carbon and reducing N_2O emission.

Keywords: food, fertility, plant nutrition, emission, sustainable agriculture

1. Introduction

By 2050, the world populace is assessed to be 9.2 billion. Horticulture may be essential for feasible advancement, destitution decrease, and nourishment security in developing countries. During this period, they have to increment rural generation by 70% to meet the expanded request for food [1]. The food supply in sub-Saharan Africa faces difficulties as the human population grows and the opportunities to expand arable land are limited. There is a constant decrease in soil fertility, which results in declining yields. The cultivation of essential crops such as rice and wheat is anticipated to be detrimentally affected by climate change, projected to affect a minimum of 22% of these areas by the year 2050 [2] and Exacerbate the phenomenon of worldwide climate change. The escalation in the amount of greenhouse gases (GHGs), primarily carbon dioxide (CO_2), methane (CH_4), and nitrous oxide (N_2O), in the atmosphere, is the leading cause of global warming. The majority of developing nations rely on the agricultural industry as their main source of economic stability [3]. Among the challenges hampering agricultural productivity in sub-Saharan Africa, the inadequacy of soil nutrients and ineffective land management techniques have been recognized as major obstacles [4].

The agricultural sector accounts for a substantial portion of greenhouse gas emissions, particularly from livestock through processes such as intestinal fermentation and the management of manure, as well as from agricultural soils due to the application of excessive nitrogen fertilizers and the decomposition of organic matter. Roughly 14% of the total worldwide emissions of greenhouse gases can be attributed to agriculture [5]. The solution to the impacts of climate change is partially found within the problem itself. With proper management of cultivated soil and implementation of efficient policies, significant amounts of carbon can be absorbed from the atmosphere and stored in the soil, which would subsequently lead to a decrease in CH_4 and CO_2 emissions [6]. Prolonged utilization of mineral fertilizers deteriorates the chemical, physical, and biological properties of soil along with its overall health [7]. The usage of organic fertilizers as a source of nutrients is gaining popularity due to the escalating costs and detrimental impacts of chemical fertilizers [7, 8]. The inadequate utilization of organic fertilizers by rural Ethiopian farmers is negatively affecting crop production in the area and the general food safety situation.

One solution suggested for the soil fertility crisis in sub-Saharan Africa is the use of organic resources, considered the most practical choice [9]. Several types of organic resources can be utilized, namely farm manure (FYM), poultry litter (PL), poultry manure (PM), as well as manure and vermicompost [9] and by up to 50% [10]. For example, organic materials such as FYM are traditionally used by rice farmers [8]. FYM provides essential nutrients, including N, P, K, Ca, Mg, and S, that are crucial for the growth of plants. It also contains micronutrients such as Fe, Mn, Cu, and Zn. Therefore, it serves as a combination of nourishment for plants [11]. It improves soil's physical, chemical, and biological properties [5]. The structure can be improved by employing organic FYM on soil, leading to better growth conditions for roots [12]. Farmyard manure enhances the ability of the soil to retain water [13]. The utilization of organic fertilizers to enhance soil texture and nutrient transfer and sustain soil well-being has generated curiosity regarding organic cultivation [5]. Chicken manure is the primary organic fertilizer of substantial importance due to its high levels of essential nutrients like nitrogen, phosphorus, and potassium. According to [14], the utilization of chicken manure led to a rise in exchangeable soil cations and nitrogen levels, elevating it from 0.09 to 0.14%. The excrement produced by chickens on farms, commonly known as chicken manure, undergoes a gradual process of decomposition and displays a considerably elevated level of phosphorus in comparison to other organic materials used for nutritional purposes. It comprises approximately 3.03% nitrogen, 2.63% phosphorus, and 1.4% potash [15].

Having a comprehensive knowledge of the impact of carbon, nitrogen, and essential soil characteristics on soil gas emissions is crucial in order to minimize the long-lasting negative consequences that may arise even after discontinuation of the use of fertilizers and composts. This is an important step toward mitigating the residual effects. This paper aims to determine how organic fertilizers can enhance soil productivity, secure food supply, and safeguard the environment against climate change. The primary aim of the review is to examine the role of organic fertilizers in enhancing the yield of crops. Examine the significance of using organic fertilizers in relation to enhancing soil fertility, promoting crop yields, mitigating climate change, and analyzing their effects on ensuring sustainable agricultural practices.

2. Concepts of organic fertilizers

Organic fertilizers refer to naturally occurring substances that possess a well-defined chemical composition and offer plant nutrients in a readily available form, thereby possessing a high analytical worth [16]. Organic fertilizers comprise fertilizers obtained from animal waste, human waste, or plant material (such as vegetables), organic matter such as compost and manure. The use of organic fertilizers involves using raw materials derived from nature, typically referring to our biodegradable apparel. Normally, compost is created through the process of breaking down materials that can be naturally broken down. These waste materials encompass paper, foliage, peels from fruits, uneaten food items, and also fruit juices. The incorporation of organic fertilizers is a beneficial supplement to enhance the quality of the soil. This facilitates the soil to attain an optimum condition for cultivation.

2.1 Organic fertilizer on crop production and soil nutrients

According to [17] the report, the most successful results in terms of grain yield and test weight were obtained through the utilization of 10 tons per hectare of PM, paired with 125 kilograms per hectare of nitrogen in the form of urea. The utilization of various levels of N, PM, and their combination did not significantly impact the concentrations of P and K in both the plant stover and grain. According to research, the utilization of poultry manure resulted in a significant enhancement of crop growth and maize yield by up to 40%. The implementation of four tons of PM per hectare led to the most substantial development and productivity of maize [18]. Indeed [19] reported the utilization of organic fertilizers may result in a substantial productivity boost of 15.2% for kiwifruit, as opposed to using mineral fertilizers. The nutrient levels in kiwifruits may be increased by using organic fertilizers. The outcomes indicated that organic fertilizers that have abundant quantities of diminutive organic matter could be more effective in stimulating crop yield.

The active small molecular organic matter in the produced organic fertilizer led to more noticeable effects compared to the conventional organic fertilizer derived from fermented pig manure. One reason for the ease of mineralization and subsequent release of mineral nutrients into the soil is the smaller size of organic molecules. Low molecular weight organic acids, which belong to the group of small organic molecules, have been extensively demonstrated to aid in the mobilization of soil P by means of dissolution and complex formation [19, 20]. The increased accessibility of metallic elements can be achieved through the creation of organic-metal complexes using low molecular weight organic acids [21]. Additionally, the employment of the AF application resulted in improved fruit characteristics such as increased firmness and higher levels of soluble solid, soluble sugar, titratable acid, and vitamin C. Other investigations had reported comparable findings. Ma et al. [19] it has been discovered that the organic matter present in soil had a direct association with the weight per fruit, levels of soluble solids, soluble sugar, and vermicompost (VC) content in jujuba fruits. Hence, organic fertilizer appears to be a valuable and environmentally friendly way to improve the mineral availability in the soil and improve fruit quality of tomato [22]. Despite the utilization of fermented pig manure containing equal levels of organic matter, the utilization of mineral fertilizer with OF (organic fertilizer) intervention did not lead to any enhancement in the yield or caliber of kiwifruits. According to Fallahi and Seyedbagheri [23] suggestion that humic substances do not have an impact on the quality of apples implies that organic matters with high molecular weights that are resistant to degradation are not significantly conducive to the growth of crops.

2.1.1 Roles of small molecular organic matters for crops

The introduction of the mineral plant nutrition theory has substantially ensured our food security since its inception [24]. It must be acknowledged that the accumulation of organic matter is an integral part of the growth of crops. Carbon is, without a doubt, the crucial factor determining the growth of crops. The crops are experiencing a lack of carbon absorption to a certain extent. As an illustration, augmenting the CO_2 levels within the greenhouse has the potential to boost crop yield by providing an ample amount of CO_2 for the process of photosynthesis [25, 26]. Moreover, plants have the ability to absorb nutrients from both their roots and leaves, including minute organic substances, in order to foster their development. Sulmon et al. [27] reported that, the utilization of external carbohydrates was employed to enhance the process of phytoremediation, given that the plants were capable of incorporating the exogenous carbohydrates.

Apart from carbon, it is possible for plants to inherently obtain nitrogen directly from organic sources without the need for microbial mineralization [28]. Hirner et al. [29] discovered a transporter in the root epidermis with a strong affinity for taking in amino acids at the cellular level. Ge et al. [30] also noted that glycine N uptake accounted for 21% of nitrogen uptake in tomatoes, following that of KNO_3-N and NH_4Cl-N. Crops can directly absorb other types of small N-molecules like urea, polyamines, and polypeptides resulting from enzymatic cleavage [29].

Utilizing small organic molecules such as amino acids, peptides, and carbohydrates directly provides a more effective means of enhancing crop growth without the need for numerous assimilation processes compared to the absorption of CO_2. Furthermore, vegetation would redistribute their natural substances in order to combat environmental stress [30, 31]. Adding live natural substances to the soil could help protect crops and safeguard their investment of carbon against environmental challenges. Utilizing organic fertilizers that contain high levels of active small molecular organic agents can enhance both the yield and quality of crops, as compared to traditional organic fertilizers.

2.1.2 Effect of compost on yield of cereals

The study conducted *adaa* district Eastern Shewa Oromia region by Bhattacharyya et al. [32] indicated that the combination of dry matter compost and inorganic fertilizers resulted in a grain yield value of 0.67 t ha^{-1} for bread wheat. The untreated region or area had the lowest crop output, quantified at 260 grams per square meter or 2.6 t ha^{-1} . Previous studies conducted have confirmed similar results. On the other hand, a study conducted by [33] in the Tigray area, it has been observed that the cultivation of teff and barley in plots treated with mineral fertilizer and 6.4 t ha^{-1} yr^{-1}

compost results in a substantial increase in grain yield compared to plots treated with 3.2 t ha^{-1} yr^{-1} compost and control plots. The plots treated with 6.4 t ha^{-1} compost and mineral fertilizer displayed no noticeable disparity.

2.1.3 Effects of crop residue

The utilization of crop residues as a soil enhancer is frequently restricted because it poses difficulties for both mechanical and manual cultivation, as well as causing harmful impacts on crop output due to the presence and continuation of pests and diseases [34] allelopathy and short-term nutrient deficiency [35].

A large proportion of crop residues are either utilized as cattle feed or incinerated due to these factors. Ocio et al. [36] evaluated that edit residue contain on normal 40, 10, and 80% of the N, P, and K right now connected as fertilizer. A ton of maize buildup contains 4–8 kg N, 1.5–1.8 kg P, 13–16 kg K, 3.8–6.6 kg Ca, and 1.5–3.4 kg Mg. Buildups of cereal crops include 60 to 75% of the whole biomass generation and have lower supplement concentrations than the grain [37]. In this manner, returning them to the soil frameworks especially, where no or moo inputs are utilized, is fundamental in abating supplement misfortunes. Be that as it may, edit buildups by themselves are not sufficient to balanced supplement mining in sub-Saharan Africa. Edit buildup administration impacts the accessibility of supplements, particularly N. Agreeing with the consideration conducted by Dejene [37] in Gozamen Woreda Eastern Gojam Amhara Local, crop residue is commonly utilized for keeping up soil richness and crop production in two ways.

2.1.4 Effects of farm yard manure on soil fertility and yield

As Tolessa and Friesen [38] detailed natural fertilizers, particularly FYM, have a noteworthy part in keeping up and moving forward the chemical, physical, and organic properties of soils and in supporting maize abdicate in the western portion of Ethiopia. They moreover detailed that 10 t ha^{-1} of FYM are measurably at proportionality with the current agronomic suggestion of inorganic fertilizers N and P for maize. Another study by Zelalem [39, 40] at Haraghe Zone Oromia Locale Eastern Ethiopia showed that 10 t ha^{-1} of FYM and 100 kg ha^{-1} N + 100 kg ha^{-1} P appeared no noteworthy distinction on maize grain abdicate but altogether vary from the control treatment. Moreover, Negassa, et al. [40] demonstrated that the direness of utilizing natural excrement has been picking up ground within the wake of expanding, taken a toll on fertilizer with each passing year, and certain other characteristic impediments with the utilization of chemical fertilizers.

2.1.5 Effects of green manure on soil fertility

The study conducted by Getu and Teshager [41] on the impact of green excrement plants on sorghum surrender and soil ripeness in eastern Amhara of Ethiopia uncovered that there was factually noteworthy ($P < 0.05$) distinction within the grain surrender of sorghum due to the impact of intercropping with the green fertilizers.

2.1.6 Effects of biogas slurry on yield

Using bio-slurry as a liquid or composted application, either on its own at 20 t ha^{-1} or combined with a full dose of chemical fertilizer at 10 t ha^{-1}, resulted in improved percentages of yield for a variety of crops (maize, soybean, wheat, sunflower, cotton, groundnut, cabbage, and potato) when compared to controls [42]. According to

Krishna [43] by utilizing bio-slurry, the crop production of rice and maize observed a rise of 34 percent while wheat saw an increase of 25 percent. The application of bio-slurry in various forms elevated both the amount and caliber of the harvest, including crops, vegetables, and fruits, in addition to enhancing the plants' resistance to diseases [42]. Indeed [44], a study was conducted that compared the impact of biogas slurry and inorganic fertilizer on soil characteristics, as well as the growth and yield of white cabbage (*Brassica oleracea* var. *capitata* f. *alba*). At Sebeta Hawas Woreda, South West Shewa zone Oromia Region, it was found that using a combination of slurry compost and a complete dose of fertilizer resulted in a 38.4% increase in crop yield compared to using a full dose of inorganic fertilizer alone. Five tons of slurry compost when it came to crop yield. Slurry compost quantity of 8 tons per hectare.

2.1.7 Effects of poultry manure on soil fertility and yield and yield-related parameters

An exceptional natural fertilizer, poultry manure is rich in essential nutrients like nitrogen, phosphorus, potassium, and others. Unlike chemical fertilizer, it contributes organic matter to the soil that enhances soil quality, increases nutrient retention, promotes aeration, boosts soil moisture retention, and improves water infiltration [45]. According to the findings, poultry waste offers a more easily accessible source of phosphorus for plants compared to other forms of organic manure [46]. Poultry waste proves to be a useful fertilizer and can potentially replace the use of synthetic fertilizers.

The utilization of poultry manure resulted in a significant rise of 53% in soil nitrogen levels, increasing from 0.09 to 0.14%. In addition, the application of manure enhanced the presence of exchangeable cations in the soil [47]. The primary motives for utilizing PM in agriculture are to improve soil quality through organic amendment and to supply crops with essential nutrients [48]. Likewise, [49] reported that the use of PM significantly impacted various aspects of maize growth and yield, including plant height, row count per cob, number of grains per row, the weight of 1000 grains, grain yield, biological yield, and harvest index. The highest possible values for each parameter were observed when 12 tonnes per hectare of PM were utilized. The author previously stated that the composition of PM includes approximately 2.04% nitrogen, 2.06% phosphorus, and 1.86% potassium). Indeed [48], according to the data, PM comprises approximately 3.03% nitrogen, 2.63% phosphorus, and 1.4% potash. The data presented indicates that the plot where 12 t ha^{-1} of poultry manure was utilized had the highest grain yield with a significant value of 5.11 t ha^{-1}. The next highest yield was recorded from the plot using 10 t ha^{-1} PM, which was statistically equivalent to the setup that used 8 t ha^{-1} PM, and produced grain yields of 4.16 and 3.60 t ha^{-1}, respectively. The plots treated with 6 t ha^{-1} poultry manure had a statistically identical grain yield to that of the control treatment. According to [47] reported that poultry manure significantly increased grain yield.

According to findings [50], it is suggested that utilizing poultry manure to replace 50% of inorganic fertilizer can effectively minimize the need for chemical fertilizers while maintaining crop productivity. The utilization of fertilizers containing 50% NPK and 100% PM and fertilizers containing pure 100% NPK resulted in the most bountiful pod and seed yields per plant. The control and 100% PM treatments exhibited the minimum amount of pods per plant and seed yield per plant. The finest origin of crop nutrients is poultry waste (**Table 1**).

Nutrient element	Values (%)
N	4.50
P_2O_5	2.50
K_2O_5	2.00
CaO	2.00
MgO	1.00
S	0.50
Fe	0.04
Mn	0.09
Zn	0.09
Other characteristics	0.50

Table 1.
Chemical composition of poultry manure [50].

2.2 Organic fertilizer on quality improvement

The primary physical reason for the decrease in food production per person in Ethiopia is acknowledged as the result of land deterioration and the resulting reduction in soil productivity [15]. The depletion of land's ability to support crops and livestock caused by the use of man-made chemicals has a direct negative impact on the production of food and animal feed. The use of chemical substances leads to a decrease in land productivity, which is worsened by inadequate land administration [51]. Nevertheless, it is possible to reduce it by utilizing organic fertilizers that are derived from excrement and urine [52].

The fertility of soil can be enhanced through the use of organic fertilizers that have an impact on its physical, chemical, and biological qualities. It enhances the movement of water and air through the soil, thereby boosting its capacity to retain moisture [53]. According to [54] According to the report, the usage of organic fertilizers also has a positive impact on the soil as it generates clay humic complexes that boost the soil's ability to adsorb vital nutrients like calcium, magnesium, and potassium. Furthermore, it enhances the activity of microorganisms that participate in the mineralization process. A study conducted by [55] stated that the soil pH incorporating organic matter into the soil could substantially elevate the pH level, as well. The elevated presence of fundamental nutrients in organic supplements and the reduction of hydrous oxides in soil with the aid of poultry manure are the reasons assigned for this outcome [56].

According to [57] findings, animal manure is the prime source of soil fertility management to improve the way for many farmers in Ethiopia. It is used as fertilizer to ameliorate soil fertility depletion in many parts of Ethiopia. Indeed [58], according to findings, approximately 87% of farmers in Kindo Koisha, a region in Southern Ethiopia, use animal manure. The reason for this is that the utilization of animal excrement leaves a lasting impact on the land [59]. The impact may differ depending on the quantities administered. The feasibility of this is reliant on the presence of domestic animals and the assistance of family workers for transportation to their farming lands [60]. Currently, biomass is widely utilized as a primary source of energy in households [51]. The utilization of PM significantly raised the levels of exchangeable cations and soil nitrogen content, elevating them from 0.09 to 0.14% [47]. The physical and chemical condition, specifically nitrogen content, of the soil was enhanced by the utilization of 10 tons per hectare of PM in conjunction with 125 kg of N [16].

2.3 Organic fertilizer on climate change mitigation

Soil microbes and the atmosphere benefit from the ecosystemic role of soils. If appropriately managed, soil can function as a significant tool for reducing the impact of climate change by sequestering carbon and reducing the discharge of greenhouse gases into the atmosphere. On the contrary, soil carbon can emanate in the form of carbon dioxide (CO_2) and add to climate change if the soil is not managed effectively or farmed using unsustainable agricultural methods. The gradual transformation of grassy and wooded areas into cultivation and pasturage lands throughout many centuries has caused extensive depletion of soil carbon on a global scale. Revitalizing depleted soil and embracing methods to conserve soil can significantly reduce greenhouse gas emissions from farming, promote carbon storage, and reinforce the ability to cope with climate change [57].

The earth's soil contains the most significant carbon reservoir on land, and the biogeochemical reactions happening in the soil regulate the release and absorption of greenhouse gases into the atmosphere [58]. When implementing sustainable methods to improve soil organic matter, it is important to also address the causes of soil deterioration and protect the current levels of soil carbon, especially in soils that have a high amount of organic carbon [61].

Soil-based carbon sequestration will aid in both adapting to and mitigating the effects of climate change. This will enhance the sustainability of agricultural production systems, improve the resilience of agricultural ecosystems as a whole, and uphold the ecosystem services that rely on soils [62].

The world's agricultural soils span across approximately 1.5 billion hectares and possess a significant ability to sequester carbon [63]. Efficiently controlling the pool of organic carbon within the soil is a crucial objective toward attaining both adaptation and mitigation of the worldwide environmental impact [64], while advancing global food security [65]. Cropland soils are significant carbon sinks that can be utilized to alleviate and adjust to worldwide climate change. The amount and speed at which soil organic carbon is stored (typically approximately 0.55×10^{-9} Pg C ha^{-1} y^{-1}) depend on factors such as how residues are managed, and organic material is recycled, the climate, nitrogen application, and properties of the soil [64]. Like cropland soil, forest and grassland soil also have the potential to play a vital role in carbon sequestration. Soil enriched with organic fertilizer can enhance its capacity to absorb carbon, boosting soil fertility management. Effective nutrient management plays a crucial role in the sequestration of soil organic carbon (SOC), which in turn emphasizes the significance of enhancing soil fertility [66]; the levels of organic carbon and nitrogen in soil are crucial factors that determine soil productivity and quality by enhancing physical, chemical, and biological processes such as nutrient cycling, water preservation, growth of root and shoot, and upkeep of ecological health [67].

2.4 Organic fertilization for sustainable agriculture

The incorporation of organic substances into the soil can enhance soil characteristics and maintain an enduring agricultural output. Furthermore, organic substances that are of low molecular weight have the potential to participate in the regulation of both soil nutrient dynamics and agricultural productivity. According to [66] study in Nigeria reported, when 10 metric tons per hectare of PM along with urea and muriate of potash fertilizer were utilized, a sustainability index (SI) of 72.5% was observed. According to the study, using 10 tons of PM alongside NK that has a 72.5% SI is a more efficient and ecologically sound approach than utilizing NK, with the highest agronomic efficiency and partial factor productivity. Using a combination of synthetic and poultry manure as a source of P can result in higher profitability and sustainability in the long run [68]. Similarly to Sainju & Good (1993), the utilization of P60SSP + 60 PM enables the easy accessibility of phosphorus nutrients present in fertilizers and manure, thereby enhancing the quality of soil constituents. As a result, it can be inferred that farmers residing in semi-moist regions ought to utilize a combined approach of adding P into the soil, with an equal division of 60 kg ha^{-1} from both single super phosphate (SSP) and poultry manure. This will not only enhance the applied fertilizer phosphorus uptake efficiency (AFPU), fertilizer phosphorus use efficiency (FPUE), and phosphorus index ratio (PIR), but also generate greater and consistent wheat yields. Furthermore, by adopting this method, there will be reduced dependence on chemical fertilizers, which in turn will lower the potential dangers connected to continuous and excessive use of synthetic fertilizers on the soil and atmosphere.

2.5 Determinant factors for organic fertilizers

According to [69], several factors significantly influenced the adoption of organic fertilizers in Ethiopia, including the age and marital status of the household head, educational level, labor availability, farming experience, farm, and livestock size, access to information and extension services, labor costs, household income, soil fertility, and distance between the farm and home.

3. Conclusions

Based on an analysis of various literary sources, it has been determined that the following points are evident. Based on the review, one can infer or deduce.

- The usage of organic fertilizer enhances the production of a variety of agricultural crops.
- Because of its residuals that contribute to crop production beyond a single season, as well as its eco-friendliness, organic fertilizer surpasses other soil fertility management tactics in benefits.
- The utilization of organic fertilizer not only enhances the quality of soil but also plays a pivotal role in reducing the impact of climate change by boosting carbon sequestration and enhancing the nutrient usage efficiency of crops.
- Moreover, the usage of organic fertilizers plays a crucial part in promoting sustainability in agriculture by enhancing the overall physical, chemical, and biological properties of the soil.
- The utilization of organic fertilizer was greatly impacted by the determinant aspects.

Acknowledgements

The authors are thankful to the anonymous reviewers for their comments and suggestions to improve the quality of this review paper. This review chapter was carried out as part of the Ph.D. research work of the first author.

Author contributions in writing review and editing

Habtamu Tadele performed the literature search and drafted and/or critically revised the work. Birtukan Amare revised the draft proposal and corrected reviewed the article after reviewers' comments.

Data availability statement

All data generated or analyzed during this review are included in this published article.

Statements and declarations

The authors declare no conflict of interest.

Author details

Habtamu Tadele Belay[1*] and Birtukan Amare Kebede[2]

1 Bahir Dar University, Bahir Dar, Ethiopia

2 Burie Agricultural College, Burie, Ethiopia

*Address all correspondence to: habtietoo@gmail.com

References

[1] I.F.I. Association. Fertilizers and their use: A pocket guide for extension officers, Food & Agriculture Org. 2000

[2] Adhikari U, Nejadhashemi A, Woznicki S. Climate change and eastern Africa: A review of impact on major crops. Food and Energy Security. 2015;**4**:110-132

[3] Saraceno E. Rural Development Policies and the Second Pillar of the Common Agricultural Policy (version 16.08.02), Documento presentado en el taller "Desirable Evolution of the CAP: A Contribution", organizado por ARL y DATAR, Bruselas, Bélgica. 2002

[4] Vanlauwe B, Descheemaeker K, Giller KE, Huising J, Merckx R, Nziguheba G, et al. Integrated soil fertility management in sub-Saharan Africa: Unravelling local adaptation. The Soil. 2015;**1**(1):491-508

[5] Khan NI, Malik AU, Umer F, Bodla MI. Effect of tillage and farm yard manure on physical properties of soil. International Research Journal of Plant Science. 2010;**1**(4):75-82

[6] Gaskell M, Smith R, Mitchell J, Koike ST, Fouche C, Hartz T. Soil fertility management for organic crops. 2007

[7] Mahajan A, Gupta R. In: Mahajan A, Gupta RD, editors. Bio-Fertilizers: Their Kinds and Requirement in India, Integrated Nutrient Management (INM) in a Sustainable Rice-Wheat Cropping System. Springer Science plus Business Media; 2009. pp. 75-100

[8] Satyanarayana V, Vara Prasad P, Murthy V, Boote K. Influence of integrated use of farmyard manure and inorganic fertilizers on yield and yield components of irrigated lowland rice. Journal of Plant Nutrition. 2002;**25**(10):2081-2090

[9] Amoding A, Stephen Tenywa J, Ledin S, Otabbong E. Effectiveness of crop-waste compost on a Eutric Ferralsol. Journal of Plant Nutrition and Soil Science. 2011;**174**(3):430-436

[10] Kiani MJ, Abbasi MK, Rahim N. Use of organic manure with mineral N fertilizer increases wheat yield at Rawalakot Azad Jammu and Kashmir. Archives of Agronomy and Soil Science. 2005;**51**(3):299-309

[11] Shirani H, Hajabbasi MA, Afyuni M, Hemmat A. Effects of farmyard manure and tillage systems on soil physical properties and corn yield in Central Iran. Soil and Tillage Research. 2002;**68**(2):101-108

[12] Prasad B, Sinha S. Long-term effects of fertilizers and organic manures on crop yields, nutrient balance, and soil properties in rice-wheat cropping system in Bihar, long-term soil fertility experiments in Rice-wheat cropping systems. In: Rice-Wheat Consortium Paper Series 6. West Africa Journal of Applied Ecology (WAJAE). 2006;**9**(12):1-11

[13] Mengistu DK, Mekonnen LS. Integrated Agronomic Crop Managements to Improve Tef Productivity Under Terminal Drought. In: Ismail Md. Mofizur Rahman, editor. Water Stress. Europe: IntechOpen; 2012;**12**:235-254

[14] Boateng SA, Zickermann J, Kornahrens M. Poultry manure effect on growth and yield of maize. West African Journal of Applied Ecology. Ghana. 2006;**9**(12):1-11

[15] Mubeen K, Wasaya A, Rehman HU, Yasir TA, Farooq O, Imran M, et al. Integrated phosphorus nutrient sources improve wheat yield and phosphorus use efficiency under sub humid conditions. PLoS One. 2021;**16**(10):e0255043

[16] Gupta PK. Handbook of soil, fertilizer and manure, Agrobios (India). 2003

[17] Yadav S, Meena R, Seema SK, Sharma D. Effect of nitrogen and poultry manure on yield and nutrients uptake by maize (Zea mays). The Indian Journal of Agricultural Sciences. 2019;**89**(11)

[18] Uwah DF, Afonne FA, Essien AR. Integrated nutrient management for sweet maize (Zea mays (L.) saccharata strut.) production in Calabar, Nigeria. Australian Journal of Basic and Applied Sciences. 2011;**5**(11):1019-1025

[19] Ma X, Li H, Xu Y, Liu C. Effects of organic fertilizers via quick artificial decomposition on crop growth. Scientific Reports. 2021;**11**(1):1-7

[20] Khademi Z, Jones D, Malakouti M, Asadi F. Organic acids differ in enhancing phosphorus uptake by Triticum aestivum L.—Effects of rhizosphere concentration and counterion. Plant and Soil. 2010;**334**(1):151-159

[21] Fink JR, Inda AV, Tiecher T, Barrón V. Iron oxides and organic matter on soil phosphorus availability. Ciencia e Agrotecnologia. 2016;**40**:369-379

[22] Mauromicale G, Longo AMG, Monaco AL. The effect of organic supplementation of solarized soil on the quality of tomato fruit. Scientia Horticulturae. 2011;**129**(2):189-196

[23] Fallahi E, Fallahi B, Seyedbagheri MM. Influence of humic substances and nitrogen on yield, fruit quality, and leaf mineral elements of 'early spur Rome'apple. Journal of Plant Nutrition. 2006;**29**(10):1819-1833

[24] Nikolskii AA, Vanisova EA. Philosophy of ecology of Justus von Liebig: Different Liebig, RUDN. Journal of Ecology and Life Safety. 2020;**28**(1):75-81

[25] Boondum S, Chulaka P, Kaewsorn P, Nukaya T, Takagaki M, Yamori W. Carbon dioxide (CO_2) enrichment in greenhouse enhanced growth and productivity of tomato (Solanum lycopersicum L.) during winter. International Forum on Horticultural Product Quality. 2018;**1245**:61-64

[26] Pan T, Ding J, Qin G, Wang Y, Xi L, Yang J, et al. Interaction of supplementary light and CO_2 enrichment improves growth, photosynthesis, yield, and quality of tomato in autumn through spring greenhouse production. HortScience. 2019;**54**(2):246-252

[27] Sulmon C, Gouesbet G, Couee I, El Amrani A, inventors; Universite de Rennes 1, assignee. Method for improving the phytoremediation of polluted sites by providing plants with exogenous carbohydrates. US: United States patent 8,222,187; 2012

[28] Change IC. Impacts, Adaptation and Vulnerability. Contribution of Working Group II to the Fourth Assessment Report of the Intergovernmental Panel on Climate Change. Cambridge, UK: Cambridge University Press; 2007

[29] Hirner A, Ladwig F, Stransky H, Okumoto S, Keinath M, Harms A, et al. Arabidopsis LHT1 is a high-affinity transporter for cellular amino acid uptake in both root epidermis and leaf mesophyll. The Plant Cell. 2006;**18**(8):1931-1946

[30] Ge T, Song S, Roberts P, Jones D, Huang D, Iwasaki K. Amino acids as a

nitrogen source for tomato seedlings: The use of dual-labeled (13C, 15N) glycine to test for direct uptake by tomato seedlings. Environmental and Experimental Botany. 2009;**66**(3):357-361

[31] Jones DL, Healey JR, Willett VB, Farrar JF, Hodge A. Dissolved organic nitrogen uptake by plants—An important N uptake pathway? Soil Biology and Biochemistry. 2005;**37**(3):413-423

[32] Bhattacharyya P, Das S, Adhya T. Root exudates of rice cultivars affect rhizospheric phosphorus dynamics in soils with different phosphorus statuses. Communications in Soil Science and Plant Analysis. 2013;**44**(10):1643-1658

[33] Araya H. The effect of compost on soil fertility enhancement and yield increment under smallholder farming. In: The Case of Tahtai-Maichew District-Tigray Region. Ethiopia: University of Hohenheim, Germany, Doctor, University of Hohenheim; 2010

[34] Genizeb A. Evaluation of Alternative Soil Amendments to improve soil fertility and response tobread wheat (Triticum aestivum) Productivity in Ada'a Disrict, Central Ethiopia, [M.Sc thesis], Addis Ababa University. 2015

[35] Larson W, Clapp C, Pierre W, Morachan Y. Effects of increasing amounts of organic residues on continuous corn: II. Organic carbon, nitrogen, phosphorus, and sulfur 1. Agronomy Journal. 1972;**64**(2):204-209

[36] Ocio J, Brookes P, Jenkinson D. Field incorporation of straw and its effects on soil microbial biomass and soil inorganic N. Soil Biology and Biochemistry. 1991;**23**(2):171-176

[37] Dejene T. Assessment of the practices and aspects of farmland management in Gozamen District, East Gojjam Zone, Ethiopia, [Msc. thesis], College Of Social Science, Addis Abeba University. 2011. pp. 40-41

[38] Tolessa D, Friesen D. Effect of enriching faryard manure with mineral fertilizer on grain yield of maize at Bako, Western Ethiopia. Integrated approaches to higher maize productivity in the New Millennium. 2002

[39] Zelalem B. Improving and sustaining soil fertility by use of enriched farmyard manure and inorganic fertilizers for hybrid maize (BH-140) production at West Hararghe zone, Oromia, eastern Ethiopia. African Journal of Agricultural Research. 2013;**8**(14):1218-1224

[40] Negassa W, Gebrekidan H, Friesen D. Integrated use of farmyard manure and NP fertilizers for maize on farmers' fields. Journal of Agriculture and Rural development in the Tropics and Subtropics (JARTS). 2005;**106**(2):131-141

[41] Getu A, Teshager A. Effect of adaptable green manure plants on sorghum yields and soil fertility in eastern Amhara region of Ethiopia. Journal of Biology, Agriculture and Healthcare. 2015;**5**:223-231

[42] Dhobighat C, Painyapani T. Physico-chemical Analysis of Bio-slurry and Farm yard Manure for Comparison of Nutrient Contents and other Benefits so as to Better Promote Bio-slurry, Yashoda Sustainable Development (P) Ltd, Final Report, Nepal. 2006. pp. 30-31

[43] Krishna P. Response to bio-slurry application on maize and cabbage in Lalitpur District, Final Report his Majesty's Government of Nepal, Ministry of Science and Technology. Alternative Energy Promotion Centre, Nepal. 2001

[44] Debebe Y, Itana F. Comparative study on the effect of applying biogas

slurry and inorganic fertilizer on soil properties, growth, and yield of white cabbage (Brassica oleracea varcapitata f. alba). Journal of Biology, Agriculture and Healthcare. 2016;**6**(19):19-26

[45] Addis Z. Organic fertilizers use and application for cereal crop production in Ethiopia. Journal of Natural Sciences Research. 2019;**9**(10):14-25

[46] Deksissa T, Short I, Allen J. Effect of soil amendment with compost on growth and water use efficiency of Amaranth. In: Proceedings of the UCOWR/NIWR Annual Conference, International Water Research Challenges for the 21st Century and Water Resources Education, Durham, NC, USA; 2008

[47] Boateng S, Zickermann A, Kornaharens M. Effect of poultry manure on growth and yield of maize, West African. Journal of Applied Ecology. 2006;**9**:1-11

[48] Farhad W, Saleem M, Cheema M, Hammad H. Effect of poultry manure levels on the productivity of spring maize (Zea mays L.). Journal of Animal and Plant Sciences. 2009;**19**(3):122-125

[49] Garg S, Bahl G. Phosphorus availability to maize as influenced by organic manures and fertilizer P associated phosphatase activity in soils. Bioresource Technology. 2008;**99**(13):5773-5777

[50] Gezahegn AM, Martini M. Effects of residual organic manure and supplemental inorganic fertilizers on performance of subsequent maize crop and soil chemical properties. International Journal of Research Studies in Agricultural Sciences,(IJRSAS). 2020;**6**(1):1-9

[51] Agegnehu G, Amede T. Integrated soil fertility and plant nutrient management in tropical agro-ecosystems: A review. Pedosphere. 2017;**27**(4):662-680

[52] Webb NP, Marshall NA, Stringer LC, Reed MS, Chappell A, Herrick JE. Land degradation and climate change: Building climate resilience in agriculture. Frontiers in Ecology and the Environment. 2017;**15**(8):450-459

[53] Bahri S, Natsir A, Hasan S, Sirajuddin S. Combination urea and compost fertilizer with different defoliation affected corn and peanut production based on integrated farming system. In: IOP Conference Series: Earth and Environmental Science. Bristol, England: IOP; 2019;**247**:1-10

[54] Abegaz A. Farm Management in Mixed Crop-Livestock Systems in the Northern Highlands of Ethiopia. Wageningen, The Netherlands: Wageningen University; 2005

[55] Webb J, Sorensen P, Velthof G, Amon B, Pinto M, Rodhe L, et al. An assessment of the variation of manure nitrogen efficiency throughout Europe and an appraisal of means to increase manure-N efficiency. Advances in Agronomy. 2013;**119**:371-442

[56] Bahri S. Syamsul Bahri: Jurnal internasional integrated agricultural system study on crops: Intercropping of corn-peanut and beef cattle fattening. International Journal of Current Research in Biosciences and Plant Biology. 2018;**5**(4):24-29

[57] Haile W, Boke S, Kena K. Integrated Soil Fertility management options for sustainable crop production: Review of Research Findings from Southern Regional State of Ethiopia. In: Improved natural Resources management Technologies for Food Security, Poverty education and Sustainable Development. Proceedings

of the 10th Conference of the Ethiopian Society of Soil Science. Addis Ababa, Ethiopia: EIAR; 2010. pp. 163-175

[58] Elias E. Farmers' Perceptions of Oil Fertility Changes and Management. Addis Ababa, Ethiopia: Institute for Sustainable Development; 2002

[59] N. Nabahungu, J. Semoka, C. Zaongo, Limestone, Minjingu phosphate rock and green manure application on improvement of acid soils in Rwanda, Advances in Integrated Soil Fertility Management in Sub-Saharan Africa: Challenges and Opportunities. Berlin, Heidelberg: Springer; 2007. pp. 703-712

[60] Hue N. Correcting soil acidity of a highly weathered Ultisol with chicken manure and sewage sludge. Communications in Soil Science and Plant Analysis. 1992;**23**(3-4):241-264

[61] Tegene B. Indigenous soil knowledge and fertility management practices of the south Wällo highlands. Journal of Ethiopian Studies. 1998;**31**(1):123-158

[62] Roy RN, Finck A, Blair G, Tandon H. Plant nutrition for food security, a guide for integrated nutrient management. FAO Fertilizer and Plant Nutrition Bulletin. 2006;**16**:368

[63] Scharlemann JP, Tanner EV, Hiederer R, Kapos V. Global soil carbon: Understanding and managing the largest terrestrial carbon pool. Carbon Management. 2014;**5**(1):81-91

[64] Lal R. Managing soils and ecosystems for mitigating anthropogenic carbon emissions and advancing global food security. Bioscience. 2010;**60**(9):708-721

[65] Smith J, Abegaz A, Matthews RB, Subedi M, Orskov ER, Tumwesige V, et al. What is the potential for biogas digesters to improve soil fertility and crop production in sub-Saharan Africa? Biomass and Bioenergy. 2014;**70**:58-72

[66] West T, Marland G. Net carbon flux from agricultural ecosystems: Methodology for full carbon cycle analyses. Environmental Pollution. 2002;**116**(3):439-444

[67] Sedjo R, Sohngen B. Carbon sequestration in forests and soils. Annual Review of Resource Economics. 2012;**4**(1):127-144

[68] Sainju U, Good R. Vertical root distribution in relation to soil properties in New Jersey Pinelands forests. Plant and Soil. 1993;**150**(1):87-97

[69] Muluneh MW, Talema GA, Abebe KB, Dejen Tsegaw B, Kassaw MA, Teka Mebrat A. Determinants of organic fertilizers utilization among smallholder farmers in South Gondar zone, Ethiopia. Environmental Health Insights. 2022;**16**:11786302221075448

Section 2

Production Techniques and Formulations

Chapter 5

A Bioaugmentation Strategy for Promoting the Humification Process during Composting by Microbial Inoculants: A Review

Xiao-Lin Lu, Kai Ding, Xiao-Xia Dong, Gang Li and Jun Ma

Abstract

Stimulating compost humification is an important method to facilitate carbon sequestration, especially against the background of carbon neutrality. However, the disadvantages of traditional composting, including long humification cycles and high environmental risks, restrict its application. Microbial inoculants markedly increase the humus content of compost, and their performance is considerably influenced by the nature of the material, the microbial species, the inoculation dosages, and the inoculation methods. So far, the effects of microbial inoculants on compost maturity and microbial diversity have been widely studied, whereas an overview of their regulatory role in humus formation is still lacking. This review summarizes the promotional effects of microbial inoculants on humification and related biological mechanisms during composting. Further research on the development of microbial inoculants and the optimization of inoculation methods will promote humification and facilitate the production of high-quality compost.

Keywords: carbon sequestration, microbial inoculants, compost humification, biological mechanisms, high-quality compost

1. Introduction

The problem of solid organic waste is becoming increasingly serious due to the worldwide expansion of different industries [1]. Until some years ago, landfill was still the most commonly used disposition of solid organic waste in many countries [2]. This not only causes bad odors but also creates greenhouse gas emissions and soil pollution [3]. Composting is a more useful way of converting various solid organic wastes into humus-like stable products by microorganisms [4].

The composting process is mainly a result of microbial metabolism [5]. Different microorganisms have different functions during composting. However, traditional composting is characterized by a long humification cycle, which is related to the poor activity of indigenous microbes. Thus, regulating microorganisms is a feasible way to facilitate nutrient transformation and humus enhancement in the composting process.

As a biological regulating method, inoculation is an effective way to improve the composting microenvironment and increase microbial activity, which can accelerate compost maturation [6]. The addition of microbial inoculants is also a main bioaugmentation strategy and can markedly increase the levels of humic substances (HS) during composting [3]. However, the efficiency of this approach is sometimes uncertain because of the differences in inoculation methods and the quantity and types of the microbial inoculant [7–12]. Therefore, the improvement of the humification process with microbial inoculants during composting is an important research field.

Given that stimulating compost humification is an important method to facilitate carbon sequestration, especially in the context of carbon neutrality [13], studies identifying the driving mechanisms of humus formation and searching for ways to improve the HS amounts during composting have been performed.

Generally, HS formation is the result of polymerization or condensation of humus precursors, such as phenols, amino acids, and reducing sugars, derived from macromolecule degradation and microbial synthesis under the actions of different metabolic pathways (**Figure 1**) [14, 15]. As one of important components of HS, humic acids (HA) are heterogeneous complexes with various molecular weights and functional groups. A higher HA content indicates a higher degree of humification, facilitating carbon sequestration and soil remediation [16]. Additionally, HA formation might be influenced by various factors, including the precursors, environmental factors, and functional microbial activities [17]. According to a previous study, thermophilic microbes play a key role in humification [18]. However, few reviews have summarized the mechanisms for improving compost humification by bioaugmentation.

The use of microbial inoculants as a bioaugmentation strategy for the green disposal of solid organic waste is widely recommended. Numerous studies reported the considerable promotion of humification by applying microbial inoculants [19]. At the same time, the mechanisms behind these phenomena have attracted increased attention. Nowadays, with the development of novel analytical techniques, we have gained a deeper understanding of how the addition of functional microorganisms

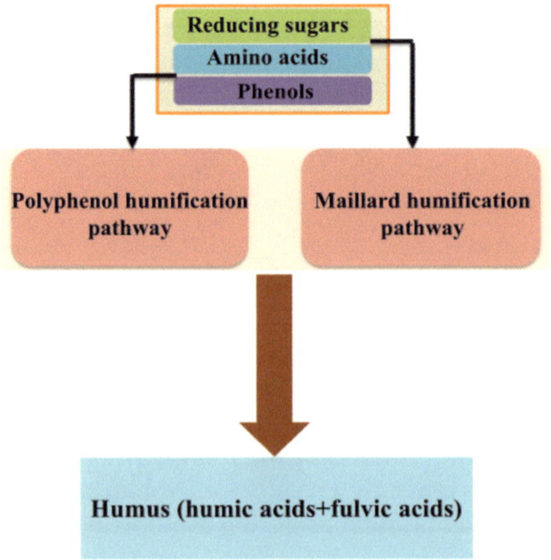

Figure 1.
Formation pathway of humus during the aerobic composting process.

affects humification during composting. In this review, the biological mechanisms of microbial inoculants in promoting compost humification are systematically summarized based on recent studies, with the aim to provide scientific guidance for the efficient application of functional microorganisms to promote the conversion of organic components into humus.

2. Development of microbial inoculants used in composting

According to the principles of classical electromagnetism, charged ions move in a closed-loop circular motion within a uniform magnetic field. Herein, we cite a similar concept ("where it comes from, go there"), namely the "golden closed-loop rule." It implies that microorganisms are to be selected and domesticated to obtain excellent strains for adapting to the native composting environment as well as facilitating the composting process [20, 21]. Accordingly, composting samples are used as the main sources of microbial inoculants. Specially, composting samples taken during the thermophilic period, as a potential microbial mine, have been favored so far [17, 22]. These samples are subjected to a series of microbiological experiments, such as dilution separation, colony purification, molecular identification, high-temperature tolerance tests, inter-strain antagonism tests, and degradative enzyme activity assays [23], to obtain candidates for developing compound microbial inoculants to promote the composting process (**Figure 2**). In high-temperature tolerance tests, temperatures of 45–55°C are frequently selected to identify thermophile microbes [24, 25]. In terms of degradative enzyme activity, cellulase, amylase, laccase, and FPase were investigated [21]. To improve the ability of the microbial inoculants to degrade the composting matrix, enrichment cultures can be adopted for selecting specialized functional strains, following the sole carbon source method. Suitable carbon sources contain wheat straw [26], coffee husk [27], pectin, or sodium carboxy-methylcellulose [28], but the screening and cultivating of microbes from compost are time consuming and laborious [29]. In this context, the efforts needed to develop microbial inoculants can

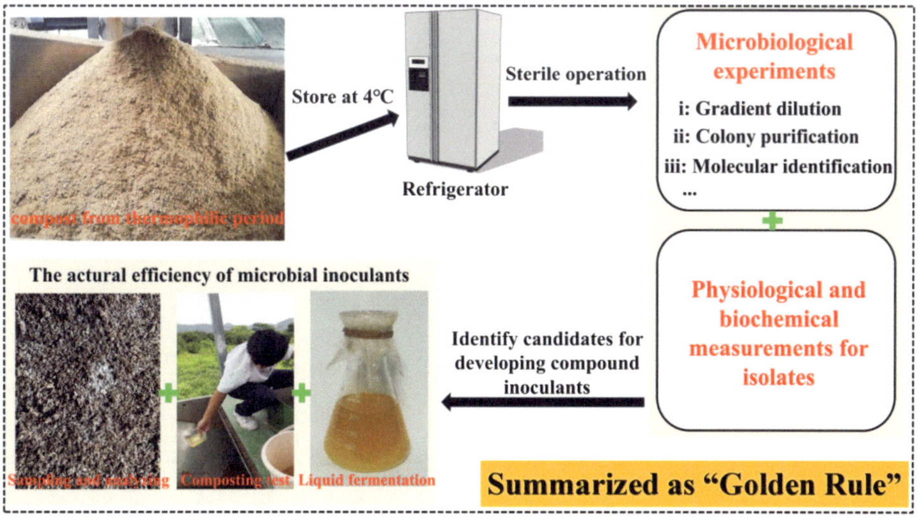

Figure 2.
Illustration of the "golden closed-loop rule" for the development of microbial inoculants used in composting.

Strains	Source	Tolerated temperature (°C)	Cultivation time (d)	Medium	Representative enzymatic activity	Reference
Phanerochaete chrysosporium and *Trichoderma longibrachiatum*	China General Microbiological Culture Collection Center	_	7	Potato dextrose medium	Ligninase, cellulases, chitinolytic, and pectic enzymes	[3]
Aeromonas caviae, *Shinella* sp., *Rhizobium* sp., *Corynebacterium pseudotuberculosis*, *Streptomyces clavuligerus*.	Rice straw samples of composting in high temperature period	45	4–5	Gaoshi No.1 medium	Cellulase activity	[17]
Bacillus licheniformis, *Aspergillus nidulans*, *Aspergillus oryzae*	Dairy manure and sugarcane leaves compost samples	50	3–5	Luria Bertani medium and potato dextrose medium	Cellulosic enzymatic	[21]
Bacillus halmapalus X-2	Cattle manure compost	55	—	Ammonia-oxidizing bacterial seed medium	—	[34]
Streptomyces sp. and *Actinobacteria bacterium*	Organic waste compost from different sources	60	5	Gaoshi No.1 medium	Cellulose activity	[35]
A. caviae, *Shinella* sp., *Rhizobium* sp., *C. pseudotuberculosis*, *S. clavuligerus*.	Rice straw compost samples from the high-temperature period	45	4–5	Gaoshi No.1 medium	Cellulase activity	[36]
Gloeophyllum trabeum	Forest litter	—	7	Potato dextrose medium	Oxidase	[37]
Lactobacillus amylophilus, *Geobacillus thermoleovorans*, and *Bacillus subtilis*	China Center of Industrial Culture Collection	—	1–2	De Man Rogosa Sharpe medium and Luria Bertani medium	Cellulase	[38]

"_": Not given.

Table 1.
Overview of the suitable microbial inoculants for improving compost humification.

be reduced *via* appropriate methods. In addition, since composting needs to be carried out throughout the year, in regions at high latitudes or during winter, the lower ambient temperature limits the composting process [30]. To overcome the adverse effects of low temperatures, it is equally necessary to develop psychrophilic functional strains to initiate composting. Based on previous studies, composting samples derived during a warm period (10°C) are ideal [31–33]. Considering the simple enzymatic performance and high environmental sensitivity of single strains, composite microbial inoculants need to be constructed to accelerate the biodegradation of organic matter in a synergistic manner [3].

According to previous studies, while microbial inoculants facilitate fermentation processes, there are drawbacks of unidirectional nutrient conversion and insignificant compost humification [34]. Above all, the development of microbial inoculants for the directional promotion of compost humification has largely been neglected. However, humification is not only critical for product quality but also facilitates the remediation of polluted soil [4]. In recent years, substantial efforts have been made to develop microbial inoculants for enhancing the humification process in composting. According to various publications, the functional microbial inoculants which could significantly promote the humification process have unique nutritional metabolic properties (**Table 1**). These functional strains were derived from bacteria, fungi, and actinomycetes and show similar enzymatic features, such as high cellulase and oxidase activities [3, 17, 21, 34–38]. Additionally, they can also grow at high temperatures (45–60°C) and under adverse environmental conditions. It is therefore expected that these functional microbial inoculants can considerably promote humification during large-scale composting comparison trials. The next step is therefore the development of commercial products using these microbes [39], which represent an innovation in the field of microbial inoculant technology.

3. Inoculation of microbial inoculants: types, amounts, and time

The inherent recalcitrant nature of raw materials hinders the degradation and transformation of organic fractions, leading to a lower composting efficiency [36]. The addition of microbial inoculants can increase microbial activity, accelerate organic matter degradation, release more precursors, and increase the humification degree of compost products [34, 35, 40]. The type, amount, and inoculation time of microbial inoculants [9, 10, 35, 38, 41, 42] have significant effects on the formation of HS during composting. Thus, it is essential to fully explore the correct use of microbial inoculants to improve the HA levels.

The types of microorganisms used for promoting compost humification are bacteria, fungi, and actinomycetes (**Table 1**). Of these, *Bacillus* sp [34, 38]., *Phanerochaete chrysosporium* [3], and *Streptomyces* sp. [17, 35, 36], can considerably accelerate HS formation. In this regard, actinomycetes have more desirable features than bacteria and fungi, such as thermo-tolerance, adaptability to harsh environments, higher amounts of hydrolytic enzymes that degrade lignocellulose, and a more pronounced response to genetic modification [35, 43]. The formation of HS mainly occurs during the curing period [37], during which actinomycetes are extremely abundant, making them ideal candidates for compost humification bioaugmentation.

The inoculum amount is also a significant factor affecting the humification process during composting [44]. Based on previous findings, inoculation at the level of 4% (in dry weight) is more conducive to the enhancement of HA than that at the level of 2% [9].

Given that various multifunctional microbe populations at different periods drive the composting process [3], it may be justified to inoculate specific microbial inoculants during different periods to reduce competition among microbes [12] and therefore accelerate the humification process. Some studies showed that composting can be considerably promoted by inoculation during the cooling period [35–43]. The multistage inoculation of composite microbial inoculants during the entire composting process also significantly increased the HA level [10, 38, 42]. Accordingly, the addition of microbial inoculants during the cooling period may be a feasible way for improving humification, especially the inoculation of actinomycetes at the level of 4% or higher.

4. Increasing the compost HA levels by adding microbial inoculants

The increase in the HA level indicates the increase in compost humification. As shown in **Table 2**, the addition of microbial inoculants can considerably promote compost humification and increases the content of HA. Different microbial inoculants have different effects on the HA content at the same inoculum level. For example, at a level of 1%, the inoculation of protein-, starch-, oil-, and lignocellulose-degrading microbes as well as ammonia-oxidating bacteria resulted in a more significant increase in the HA content, with levels being 216.88% higher compared to those

Composting materials	Main inoculant types	Inoculation time	Dosage	Effects on HA content	Reference
Chicken manure biogas residue, spent mushroom straw, and rice straw	*Phanerochaete chrysosporium* and *Trichoderma longibrachiatum*	Day 0	5% (w/w, in fresh weight)	Increase by 23.6%	[3]
Medicinal herbal residues	*B. subtilis*, *Aspergillus niger*, *Myceliophthora thermophila*, *Saccharomyces cerevisiae*, *Streptomyces pratensis*, and *S. violascens*	Days 0, 25, and 55, respectively	0.1% (w/w, in dry weight)	Increase by 12.44%	[10]
Wheat straw and fresh cattle manure	*B. clarkii* and *B. halmapalus*	Day 0	5% (v/w, in fresh weight)	Increase by 27.58%	[34]
Pig manure and wheat straw	*Gloeophyllum trabeum*	Prior to composting	—	Increase by 17.2%	[37]
Biogas residue, sawdust, and food waste	*Lactobacillus amylophilus*, *Geobacillus thermoleovorans*, and *B. subtilis*	Days 0, 2, and 14, respectively	0.5% (w/w, in fresh weight)	Increase by 77.78% in humus index	[38]
Food waste and sawdust	Protein-, starch-, oil-, and lignocellulose-degrading microbes, ammonia-oxidating bacteria	Day 0	1% (w/w, in fresh weight)	Increase by 216.88%	[40]

Composting materials	Main inoculant types	Inoculation time	Dosage	Effects on HA content	Reference
Corn straw dairy manure	*Streptomyces* sp. and Actinobacteria	Day 6	2% (w/w, in dry weight)	Increase by 18.75%	[43]
Fresh swine manure and naturally dried corn straw	*Acinetobacter pittii*, *B. subtilis* subsp. stercoris, and *Bacillus altitudinis*	Day 0	1% (v/w, in fresh weight)	Increase by 5.92%	[45]
Fresh cattle manure and dry corn straw	*B. subtilis*, *Bacillus licheniformis*, *B. cereus*, and *Streptomyces nogalater*	Day 0	5% (v/w, in fresh weight)	Increase by 46.12%	[46]

"_": Not given.

Table 2.
Overview of the promotion effects of microbial inoculants on compost humification.

of the control [40]. At an inoculation level of 5%, *Bacillus clarkii* and *B. halmapalus* significantly increased the compost HA content by 27.58% compared to the control in the composting of wheat straw and cattle manure [34]. As a lower inoculation amount generally results in a greater potential increase in HAs, the development of specific microbial inoculants is more important than the optimization of inoculation amounts in the improvement of the HA content. In the following section, the promotion of biotic processes by microbial inoculants is discussed.

5. Bioaugmentation mechanisms and factors related to an increase in compost HA by adding microbial inoculants

The formation of HS is a complex process [21]. Many hypotheses regarding the information of HS include the lignin-protein, phenol-protein, and sugar-amine condensation theories [47]. The core of these theories is that precursors are polymerized to form HS *via* biotic and abiotic ways [15]. In addition to regulating environmental factors and improving the native microbial community in compost, inoculating exogenous microbial inoculants to increase the humification degree of compost products is the main biological method [34]. Due to the complex interactions among microorganisms, advanced mathematical statistics are needed to accurately reflect the ecological interactions of the humification-associated community in the composting environment. Recent studies revealed the bioaugmentation mechanisms and influencing factors of compost humification, which are summarized in **Figure 3**.

The variations in precursor types and concentrations during composting are related to the efficiency of humus synthesis. In addition, the humification process includes the production of precursors and the polymerization of HS, both of which occur sequentially during the whole composting process [17, 36, 47]. Therefore, the deep resolution of bioaugmentation mechanisms about microbial inoculants to composting relies on elucidating the dynamic changes in the precursors, microbial activity, and community structure during different composting periods [48]. Many studies have shown that the inoculation of exogenous microbial inoculants can

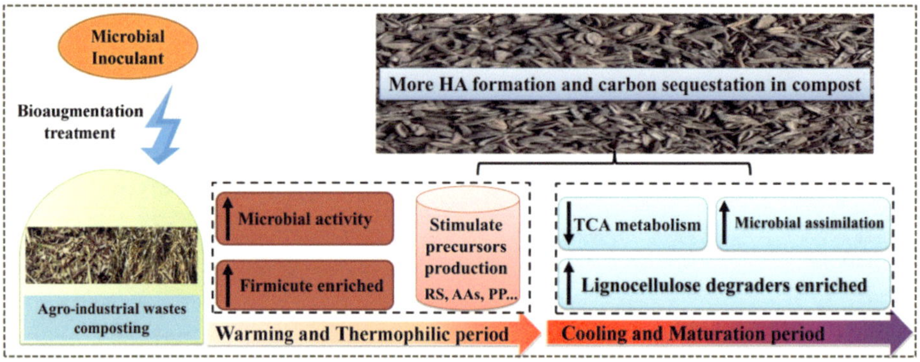

Figure 3.
Conceptual framework summarizing the bioaugmentation mechanisms for the application of microbial inoculants to composting systems. RS, reducing sugars; AAs, amino acids; PP, polyphenol; TCA, tricarboxylic acid.

significantly facilitate the enhancement of functional microbial activity during the warming and thermophilic periods of the composting process, accelerate the degradation of cellulose, proteins, and carbohydrates, and release large amounts of precursors [3, 18, 34, 37, 40, 46]. During these periods, the richness and diversity of the microbial community are markedly increased by the addition of microbial inoculants [21]. Moreover, Firmicutes, involved in the degradation of available organic components, are enriched [45]. The higher relative abundances of lignocellulose-degrading microbes (laccase producing microbes, Chloroflexi, Actinomycetes, fungi, and *Luteimonas*) go along with a decline in pile temperature during the cooling and maturation periods [3, 37, 38, 43], with the synergistic decomposition of lignin [17, 49]. The effective degradation of lignin can provide more carbon skeletons for humus formation [48]. Additionally, microbial inoculants can also considerably reduce the level of carbon metabolism, which is linked to the tricarboxylic acid cycle [44], and stimulate the microbial assimilation of precursors (polyphenols and quinones) [16, 50]. These processes facilitate HA formation and improve carbon sequestration, resulting in a high-value compost.

Generally, HA and HS formation is affected by environmental factors [46]. Specifically, the suitable C/N ratio and the total organic carbon (TOC) level can increase the richness and activity of microbes, resulting in the accumulation of soluble sugar and amino acids in the early period of composting [36]. The increase in the pH and the decline in the TOC results in the enhanced microbial synthesis of HA at the later period of the composting process [17, 34, 36, 40]. Thus, controlling the environmental factors and functional microbes can accelerate the formation of HAs based on the addition of microbial inoculants.

6. Future perspectives

As described above, the addition of microbial inoculants promotes solid organic waste decomposition, stimulates precursor production, increases the HA content, and alleviates several drawbacks of noninoculated composting systems. However, there are still some uncertainties. We highlight some perspectives for the further exploration of the use of functional microbial inoculants:

1. The instability effect is one of the main restrictions in the promotion of microbial inoculant use. Some investigations have confirmed the positive effects of microbial inoculants on humification only at a suitable inoculation level [44]. Further works will have to focus on the inoculation levels for functional microbes used to promote humus formation and related mechanisms, especially regarding the interactions among microbial inoculants, humification, and compost microbiota during composting.

2. Bioaugmentation using multistage inoculation has been well estimated in small-scale composting systems [10]. However, to guide real production, the effectiveness of multistage inoculation remains to be explored, especially regarding the effects of multistage inoculation on humus formation in large-scale composting systems [51]. Efforts will also be made to identify the carbon and nitrogen cycle functional genes related to HA production pathways, using novel analytical methods.

3. Microbial inoculants are generally cheaper than other compost additives. However, the acquisition of high-efficiency functional strains requires more time. In the near future, *in situ* selection and omics methods should be used to rapidly develop simple and stable compositive microbial inoculants (including yeast) that can not only adapt to the native compost environment but also degrade novel pollutants.

7. Conclusions

The addition of microbial inoculants to various solid wastes is a potential method for improving the humification process and increasing carbon sequestration. However, further research is needed to gain a deeper understanding of the biotic mechanisms underlying the microbial inoculants' effect on composting and to explore the relationships among functional microbes, the humification process, and the compost microbiota in large-scale composting systems. A comparison of the influences of different inoculation methods on composting processes and compost quality is also needed. Moreover, the development of multifunctional strains and the optimization of inoculation amounts would further justify the use of microbial inoculants in the compost industry.

Acknowledgements

This work was supported jointly by the research and development project of company's inner key technologies (Optimization of the sequestration capacity of beneficial microbes on organic fertilizer medium). The authors would like to express their gratitude to MogoEdit (https: //www.mogoedit.com/) for the expert linguistic services provided.

Conflict of interest

The authors declare no conflict of interest.

Author details

Xiao-Lin Lu[1], Kai Ding[1,2]*, Xiao-Xia Dong[3], Gang Li[1] and Jun Ma[1]

1 Yangtze Delta Region Healthy Agriculture Institute (Zhejiang) Co., Ltd, Tongxiang, P.R. China

2 Key Laboratory of Urban Environment and Health, Institute of Urban Environment, Chinese Academy of Sciences, Xiamen, P.R. China

3 Zhongke Jiaci (Kunshan) Environmental Protection Technology Co., Ltd, Kunshan, Jiangsu Province, P.R. China

*Address all correspondence to: kding@ydrhai.com

References

[1] Li W, Liu Y, Hou Q, Huang W, Zheng H, Gao X, et al. *Lactobacillus plantarum* improves the efficiency of sheep manure composting and the quality of the final product. Bioresource Technology. 2020;**297**:122456. DOI: 10.1016/j.biortech.2019.122456

[2] Song C, Li M, Qi H, Zhang Y, Liu D, Xia X, et al. Impact of anti-acidification consortium on carbohydrate metabolism of key microbes during food waste composting. Bioresource Technology. 2018;**259**:1-9. DOI: 10.1016/j.biortech.2018.03.022

[3] He J, Zhu N, Xu Y, Wang L, Zheng J, Li X. The microbial mechanisms of enhanced humification by inoculation with *Phanerochaete chrysosporium* and *Trichoderma longibrachiatum* during biogas residues composting. Bioresource Technology. 2022;**351**:126973. DOI: 10.1016/j.biortech.2022.126973

[4] Wu J, Zhao Y, Yu H, Wei D, Yang T, Wei Z, et al. Effects of aeration rates on the structural changes in humic substance during co-composting of digestates and chicken manure. Science of the Total Environment. 2019;**658**:510-520. DOI: 10.1016/j.scitotenv.2018.12.198

[5] Zhang Y, Zhao Y, Chen Y, Lu Q, Li M, Wang X, et al. A regulating method for reducing nitrogen loss based on enriched ammonia-oxidizing bacteria during composting. Bioresource Technology. 2016;**221**:276-283. DOI: 10.1016/j.biortech.2016.09.057

[6] Sun Q, Wu D, Zhang Z, Zhao Y, Xie X, Wu J, et al. Effect of cold-adapted microbial agent inoculation on enzyme activities during composting start-up at low temperature. Bioresource Technology. 2017;**244**:635-640. DOI: 10.1016/j.biortech.2017.08.010

[7] Chen Y, Wang Y, Xu Z, Liu Y, Duan H. Enhanced humification of maize straw and canola residue during composting by inoculating *Phanerochaete chrysosporium* in the cooling period. Bioresource Technology. 2018;**247**:190-199. DOI: 10.1016/j.biortech.2019.122075

[8] Jiang J, Liu X, Huang Y, Huang H. Inoculation with nitrogen turnover bacterial agent appropriately increasing nitrogen and promoting maturity in pig manure composting. Waste Management. 2015;**39**:78-85. DOI: 10.1016/j.wasman.2015.02.025

[9] Liu L, Wang S, Guo X, Zhao T, Zhang B. Succession and diversity of microorganisms and their association with physicochemical properties during green waste thermophilic composting. Waste Management. 2018;**73**:101-112. DOI: 10.1016/j.wasman.2017.12.026

[10] Lu XL, Wu H, Song SL, Bai HY, Tang MJ, Xu FJ, et al. Effects of multi-phase inoculation on the fungal community related with the improvement of medicinal herbal residues composting. Environmental Science Pollution Research. 2021;**28**:27998-28013. DOI: 10.1007/s11356-021-12569-7

[11] Wei Y, Zhao Y, Shi M, Gao Z, Lu Q, Yang T, et al. Effect of organic acids production and bacterial community on the possible mechanism of phosphorus solubilization during composting with enriched phosphate-solubilizing bacteria inoculation. Bioresource Technology. 2018;**247**:190-199. DOI: 10.1016/j.biortech.2017.09.092

[12] Xi B, He X, Dang Q, Yang T, Li M, Wang X, et al. Effect of multi-stage inoculation on the bacterial and fungal

community structure during organic municipal solid wastes composting. Bioresource Technology. 2015;**247**:190-199. DOI: 10.1016/j.biortech.2015.07.069

[13] Jiang J, Wang Y, Yu D, Hou R, Ma X, Liu J, et al. Combined addition of biochar and garbage enzyme improving the humification and succession of fungal community during sewage sludge composting. Bioresource Technology. 2022;**346**:126344. DOI: 10.1016/j.biortech.2021.126344

[14] Stevenson FJ. Humic Chemistry: Genesis, Composition, Reactions. New York: Reactions. John Wiley and Sons; 1994. Available from: http://journals.lww.com/00010694-198302000-00014

[15] Wu J, Zhao Y, Wang F, Zhao X, Dang Q, Tong T, et al. Identifying the action ways of function materials in catalyzing organic waste transformation into humus during chicken manure composting. Bioresource Technology. 2020;**303**:122927. DOI: 10.1016/j.biortech.2020.122927

[16] Wei Z, Mohamed TA, Zhao L, Zhu Z, Zhao Y, Wu J. Microhabital drive microbial anabolism to promote carbon sequestration during composting. Bioresource Technology. 2022;**346**:126577. DOI: 10.1016/j.biortech.2021.126577

[17] Wu D, Wei Z, Qu F, Mohamed TA, Zhu L, Zhao Y, et al. Effect of Fenton pretreatment combined with bacteria inoculation on humic substances formation during lignocellulosic biomass composting derived from rice straw. Bioresource Technology. 2020;**303**:122849. DOI: 10.1016/j.biortech.2020.122849

[18] Ma F, Zhu T, Yao S, Quan H, Zhang K, Liang B, et al. Coupling effect of high temperature and thermophilic bacteria indirectly accelerates the humifcation process of municipal sludge in hyperthermophilic composting. Process Safety and Environmental Protection. 2022;**166**:469-477. DOI: 10.1016/j.pesp.2022.08.052

[19] Rastogi M, Nandal M, Khosla B. Microbes as vital additives for solid waste composting. Heliyon. 2020;**6**:e03343. DOI: 10.1016/j.heliyon.2020.e03343

[20] Kang J, Yin Z, Pei F, Ye Z, Song G, Ling H, et al. Aerobic composting of chicken manure with penicillin G: Community classification and quorum sensing mediating its contribution to humification. Bioresource Technology. 2022;**352**:127097. DOI: 10.1016/j.biortech.2022.127097

[21] Xu J, Jiang Z, Li M, Li Q. A compost-derived thermophilic microbial consortium enhances the humification process and alters the microbial diversity during composting. Journal of Environmental Management. 2019;**243**:240-249. DOI: 10.1016/j.jenvman.2019.05.008

[22] Li C, Li H, Yao T, Su M, Ran F, Han B, et al. Microbial inoculation influences bacterial community succession and physicochemical characteristics during pig manure composting with corn straw. Bioresource Technology. 2019;**289**:121653. DOI: 10.1016/j.biortech.2019.121653

[23] Jurado M, Lopez MJ, Suarez-Estrella F, Vargas-Garcia MC, Lopez-Gonzalez JA, Moreno J. Exploiting composting biodiversity: Study of the persistent and biotechnologically relevant microorganisms from lignocellulose-based composting. Bioresource Technology. 2014;**162**:283-293. DOI: 10.1016/j.biortech.2014.03.145

[24] Chandna P, Nain L, Singh S, Kuhad RC. Assessment of bacterial

diversity during composting of agricultural byproducts. BMC Microbiology. 2013;**13**:99. Available from: http://www.biomedcentral.com/1471-2180/13/99

[25] Lopez-Gonzalez J, Suarez-Estrella F, Vargas-Garcia M, Jurado M, Moreno J. Dynamics of bacterial microbiota during lignocellulosic waste composting: Studies upon its structure, functionality and biodiversity. Bioresource Technology. 2015;**175**:406-416. DOI: 10.1016/j.biortech.2014.10.123

[26] Alessi AM, Bird SM, Oates NC, Li Y, Dowle AA, Novotny EH, et al. Defining functional diversity for lignocellulose degradation in a mcirobial community using multi-omics studies. Biotechnology for Biofuels. 2018;**11**:166. DOI: 10.1186/s13068-018-1164-2

[27] Cerda A, Mejias L, Gea T, Sanchez A. Cellulase and xylanase production at pilot scale by solid-state fermentation from coffee husk using specialized consortia: The consistency of the process and the microbial communities involved. Bioresource Technology. 2017;**243**:1059-1068. DOI: 10.1016/j.biortech.2017.07.076

[28] Wang J, Liu Z, Xia J, Chen Y. Effect of microbial inoculation on physicochemical properties and bacterial community structure of citrus peel composting. Bioresource Technology. 2019;**291**: 121843. DOI: 10.1016/j.biortech.2019.121843

[29] Qi H, Zhao Y, Wang X, Wei Z, Zhang X, Wu J, et al. Manganese dioxide driven the carbon and nitrogen transformation by activating the complementary effects of core bacteria in composting. Bioresource Technology. 2021;**330**:124960. DOI: 10.1016/j.biortech.2021.124960

[30] Gou C, Wang Y, Zhang X, Lou Y, Gao Y. Inoculating with a psychrotrophic-thermophilic complex microbial agent accelerates onset and promotes maturity of dairy manure-rice straw composting under cold climate conditions. Bioresource Technology. 2017;**243**:339-346. DOI: 10.1016/j.biortech.2017.06.097

[31] Abdellah YAY, Li T, Chen X, Cheng Y, Sun S, Wang Y, et al. Role of psychrotrophic fungal strains in accelerating and enhancing the maturity of pig manure composting under low-temperature conditions. Bioresource Technology. 2021;**320**:124402. DOI: 10.1016/j.biortech.2021.124402

[32] Jiang G, Chen P, Bao Y, Wang X, Yang T, Mei X, et al. Isolation of a novel psychrotrophic fungus for efficient low-temperature composting. Bioresource Technology. 2021;**331**:125049. DOI: 10.1016/j.biortech.2021.125049

[33] Shi W, Dong Q, Saleem M, Wu X, Wang N, Ding S, et al. Microbial-based detonation and processing of vegetable waste for high quality compost production at low temperatures. Bioresource Technology. 2022;**369**:133276. DOI: 10.1016/j.biortech.2022.133276

[34] Xu Z, Li R, Wu S, He Q, Ling Z, Liu T, et al. Cattle manure compost humification process by inoculation ammonia-oxidizing bacteria. Bioresource Technology. 2022;**344**:126314. DOI: 10.1016/j.biortech.2022.126314

[35] Zhao Y, Zhao Y, Zhang Z, Wei Y, Wang H, Lu Q, et al. Effect of thermo-tolerant actinomycetes inoculation on cellulose degradation and the formation of humic substances during composting. Waste Management. 2017;**68**:64-73. DOI: 10.1016/j.wasman.2017.06.022

[36] Wu D, Xia T, Zhang Y, Wei Z, Qu F, Zheng G, et al. Identifying driving

factors of humic acid formation during rice straw composting based on Fenton pretreatment with bacterial inoculation. Bioresource Technology. 2021;**337**:125403. DOI: 10.1016/j.biortech.2021.125403

[37] Zhu N, Zhu Y, Liang D, Li B, Jin H, Dong Y. Enhanced turnover of phenolic precursors by *Gloeophyllum trabeum* pretreatment promotes humic substance formation during co-composting of pig manure and wheat straw. Journal of Cleaner Production. 2021;**315**:128211. DOI: 10.1016/j.jclepro.2021.128211

[38] Xu M, Yang M, Sun H, Meng J, Li Y, Gao M, et al. Role of multistage inoculation on the co-composting of food waste and biogas residue. Bioresource Technology. 2022;**361**:127681. DOI: 10.1016/j.biortech.2022.127681

[39] Wang G, Kong Y, Yang Y, Ma R, Shen Y, Li G, et al. Superphosphate, biochar, and a microbial inoculum regulate phytotoxicity and humification during chicken manure composting. Science of the Total Environment. 2022;**824**:153958. DOI: 10.1016/j.scitotenv.2022.153958

[40] Zhou X, Li J, Zhang J, Deng F, Chen Y, Zhou P, et al. Bioaugmentation mechanism on humic acid formation during composting of food waste. Science of the Total Environment. 2022;**830**:154783. DOI: 10.1016/j.scitotenv.2022.154783

[41] Zhang T, Wu X, Shaheen SM, Rinklebe J, Bolan NS, Ali E, et al. Effects of microorganism-mediated inoculants on humification processes and phosphorus dynamics during the aerobic composting of swine manure. Journal of Hazardous Materials. 2021;**416**:125738. DOI: 10.1016/j.jhazmat.2021.125738

[42] Lu X, Bai H, Zhang A, Dai C, Ma Y, Jia Y. Effect of multi-stage inoculation of materity and microbial diversity during Chinese medicinal herbal residue composting. Research of Environmental Sciences. 2021;**34**:439-449. DOI: 10.13198/j.issn.1001-6929.2020.08.03

[43] Zhao Y, Lu Q, Wei Y, Cui H, Zhang X, Wang X, et al. Effect of actinobacteria agent inoculation methods on cellulose degradation during composting based on redundancy analysis. Bioresource Technology. 2016;**219**:196-203. DOI: 10.1016/j.biortech.2016.07.117

[44] Duan M, Zhang Y, Zhou B, Qin Z, Wu J, Wang Q, et al. Effects of *Bacillus subtilis* on carbon components and microbial functional metabolism during cow manure–straw composting. Bioresource Technology. 2020;**303**:122868. DOI: 10.1016/j.biortech.2020.122868

[45] Li C, Li H, Yao T, Su M, Ran F, Li J, et al. Effects of swine manure composting by microbial inoculation: Heavy metal fractions, humic substances, and bacterial community metabolism. Journal of Hazardous Materials. 2021;**415**:125559. DOI: 10.1016/j.jhazmat.2021.125559

[46] Wang X, Tian L, Li Y, Zhong C, Tian C. Effects of exogenous cellulose-degrading bacteria on humus formation and bacterial community stability during composting. Bioresource Technology. 2022;**359**:127458. DOI: 10.1016/j.biortech.2022.127458

[47] Wu J, Zhao Y, Zhao W, Yang T, Zhang X, Xie X, et al. Effect of precursors combined with bacteria communities on the formation of humic substances during different materials composting. Bioresource Technology. 2017;**226**:191-199. DOI: 10.1016/j.biortech.2016.12.031

[48] Wang YY, Zhao B, Ma LT, Li L, Deng YQ, Xu Z. Humification process and microbial driving mechanism of composting. Biotechnology Bulletin. 2022;**38**:22-28. DOI: 10.13560/j.cnki.biotech.bull.1985.2022-0134

[49] Qu F, Wu D, Li D, Zhao Y, Zhang R, Qi H, et al. Effect of Fenton pretreatment combined with bacterial inoculation on humification characteristics of dissolved organic matter during rice straw composting. Bioresource Technology. 2022;**344**:126198. DOI: 10.1016/j.biortech.2022.126198

[50] Zhang Z, Zhao Y, Yang T, Wei Z, Li Y, Wei Y, et al. Effects of exogenous protein-like precursors on humification process during lignocellulose-like biomass composting: Amino acids as the key linker to promote humification process. Bioresource Technology. 2019;**291**:121882. DOI: 10.1016/j.biortech.2019.121882

[51] Chen L, Chen Y, Li Y, Liu Y, Jiang H, Li H, et al. Improving the humification by additives during composting: A review. Waste Management. 2023;**158**:93-106. DOI: 10.1016/j.wasman.2022.12.040

Chapter 6

A Holistic Approach of Organic Farming in Improving the Productivity and Quality of Horticultural Crops

Emrul Kayesh, Joydeb Gomasta, Nadira Bilkish, Khadiza Akter Koly and Sharmila Rani Mallick

Abstract

Horticultural crops take into account fruits, vegetables, medicinal, aromatic, and ornamental plants. These crops play a vital role in dietary nutritional components and sources of medicines and aroma along with extensive aesthetic values for human beings. Horticulture is also becoming essential to meet the demand for fruits, vegetables, and other horticultural products for the fast-growing global population. With the rise of population, industrialization, and globalization, the arable soil resource is abated rapidly. Again, as a result of green revolution in the post-independence age, the "resource degrading" chemical or inorganic agriculture has given way to "resource protective" biological or organic farming as a means of preserving agricultural production against the demands placed on the earth's limited natural resources in the many developing nations. Organic farming is a holistic approach that promotes environmentally, socially, and economically sound production of food. During the last two decades, there has also been a significant sensitization of the global community toward environmental preservation and assuring food quality. This chapter aims to provide an updated knowledge of organic agriculture and its potential uses for enhancing productivity and quality of horticultural crops, saving the soil from chemical contamination and environmental preservation to ensure safe food for human beings.

Keywords: organic agriculture, holistic approach, productivity, quality, horticultural crops

1. Introduction

Holistic Management is a whole farm planning system that helps farmers, ranchers, and land stewards better manage agricultural resources in order to reap sustainable environmental, economic, and social benefits [1]. A holistic approach encompasses food security and environmental and social goals. It helps restore the health of agricultural ecosystems and increases the resilience of farms to future challenges. Roger M. Savory [2] is originally credited with the development of the term Holistic Management in agriculture that is designed to restore degraded grasslands using a method that integrates economic, social, and environmental variables (particularly movements of grazing livestock) into land management. Again, agriculture is facing difficulty feeding the vast population that is expanding quickly while still having enough food to spare for future generations. But sustainable human agricultural activities directly or indirectly responsibly cause climate change, the depletion of nonrenewable resources, and water contamination, which altogether has been putting the future food supply in danger.

Holistic agricultural systems that ascertain enhanced productivity by making the best utilization of natural resources and ecological processes are better suited to tackle these difficulties rather than using the reductionist approaches that prioritize output maximization only [3]. Organic agriculture is a holistic system considered to sustain and enhance the profitability of organic yield [4]. Organic farming is a sustainable approach that positively impacts the environment and health of human beings and wildlife because no agrochemicals such as pesticides, insecticides, herbicides, and synthetic fertilizers are used compared to conventional farming [5]. Sustainable agricultural systems also rely on the traditional knowledge and entrepreneurial skills of farmers and include both organic farming and agroecological methods. Organic farming (OF), also known as ecological farming or biological farming, can be defined as an integrative farming technology that is ecologically, economically, and socially acceptable and that ensures sustainable supply of safe and healthy foods and fibers with the least possible amount of resource use and the least amount of ecological harm.

The IFOAM General Assembly organized in June 2008 in Italy defined organic agriculture as "a production system that sustains the health of soils, ecosystems, and people through depending on ecological processes, biodiversity, and cycles adapted to local conditions, rather than the use of inputs with adverse effects." United States Department of Agriculture (USDA) and UN-Food and Agriculture Organization (FAO) also termed organic farming as a method that, to the greatest possible extent, relies on crop rotation, crop residues, animal manures, off-farm organic waste, mineral grade rock additives and biological systems of nutrient mobilization and plant protection instead of avoiding or largely excluding the use of synthetic inputs (such as fertilizers, pesticides, hormones, feed additives). Environmental preservation, livestock production, and animal care are prioritized in organic farming [6] and discourage the application of chemical fertilizers, pesticides, and herbicides [7]. Organic farming also forbids the creation of genetically modified organisms (GMOs) and their usage in animal feed. It is distinguished by the use of regulated standards (production regulations), compelled control programs, and particular labeling strategies compared to other agricultural production techniques [8]. Organic farming is, therefore, economic, environmentally safe and produces hygienic foods with no or less pesticide residue than those made by conventional agriculture [9, 10]. The concepts of organic farming are based on the idea that everything in a living system—soil, plants, farm animals, insects, the farmer, etc.—is interconnected. Therefore, it must be based on a thorough understanding and clever management of these interactions and processes. Dependence on extraneous inputs, whether synthetic or natural, is lessened as far as possible. Organic farming is promoted as one of the sustainable agricultural systems augmenting human nutrition by producing a variety of crops including fruits, vegetables, and livestock in addition to reducing the negative effects of high-input-based agriculture [11]. More recently, organic agriculture production has been rapidly increasing in all parts of the universe [9, 12] because of people's enhanced willingness to consume organic products even at a premium price.

Moreover, healthy soils are essential for biodiversity, ecological safety, and food security, which combat climate change issues. By encouraging the switch to organic farming, we are helping to repair our planet's soil by preventing chemically induced deterioration and enhancing their potential as carbon sinks. Therefore, there needs to be a permanent shift toward ecologically sound land use, which incorporates holistic techniques like organic farming and agroecology together with the preservation and restoration of natural ecosystems like peatlands and forests. But in order to ensure mitigation and adaptation in the face of the current climate crisis, soil preservation is essential. Organic farming is, therefore, a holistic system for quality crop production, not even endangering the soil as well as the climate.

2. Global history, status, and economics

2.1 Background

The origin of the concept of organic farming is actually primitive. Organic farming first became popular at the turn of the twentieth century (**Table 1**). It began as a response to opposition to the industrialization of agriculture and worries regarding the use of chemical and mineral pesticides [14]. The "Life Reform Movement" (Lebensreform Bewegung) in Germany in the 1920s, which opposed modernization and industrialization and idealized vegetarian food, self-sufficiency, natural medicine, allotment gardens, outdoor physical work, and all types of nature conservation, was one of the early pioneers of organic agriculture [15]. But the first distinct form of organic agriculture was introduced by the Austrian Rudolf Steiner, who delivered a series of lectures in 1924 and later published the series as "Spirituals Foundations for Renewal of Agriculture" coining the term—biodynamic agriculture [16, 17]. However, in the late twentieth century, the use of chemicals in agriculture began to be widespread. Since 1990, the market for organic products has been growing rapidly. Today, organic agriculture is a mainstream interest in Western societies, although it has been criticized for not considering contradictory evidence regarding some of its claims [18, 19]. As the demand for organic produce increases, so does the area under organic cultivation.

2.2 Global organic agriculture status

Almost 38% of global land area is covered by agricultural production [20]. Even though just 1.6% of all agricultural land in the world is employed in organic agriculture [21], the proportion of organic farms and agricultural land is steadily increasing. According to the Research Institute of Organic Agriculture (FiBL) and IFOAM's (2022) latest survey, in 2020, about 74.9 million hectares of agriculture land, which was merely 11 Mha in 1999, are managed organically on a continental basis involving more than 190 countries of the world. Oceania accounts for almost half of the total organic agriculture land worldwide with a land area of 35.9 million hectares (Mha), followed by Europe (17.1 Mha), Latin America (9.9 Mha), Asia (6.1 Mha), North America (3.7 Mha), and Africa (2.1 Mha), as shown in **Table 2** and **Figure 1**, as reported by Willer et al. [21]. About one-third of the world's organically managed land is located in the developing countries [22], involving around 65%

Year	Country	Historical focuses
1911	USA	Franklints King's "Farmers for Fourties Centuries" acknowledged the Asian soil management practices, and recommended other agriculturists.
1924	Germany	Rudolf Steiner's lecture series, later published as "Spiritual Foundations for the Renewal of Agriculture" coined—biodynamic agriculture.
1927	Germany	"The Natural Farming" and "Back-to-back Land Association" movement.
1931	Germany/UK	Germany: Eward Konemann "Biological Soil Culture and Manure Economy," Vol 1 UK: Sir Albert Howard "The Waste Product of Agriculture"; often refereed as "Father of modern organic agriculture."
1932	Germany	Eward Konemann "Biological Soil Culture and Manure Economy," Vol 2.
1937	Germany	Eward Konemann "Biological Soil Culture and Manure Economy," Vol 3.
1938	Germany/UK	Germany: Ehrenfried Pfeiffer "Biodynamic Farming and Gardening" UK: Sir Robert McCarrison inspired GT Wrench's "The Wheel of Health."
1940	USA/UK	USA: Rodale Organic Gardening and Experimental Farm (Rodale Institute today), 2nd longest experimental farm on organic vs. conventional; UK: [a] Sir Albert Howard's "An Agricultural Testament," [b] Lord Walter Northbourne: "Look to the Land"—first spell out "organic farming."
1942	USA	Jerome Rodale's "Organic Farming and Gardening."
1943	UK	[a] Lady Eva Balfour, founder and the first president of Soil Association in Britain, "The Living Soil" and started [b] "Haughley Experiment"—the first longest experimental farm on organic versus nonorganic.
1945	USA	Jerome Rodale "Pay Dirt."
1947	UK	Sir Albert Howard "The Soil and Health: A Study of Organic Agriculture."
1962	USA	Rachel Carson "Silent Spring" brought about an environmental and social movement. She is often referred as "mother of environmental movement."
1970	France	Claude Aubert "L'Agriculture Biologique"—a popular book, helped to form the Frenche association Nature et Progres.
1972	France	Formed the International Federation of Organic Agriculture Movement (IFOAM).
1973	Germany	Formed Research Institute of Organic Agriculture (FiBL).
1978	Germany	FiBL started the DOK trial—the longest experimental trial among biodynamic (B), organic (O), and conventional (K).
1984	USA	First spell out "organic agriculture" in the policy document.
1989	USA	The National Research Council report entitled "Alternative Agriculture."
1990	USA	[a] Endorsed "Organic Food Production Act" that established USDA National Organic Program; [b] Nicolas Lampkin "Organic Farming," a very popular publication.

Source: [13].

Table 1.
Global organic agriculture history in the twentieth century.

of the developing countries. According to the FiBL-IFOAM report [21], organic farmland increased by 3.0 million hectares (4.1 percent) in 2020 since 2019, while in the past 10 years (from 2011 to 2020), world organic farmland increased by 104.3% (**Figure 2**). Compared with 1999, when 15 million hectares were organic, organic

Region	Producers (no.)	Retail sales	Per capita consumption
Africa	833,986	16	0.01
Asia	1,808,464	12,540	2.7
Europe	417,977	52,000	63.2
Latin America	270,472	778	1.2
North America	22,448	53,717	147.5
Oceania	15,930	1594	38.4
World	3,368,254	120,647	15.8

Source: [21].

Table 2.
Organic producers, retail sales, and consumption (million €) by region in 2020.

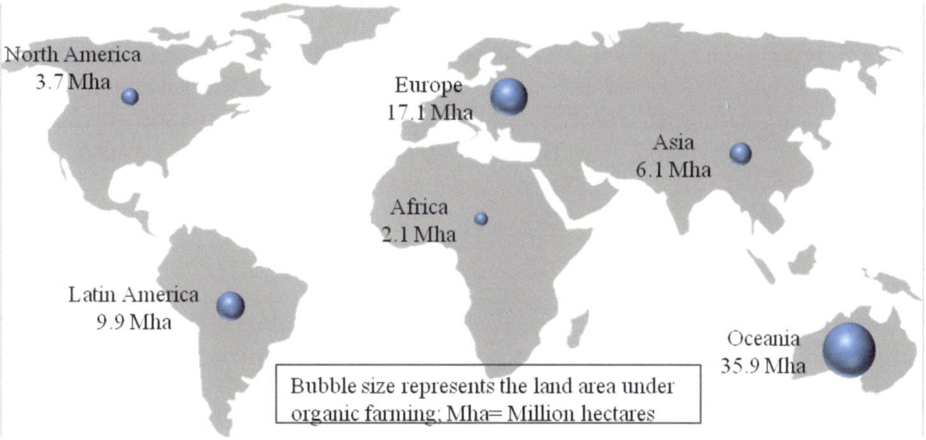

Figure 1.
Distribution of organic farming area by region in 2020 (source: [21]).

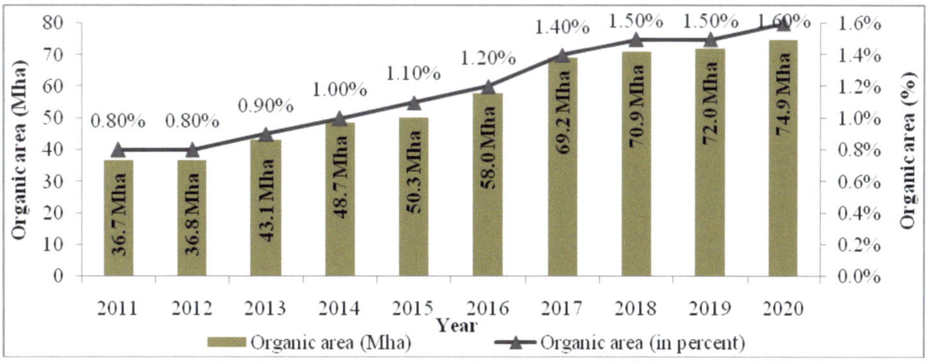

Figure 2.
Growth of the organic agricultural land and organic share in past 10 years (during 2011–2020) (source: [21]).

agricultural land has increased fivefold by 2020. The highest absolute growth was in Latin America (+19.9 percent, +1.7 million hectares), followed by Europe (+3.7 percent, +0.60 million hectares) and Asia (+7.6 percent, +0.43 million hectares). Many countries reported a significant increase; Chile and Papua New Guinea showed 650 percent and 322 percent more organic farmland, respectively. Argentina, Uruguay, and India saw the largest gains in terms of absolute hectares: in Argentina, organic farmland expanded by 781,000 (+21.3%), in Uruguay by more than 589,000 (+27.9%), and in India by over 359,000 (+15.6%). The latest FiBL survey on organic agriculture revealed that Australia occupies the largest individual land shares having 35.7 Mha of organic farmland followed by Argentina (4.5 Mha) and Uruguay (2.7 Mha). In terms of countries organic land shares to its total agricultural land, Liechtenstein has 41.6% of its agricultural land that is cultivated organically, followed by Austria (26.5%) and Estonia (22.4%). The survey also showed that 88 countries (54%) have less than 1% organic land, while another 46 countries (28%) have 1–5% organic share, and among the rest, 18 countries have 5–10% and 11 countries have more than 10% organic land (**Figure 3**). Australia had the largest organic land area in the Oceania region as well as the country became the top of the world, having had the largest individual organic area of 35,687,799 hectares representing 47.63% of the global organic farmland, whereas USA, Argentina, India, France, and Tunisia were the toppers in North America, Latin America & the Caribbean, Asia, Europe, and Africa, respectively, regarding organic agricultural land shares [21]. Again, the leading 10 countries constitute about 78.86% of the world's organic agricultural land: Australia, Argentina, Uruguay, India, France, Spain, China, USA, Italy, and Germany [21].

2.3 Economics

Over the last 20 years, the global market for organic products has grown dramatically, notably in developing nations. Organic food sales increased steadily, especially from the late twentieth century. According to the FiBL-IFOAM survey [21], at least 3.4 million organic producers existed globally in 2020. Asia accounts for 56% of the world's organic producers, followed by Africa (24%), Europe (12%), and Latin America (8%). India had the majority of the producers (1,599,010), followed by Ethiopia (219,566) and Tanzania (148,607) (**Table 2**). From 2019 to 2020, a 7.6% increment in the number of organic producers was noted. According to the survey, revenues from organic foods and beverages exceeded 120 billion euros in 2020. The United States (49.5 billion euros), Germany (15.0 billion euros), and France (12.7 billion euros) had the largest organic markets in 2020. The United States accounted for 41% of the global market, followed by the European Union and China. Switzerland had the highest per capita organic food expenditure (418 euros) in 2020. The countries with the biggest market shares for organic products were Denmark (13.1%), Austria (11.3%), and Switzerland (10.8 percent) (**Figure 4**). The market for organic agri-food goods in the EU kept expanding; however, between 2019 and 2020, imports of these products marginally dropped. The Netherlands, Germany, and Belgium were the top three EU member states for imports in 2020.

3. Basic concepts and principles followed in organic farming

As a result of various analyses of financial, environmental, and sociocultural goals, organic farming may improve conventional agriculture to the point where it may seem unnecessary to strictly forbid pesticides and mineral fertilizers as required

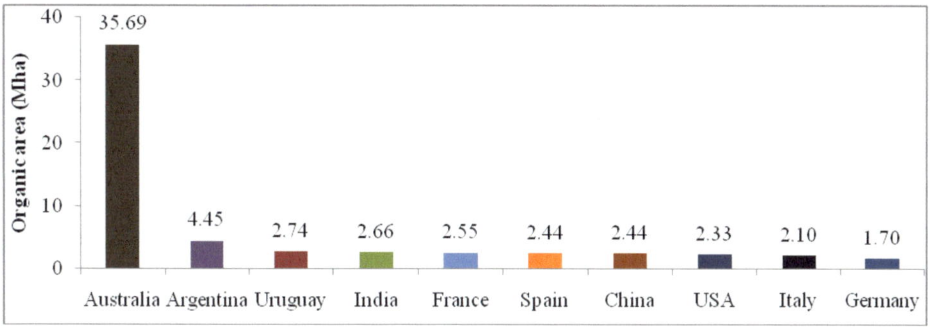

Figure 3.
Top 10 countries with the largest areas of organic agricultural land in 2020 (source: [21]).

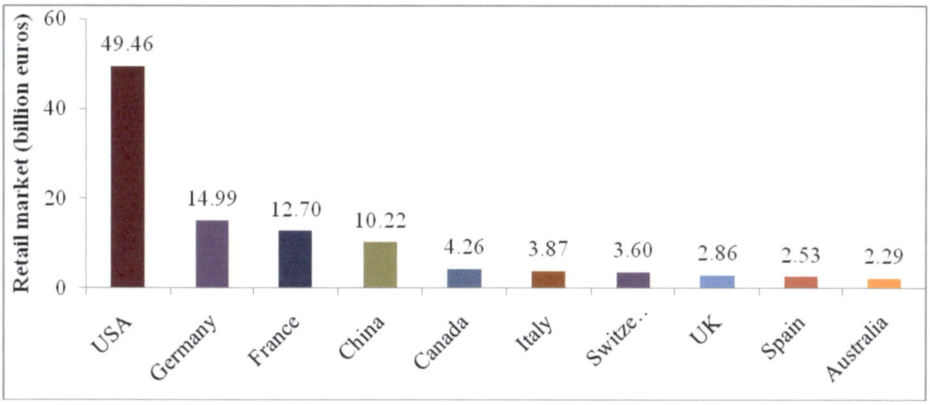

Figure 4.
The top 10 countries with the largest markets for organic food in 2020 (source: [21]).

by the organic standard (**Figure 5**). Organic farming encourages the following basic issues [23]:

- To utilize available resources and operate as much as feasible within a closed system.
- To keep soils fertile for the long run.
- To avoid all forms of pollution that may result from agricultural techniques.
- To generate sufficient amounts of food that are high in nutrients.
- To utilize as little fossil energy as possible in agricultural practices.
- To provide animals with living conditions consistent with their physiologic requirements.
- To help farmers achieve financial security through their job and realize their full potential as people.

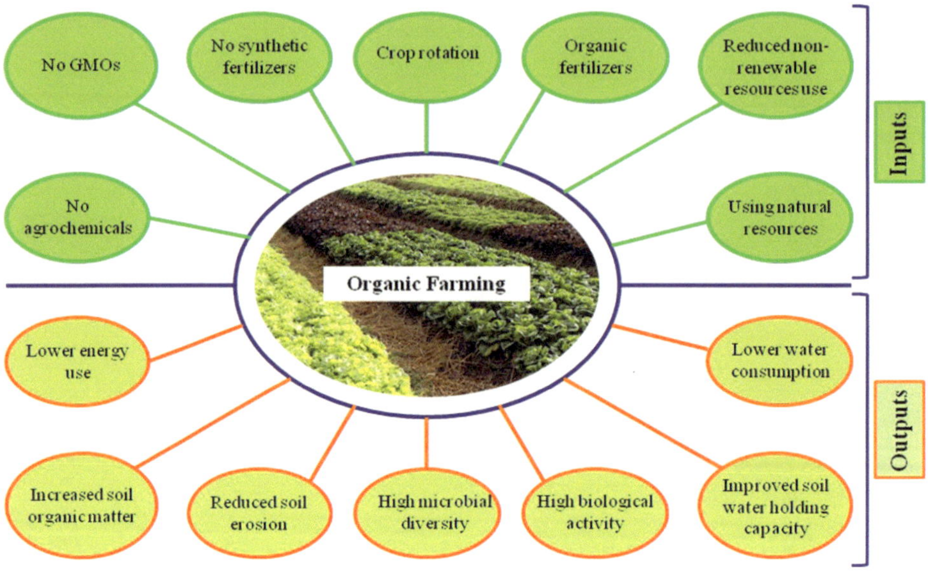

Figure 5.
Generalized input and output features of organic farming system.

Thus, organic farming methods accord with the four ethical principles [24–26].

a. Principle of health.

"Organic agriculture should sustain and enhance the health of soil, plant, animal, humans and planet as one and indivisible".

b. Principle of ecology.

"Organic agriculture should be based on living ecological systems and cycles, work with them, emulate them and help sustain them".

c. Principle of fairness.

"Organic agriculture should build on relationships that ensure fairness with regard to the common environment and life opportunities".

d. Principle of care.

"Organic agriculture should be managed in a precautionary and responsible manner to protect the health and well-being of current and future generations and the environment".

4. Strengths/blessings of organic farming

The most important benefits of organic farming are the preservation of the environment and increased resistance to ecological change, as well as the improvement of social capacity and the expansion of employment prospects. It is an ecologically safe and environmentally friendly production system spreading worldwide as the demand for sustainability increases [21, 27]. Although farm yield is less in organic system compared to conventional systems [28, 29], they are more profitable, pollinator friendly, environmentally safe and produce equally or even more healthy foods with fewer pesticide residues (**Figure 6**) [9, 31, 32]. Therefore, the benefits or overall gains from organic farming can be summarized as follows:

4.1 Profitable

In many cases, organic agriculture is significantly more profitable than conventional agriculture when premium prices are considered. Crowder and Reganold [33] stated after investigating 55 crops grown on five continents that organic agriculture was significantly more profitable (22–35%) and had higher benefit/cost ratios (20–24%) than conventional agriculture. But when organic premiums were taken away, net present values (−27 to −23%) and benefit/cost ratios (−8 to −7%) of organic agriculture were significantly lower than conventional agriculture. According to a recent global comparison research, organic farming is 13% more profitable than conventional farming on average [30]. Generally, organically produced goods fetch 10 to 50% premium price over conventional production and also possess a faster marketing rate [34, 35].

4.2 Multifunctional use and resilience

Organic food and farming practices typically increase the resilience of agroecosystems in addition to generating food by supplying a variety of ecosystem goods and services, some of which are listed below. By doing so, they might achieve social and environmental policy objectives. For instance, they cover both animal welfare and the means of subsistence for farmers and farmworkers. Grazing animals are a crucial component of the utilization of the land [6, 36]. Organic agriculture produces

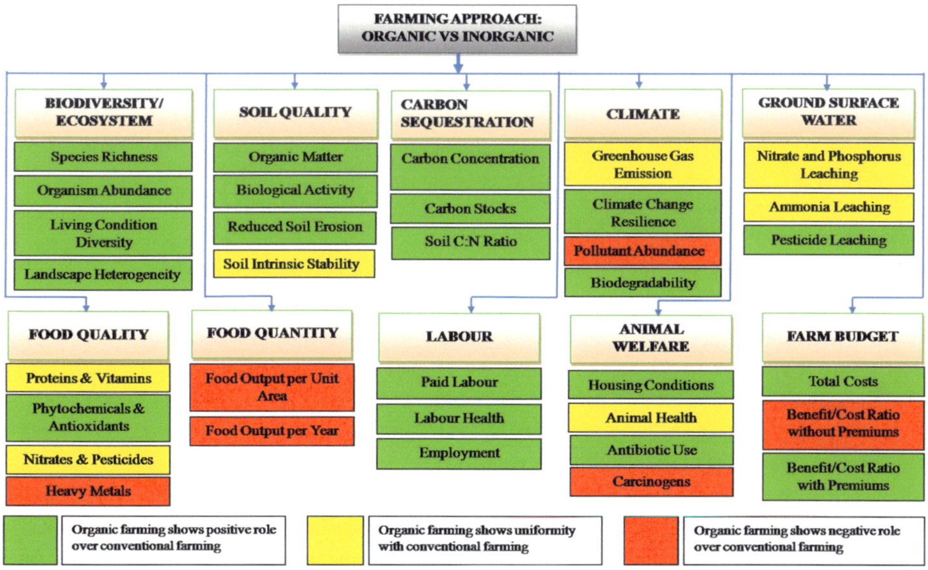

Figure 6.
Key variable indicators deferring the impacts of organic and inorganic farming for agricultural sustainability (source: [30]).

both commodity and noncommodity outputs and addresses ethical concerns such as animal welfare and the livelihoods of farmers (fair trade). According to a decades-long study on organic farming, in years of drought, organic yields can be up to 40% greater than nonorganic farms [37]. Organic farmers are more resilient and adaptable to stresses connected to climate change as well as other disruptive global stressors since they avoid the majority of fossil fuel-based inputs.

4.3 Ecosystem balance and biodiversity conservation

In most cases, organic food and farming systems increase overall biomass abundance and conserve biodiversity both within and between species, which in turn may enhance the pollination of crops and natural pest regulation [6, 36]. Comparative biodiversity assessments on organic and conventional farms reveal a 30% higher species diversity and a 50% greater abundance of flora and fauna in organic fields [38, 39]. The diversity and richness of bees significantly increased in places where the number of organic farms increased, which helped pollinate crops and wild plants over wider areas [40]. Organic farming has beneficial effects on species abundance/richness for a wide variety of taxa. Of the 99 studies reviewed by Hole et al. [39], only 8 found negative effects of organic farming on diverse individual taxon.

4.4 Soil health conservation and carbon sequestration

Organic agriculture preserves healthy soils by enhancing soil fertility, maintaining and creating a fertile living soil through the use of organic inputs in the form of green manures, farm yard manures, and compost, as well as by adopting cover crops, crop rotations, and intercropping and also by practicing minimum or no soil disturbance tillage. Crops and animals are integrated, which reduces overgrazing and makes nutrient recycling on farms easier. According to a review report, organic farming uses more organic fertilizers (such as manure, compost, and fertility-building/green manure crops) than conventional farming, with the median soil organic matter being 7% greater than conventional farming. The soil organic carbon concentrations and stocks of C per hectare are higher in top soils managed organically [30]. As a result, organic food and farming practices typically retain soil fertility in a sustainable manner, which may also lessen soil erosion and allow for the storage of carbon in organic matter.

4.5 Environmental protection

Traditional farming's heavy reliance on chemical pesticides, herbicides, and fertilizers has had a negative impact on the environment. Almost 35–65% less nitrogen leaks from arable fields into soil zones where it could harm the quality of the ground and drinking water as a result of the ban on chemical fertilizers on organic farms [41]. Leaching and run-off impacts are probably not a problem in organic farms because synthetic pesticides and herbicides are avoided. EU organic research places high emphasis on replacing copper fungicides with the breeding of disease-resistant cultivars and with easily biodegradable botanicals [6]. Nitrate leaching and greenhouse gas emissions per ha are up to 60% lower in organic farming. However, when assessed by the unit of product, the impacts of both organic and conventional farming on greenhouse gas emissions are very similar [30]. As per a meta-analysis, the area-scaled nitrous oxide emissions from organically managed soils were 492 kg CO_2 equivalents/ha lower per year than those from nonorganically managed soils [42].

4.6 Climate change adaptation

Organic food and farming systems emit fewer greenhouse gases under best farm practices, show higher yield stability in climatically extreme years, and reduce the risk of floods. Organic agriculture improves the capacity of agroecosystem to function in the face of unanticipated occurrences like climate change by boosting ecosystem resilience [43]. Organic agricultural practices reduce the need for fossil fuels, as well as their emissions of carbon dioxide and nitrous oxide, soil erosion, and carbon stocks. In comparison with high-input systems, energy consumption in organic systems has been reported to be lowered by 10 to 70% in EU nations and by 28 to 32% in the United States. Compared to conventional soils, organic systems in temperate regions almost double the effectiveness of carbon sequestration (575–700 kg carbon per ha/year) [44]. Thus, organic farming can potentially contribute to mitigating threats from climate change on crop production.

4.7 Safe and quality production

In some cases, organic food contains higher concentrations of secondary plant metabolites, antioxidants, and vitamins, as well as polyunsaturated fatty acids. Furthermore, organic food is often less contaminated with cadmium, nitrate, nitrite, and other residues. Organic food is considered healthier as compared to the food obtained by conventional farming [45, 46]. According to Baranski et al. [47], phenolic acids, flavanones, stilbenes, flavones, flavonols, and anthocyanins were significantly more concentrated in organic crops and crop-based foods.

5. Management approaches in organic farming

Effective management of nutrients, weeds, insect pests, and diseases is the major challenge for successful organic farming. Integrated management comprising cultural, mechanical, and biological practices is warranted for managing nutrients, weeds, pests, and diseases in an eco-friendly way in organic farms.

5.1 Nutrient management

Soils are a nonrenewable resource on which 95% of our food supply depends. Short-sighted chemical fertilizer applications in industrial farming are depleting soils at an alarming rate. Organic farming systems require effective nutrient management. Recycling, controlling biologically related processes like nitrogen fixation, and the sparing use of unprocessed, slowly soluble off-farm items that disintegrate in the same way as soil minerals or organic matter all promote the provision of nutrients to crop plants.

5.1.1 Green manure

Many fast-growing crops such as dhaincha (*Sesbania* sp.), sunhemp, and cowpea can fix atmospheric nitrogen at the rate of 60–100 kg/ha and can be utilized as green manure to the land. Dhaincha (*Sesbania esculenta* and *S. rostrata*) and sunhemp (*Crotalaria juncia*) are often plowed into the soil 6 to 8 weeks after being sown once sufficient vegetative development has been achieved. The use of green manure is very advantageous for organic production and preserving the health of the soil. Green manures enhance the physical and microbiological qualities of the soil in addition to adding nutrients.

5.1.2 Farm yard manure

The manure prepared using cow urine, dung, and farm waste in the backyard is called farm yard manure (FYM). This method has been followed since old times. The preparation of FYM can be by the use of any one of the methods including the sealed pit method, open pit method, and Japanese method. The soil physical property, microbial activity, and yield have been increased considerably using FYM. It is possible to recover between 70 and 80 percent of the energy provided to cattle as agricultural leftovers if the manure and urine from the animals are correctly collected.

5.1.3 Enriched compost

One of the traditional crop nutrient sources is composting organic residues. Though nutrient concentration is less, apart from NPK, it also provides the required micronutrients to the areas cultivated. Micronutrient supply satisfies the hidden hunger in the plants particularly and safeguards them against injury and toxicity. It also improves chemical, physical, and biological properties of the soil. In addition, compost is enriched externally through microbial inoculants, biofertilizers, etc. It is found that in cucumbers, the application of compost increases the yield [48].

5.1.4 Vermicompost

The technology uses earthworms as natural bioreactors for recycling nontoxic organic waste into soil. Vermicompost refers to the manure generated through rearing earthworms on a large scale in natural or artificial pits. This method is generally adopted when there is a huge quantity of undecomposed organic matter [49, 50]. Many forms of organic material can be used to prepare vermicompost; it includes manure of animals, wastes of manufacturing industries like paper waste, sugar waste of cane or cotton residues, kitchen waste, agricultural wastes, and municipal wastes having an organic origin. Higher concentrations of vermicast and vermitea improve the health of the plant, provide protection, improve growth, and also provide optimum production of crops.

5.1.5 Concentrated organic manure (oil cakes)

The oil cakes are applied in the granular form before the fertilizer use, so that nutrients that are contained in them are available for the crops. This enriches the soil organic carbon to soil, which in turn increases microbial activity. Castor cake, neem cake, and linseed cakes are few examples of nonedible cakes. As most of the edible cakes are fed to cattle as concentrates, the use of it as a nutrient source is limited in the Indian scenario.

5.1.6 Biofertilizers

Biofertilizers are the cultures of the appropriate microbial species that can fix the atmospheric nitrogen such as *Azospirillum* and *Azotobacter* in nonleguminous and *Rhizobium* species in the leguminous crops. The phosphate-mobilizing fungi (VAM) and phosphate-solubilizing bacteria are found to be more efficient in making the unavailable soil phosphorous available for the plants. It is found that the legume-rhizobium association could fix 40–120 kg/ha of nitrogen under optimum conditions. The crops inoculated with Mycorrhizal fungi are found resistant to *Fusarium oxysporum, Rhizoctonia solani*, *Phythium,* and nematode. It has been discovered that biofertilizers, such as *Rhizobium*, *Azotobacter*, *Azospirillum*, PSB Azolla, VAM, and *Pseudomonas*, are particularly powerful tools for managing fertility and biological nutrient mobilization. Use of such inputs must be ensured in all cropping scenarios because the efficacy of such microbial formulations is significantly better in no-chemical use situations.

5.2 Weed Management

The main objective of the organic system's weed management technique is to lower weed competition and reproduction to a level the farmer can tolerate. In many instances, not all weeds will be totally removed. By inhibiting the generation of weed seeds and perennial propagules, the portions of a plant that can produce a new plant, weed management should lessen competition. Regular weed control can lower weed control expenditures and help create a crop production system that is less expensive. Weeds in organic farming systems are controlled in the following ways.

5.2.1 Cultural practices

Cultural weed control includes nonchemical crop management practices ranging from variety selection to land preparation to harvesting and postharvest level. The management practices that are included in cultural weed control are (a) crop rotation, (b) cover crops, (c) intercropping, (d) mulching, (e) stale seedbed preparation, (f) soil solarization.

5.2.2 Mechanical control

The best way to manage weeds, especially on an organic farm, is through mechanical eradication, which takes time and effort intensively. One of the earliest weed management techniques is mechanical weed control, which calls for the actual removal of weeds by mechanical equipment either prior to the main crop planting or during the crop growth season. Mechanical weeders include cutting and cultivating instruments like mowers and stimmers, as well as multifunctional implements like hoes, harrows, tines, and brush weeders.

5.2.3 Biological control

Using live creatures to eradicate weeds or to stop their growth and capacity to compete with crops is known as biocontrol of weeds. The introduction of traditional biocontrol agents, frequently insects, and the expansion and widespread usage of organisms, frequently disease agents, are two categories into which biocontrol is often divided – (i) allelopathy and (ii) beneficial organisms.

5.3 Pest and disease management

The pest control strategies in organic farming aim to reduce and prevent the insect population's aggregation. The risks of pest outbreaks are minimized by enriching the soil with compost, crop rotation, intercropping, and conservation tillage [51]. Strategy for pest control in organic farming limits the use of chemical pesticides and promotes the use of organically derived pesticides. The effective control of pest population is achieved through field scouting, trap crops, insect trapping, and application of some biological control methods like introducing beneficial insects and using natural enemies to reduce the pest population. Onion thrips incidence was similar between the mineral fertilized and organic fertilized fields. Simmons et al. [52] suggested that the combination of host plant resistance and the reflective mulch could suppress the white-fly infestation that mainly affects organic vegetable production.

6. Organic farming in vegetable crop improvement

6.1 Productivity

Application of commercial organic fertilizer at recommended level results in higher vegetable yield than application of inorganic fertilizer. Alimi et al. [53] opined that the higher yield in organically managed fields might be the inclusion of Ca and Mg from organic fertilizer, which are missing from chemical fertilizer. Comparing organic treatments to conventional ones, the output of carrot roots increased [54]. In comparison with a supply of mineral nutrients, organic fertilization dramatically increased lettuce yield [55]. The organic method produced greater yields of tomatoes and cabbage than the conventional system did [56]. The use of compost resulted in a high yield of marketable cucumbers [48]. A study on organic farming in vegetable crops at IIVR, Varanasi, found that while yields under organic production gradually engrossed to yields under conventional inorganic farming in 4–5 years, the productivity of vegetable crops in organic farming was lower in the early years [57, 58]. Okra and cowpeas were produced during the summer, and tomatoes and cabbage were grown during the winter. By the fourth year, the yields from organic cultivation and conventional cultivation were equivalent (**Table 3**). However, the equivalent yield for cowpea and pea during the winter wet season was finally discovered after 3 years of continuous organic farming. In another study, Rembialkowska [60] noted that the average yield of carrots was higher by 33% on organic than conventional farms with a nonsignificant statistical variation. The average yield of the organic potatoes was significantly lower than the conventional ones [60]. Maggio et al. [61] registered lower yield in cauliflower, broad-leaved endive, and zucchini in organic cultivation

Items	Cabbage	Tomato	Okra	Cowpea (S)	Cowpea (K)	Pea
Conventional yield (t/ha)	41.00	37.40	9.26	8.00	10.26	7.30
Organic farming						
First-year yield (t/ha)	25.42	23.63	5.27	4.64	7.50	4.96
Second-year yield (t/ha)	29.54	27.75	6.57	5.76	8.97	6.28
Third-year yield (t/ha)	34.75	33.00	83.23	7.04	9.40	7.15
Fourth-year yield (t/ha)	38.83	36.84	9.16	7.84	—	—

Source: [59].

Table 3.
Average yield of different vegetables under organic farming against conventional yield.

Treatment	Yield increase over control (%)
Control/no organic matter	—
100% Cowdung	140.61
100% Farm yard manure	138.46
50% Soil +50% Recommended dose	126.49
50% Soil +50% Water hyacinth	144.03
50% Soil +50% Cowdung	164.68
50% Soil +50% Farm yard manure	169.02
50% Water hyacinth +50% Cowdung	133.33
50% Water hyacinth +50% Farm yard manure	152.28
50% Farm yard manure +50% Cowdung	155.84

Source: [62].

Table 4.
Effect of organic amendments on yield increase over control in eggplant.

Year	Status	Yield (q/ha)
Conventional		10.00
2 year	Year of conversion	5.75
4th year	Organic	7.50
5th year	Organic	8.75
6th year	Organic	10.00

Source: [64].

Table 5.
Yields of organic farming Vis-a-Vis conventional farming.

compared to conventional. Sultana et al. [62] noted a more than 150% yield increase in eggplant after organic amendments (**Table 4**).

Though for one or few years organic transition from the conventional system might give lower yield, long-term organic culture produces significantly higher yield in vegetables. According to Singh et al. [59], the yield of vegetables produced by organic farming is either equal to or greater than that produced by conventional farming after 5–6 years of organic farming practices, with the soil fertility sufficiently recovered. According to Ramesh et al. [63], organic farming has the potential to boost productivity in irrigated areas. Rajendran et al. [64] too observed that although organic farming may have lower productivity in the first few years, by the 6th year, yields were comparable to those from inorganic farming (**Table 5**).

6.2 Organic sources and vegetable productivity

Organic nutrient sources greatly influence vegetable yield. Soil amendment treatments consisted of combination of poultry compost, poultry litter, dairy compost, dairy manure, blood meal, feather meal, and Fertrell™ 5–5-3; poultry litter resulted in the highest yield in all the trials [65]. Thamburaj [66] found that applying oil cakes of margosa, castor, and groundnut (@0.2% W/W) reduces the intensity of root gall development in tomatoes. Studies showed that organically cultivation yielded 28.18 t/ha tomato, consonance with the recommended application of FYM and NPK (120:100:100 kg/ha). Reports also expressed that poultry manure and FYM in 50 kg N/ha produced maximum brinjal yield [67]. Combined manuring with FYM @ 25 t/ha + Biofertilizer (PSB + *Azotobacter/Rhizobium*) enhanced the yield of okra, cowpea, and bottle gourd by 27.5, 40.1, and 8.33%, respectively, in summer season compared to conventional system [59]. Okra responded with a greater yield to poultry manure @ 20 kg N/ha [68]. Singh et al. [59] noticed that the use of 20–30 t/ha FYM/NADEP compost, 7.5–10 t/ha vermicompost, or 7.5–10 t/ha chicken manure combined with bioinoculation of *Azatobacter* and PSB could guarantee a yield that is 20–35% higher than that of a conventional system. Even a combination of different biological fertilizers such as FYM @ 10 t/ha + vermicompost @ 3.5 t/ha or Farm yard manure @ 10 t/ha + chicken manure @ 2.5 t/ha or NADEP compost@ 10 t/ha + vermicompost @3.5 t/ha in addition to phyto-inoculation of Azatobacter and PSB was extremely efficient and generated yield comparable to the traditional inorganic method in cabbage, brinjal, broccoli, cauliflower, pea, bottle gourd, tomato, cowpea, okra crop, etc. It was noted, nonetheless, that various organic inputs behaved differently on various vegetable crops over various seasons. Howlader et al. [69] opined that poultry manure or cowdung at 5 t/ha contributed to superior yield in tomatoes (**Table 6**).

6.3 Soil fertility status

Accelerated accumulation of organic-C in organically manured fields is evident. Persistent application of organic manure enhances soil health and texture. During just 3 years, organic carbon and soil carbon stock in organic fields increased by 39 and 22.3%, respectively, compared to traditional systems [58]. With organic farming, 301.1 kg/ha/year of carbon was sequestered annually, compared to 42.6 kg/ha/year with conventional farming for cabbage. Because it enhances the physical and biological characteristics of the soil and serves as a nutrient store, organic carbon is a useful indication of soil quality. Singh and Upadhyay [70] opined that organic fertilization resulted in higher stock of organic-C as well as C sequestration rate in potato-based cropping systems. In addition, the available soil P and K levels markedly improved after organic fertilization [71]. Manjunath et al. [72] and Amanalluah [73] noted higher

Treatment	Fruits plant^{-1}		Fruit yield (t ha^{-1})	
	2016–2017	2017–2018	2016–2017	2017–2018
100% Recommended dose	41.7	34.27	54.3	89.75
75% Recommended dose	38.9	31.98	51.9	86.30
100% Recommended dose + Cowdung @ 5 t ha^{-1}	45.0	35.54	65.2	94.93
100% Recommended dose + Poultry manure @ 5 t ha^{-1}	42.8	36.38	62.5	89.65
75% Recommended dose + Cowdung @ 5 t ha^{-1}	43.5	36.58	60.9	95.46
75% Recommended dose + Poultry manure @ 5 t ha^{-1}	43.2	34.91	57.1	91.02

Source: [69].

Table 6.
Influence of organic amendments on tomato productivity.

Depth (cm)	Farming system	Soil texture (%)			Carbonates (%)	C:N ratio	Organic matter (g/kg)	C mineralization (mg/kg)	Net N mineralization (mg/kg)
		Sand	Silt	Clay					
0–10	Conventional	37.8	40.2	22.0	9.5	8.8	16.2	282	22.8
	Organic	29.7	43.6	26.7	9.1	9.2	24.9	490	32.8
10–20	Conventional	37.9	40.0	22.1	9.5	8.5	14.6	245	23.4
	Organic	29.6	43.7	26.7	9.6	8.7	23.2	437	47.2

Source: [74].

Table 7.
Comparative soil characteristics in the organic and conventional arable system in the Netherlands.

organic carbon accumulation and better uptake of nutrients under FYM-applied fields under the organic production system. Pulleman et al. [74] stated that soil organic matter content, C:N ratio, soil carbonates, N mineralization, etc., are positively influenced by organic farming along with other soil characters in the Netherlands (**Table 7**).

6.4 Soil health and microbial population

The elevated proportion of organic matter in organic farms transforms the soil into a living substrate by sustaining the micro-, meso-, and macro-fauna. Microbial performance such as dehydrogenase activity, alkaline phosphatase, and microbial biomass carbon was noted to be higher in organic matter-applied soils by 32, 26.8, and 22.4%, respectively, compared to inorganic farms [59]. Organically treated plots have more microbial populations, which aids in nutrient breakdown and boosts the availability of these nutrients to the vegetation. However, a direct impact from the microorganisms added through the manure is also feasible. In general, the increase in microbial biomass carbon in soils with organic manure addition was caused by an increase in the presence of carbon substrate that drives microbial development [75]. Likely, Manjunath et al. [72] recorded a higher growth of bacterial and actinomycetes in organic farm. In addition, Kumari et al. [76] found that the administration of organic manures increased the microbial community more than the recommended chemical fertilization. According to Singh et al. [77], an organic source of nutrients enhanced microbial activity and significantly improved dehydrogenase function. Hence, reintroducing useful microbes into the soil through organic farming enhances soil quality and vitality.

6.5 Quality characteristics of organic vegetables

Organically cultivated crops are well accepted for their higher content of vitamins, minerals, and phytochemicals (**Table 8**). Organically grown vegetables have improved quality, taste, and flavor mostly because of elevated dry matter, vitamin C, protein content, and quality, decreased free nitrates in vegetables, and reduced disease and storage losses. According to studies, the vitamin C content of organically grown cabbage, tomatoes, and cowpea grew by 17, 35, and 36%, respectively [59]. Vegetables grown organically also have superior physical characteristics. In comparison with an inorganic approach, the ascorbic acid, total phenol, and antioxidant content of peas improved by 31.8, 48.8, and 4.96%, respectively, under biological farming [70, 78]. Furthermore, organically produced vegetables have higher vitamin C, total carotenoids, flavonoids, phenolics, and higher nutrient concentrations, boosting

Vegetables	Vit-C	Fe	Mg	P
Spinach	+52	+25	-13	+14
Carrot	−6	+12	+69	+13
Lettuce	+17	+17	+29	+14
Cabbage	+43	+41	+40	+22
Potato	+22	+21	+5	0

Here, (+) and (−) indicate percent increase and decrease in nutrients in organic over conventional, respectively.

Table 8.
Vitamin and mineral nutrient contents of some organically produced vegetables.

human immune health [79]. Organically grown cucumbers had higher dry matter, sugar, and vitamin C [56]. Regarding vegetables (carrot, beetroot, lettuce, kale, leek, turnip, celeriac, and tomato), a trend has been observed for higher levels of iron and magnesium expressed in the nutritional quality and safety of organic food [80]. Additionally, β-carotene levels in organic and conventional foodstuffs have no noticeable differences [81]. Organically grown crops possess higher antioxidant properties, with individual antioxidant activity ranging between 18 and 69% higher in organic vegetables than in conventional ones [82]. Vegetables cultivated under compost fertilizing had higher Ca, Zn, and Fe contents on a fresh mass basis than plants produced using inorganic fertilizers and, finally, chicken manure [55, 83]. Hadayat et al. [84] found that potato, lettuce, tomato, carrot, and onion metal contents in conventional produce were slightly greater than in organic produce, especially for Cd and Pb.

7. Organic farming in fruit crop improvement

7.1 Fruit yield and productivity

All living beings have the natural tendency to produce their off-springs. With the go of nature, plants send and preserve their formulated and uptaken nutrients to fruits as sink after use for normal growth. Organic nutrient supply significantly improves yield over control, but the organic yield is to some extent lower than conventional yield in most cases. The relative yield of fruit crops is about 72% (28% lower than) of conventional yield [85], and this phenomenon is comparatively worse than other crops. Among the temperate fruit crops, organically grown apples and strawberries yielded 69 and 59% of the synthetic chemical used on a farm, while the other fruit crops, namely grapes, melons, apricot, blackcurrant, cherry, kiwi, peach, and pear from Europe and Turkey, had 78% productivity in organic cultivation than that of conventional farming [85]. In mango, application of FYM (50 kg/plant) + *Azospirillum culture* (250 g/tree) + PSB @ 250 g/tree produced superior yield (52.00 fruits/tree, 14.70 kg/tree, 1.47 t/ha) over other organic treatment along with control [86], while Sau et al. [87] reported that different treatments of biofertilizer showed maximum fruit weight (237.12 g) and yield (42.14 kg $plant^{-1}$) (**Figure 7**). Again, a combination of FYM (10 kg), neem cake (1.25 kg), vermicompost (5 kg), wood ash (1.75 kg), triple green manuring with cowpea, and biofertilizers (AMF @ 25 g + *Trichoderma harzianum* @ 50 g + PSB @ 50 g + *Azospirillum* @ 50 g $plant^{-1}$) recorded the highest bunch weight (9.60 kg) and yield (23.99 t/ha) in banana [88]. In guava, among recommended fertilization (600 g urea, 2000 g superphosphate, and 1000 g muriate of potash), *Jeevamrit* @ 10 liter/tree, *Azotobacter,* and *Azospirillum* @ 100 g/tree treatments, the best yield was noted from trees fertilized with recommended dose of NPK (222.43 g/fruit, 62.73 kg/plant). In another experiment, 90% RDF + 10% FYM/Tree was recorded as the best treatment in terms of better growth and yield of guava among eight organic–inorganic combinations [89]. Raghavan et al. [90] observed the highest number of fruits (1281/tree), yield (30.01 kg/tree) having extended levels of total, and reducing sugar content (26.14 and 14.51%, respectively) after organic treatment in litchi.

7.2 Fruit mineral content

Organic fertilizers largely add a complex combination of nutrient elements in soil, so as in fruits of organically treated plants. Mineral contents of fruits were found

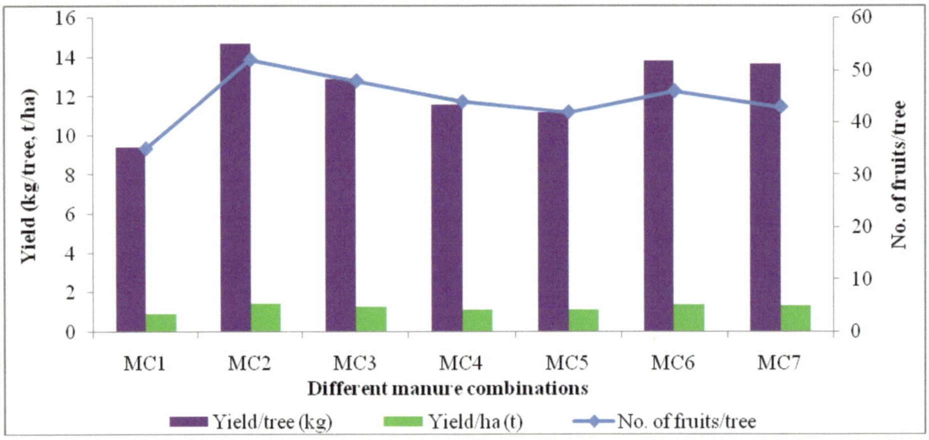

Figure 7.
Effect of organic package on fruit yield of mango cv. Alphanso. Here, MC1: FYM @ 50 kg/plant, MC2: FYM @ 50 kg/plant + Azospirillum (250 g/tree) + PSB @ 250 g/tree, MC3: FYM @ 50 kg/plant + Azatobacter (250 g/tree) + PSB @ 250 g/tree, MC4: Vermicompost @ 50 kg/plant, MC5: Vermicompost @ 50 kg/plant + Azospirillum culture (250 g/tree) + PSB @ 250 g/tree, MC6: Vermicompost @ 50 kg/plant + Azatobacter (250 g/tree) + PSB @ 250 g/tree and MC6: Vermicompost @ 50 kg/plant + Azospirillum (250 g/tree) + PSB @ 250 g/tree + vermiwash.

to be higher in fruits produced under conventional systems in comparison with the fruits produced under organic systems [91]. Harhash and Ahmed [92] analyzed that the NPK content of mango fruits differed in different organic and chemical fertilizer treatments (**Table 9**). Easmin et al. [93] observed an increased mineral content of fruit with organic matter application in the papaya field.

7.3 Fruit biochemical attributes

In general, organically produced fruits possess significantly higher total soluble solids (TSS) and lower titratable acidity (TA) in comparison with the conventionally produced fruits [94]. Compared to conventionally produced strawberries, which had 6.6% TSS and 0.99% TA, strawberries cultivated organically had a much higher TSS (7.1%) and lower TA level (0.93%) [95]. According to Leskinen et al. [96], levels of ascorbic acid in organically produced fruits were consistently higher than the levels in the conventionally grown ones. Nevertheless, Cayuela et al. [97] found no appreciable difference between strawberry fruits cultivated conventionally and organically in terms of ascorbic acid content. Again, organically grown fruits developed a significantly stronger color than conventionally grown ones [97]. The 6 kg OM/m^2 treatment produced strawberry fruits with the highest anthocyanin concentration (42.88 mg 100 g^{-1} fruit fresh weight). Despite this, strawberry plants receiving the control treatment still had anthocyanin contents that were between 17.8 and 41.8 mg 100 g^{-1}, and values lower or higher than that should not be considered acceptable [98]. *Azotobacter chorococcum* + *Azospirillum brasilense* + AM (*Glomus musseae*) + Panchagavya [3%] exhibited superiority in fruit biochemical qualities like TSS (19.70° Brix) and total sugars (13.41%) along with prolonged shelf life of 10 days in mango [87]. Easmin et al. [93]; Sharma and Negi [99], and Rahman et al. [100] observed similar trends in fruit biochemical properties in papaya, strawberry, and banana, respectively (**Table 10**).

Treatment	N (%)	P (%)	K (%)
100% mineral fertilizers (NPK) as control	1.93	0.26	1.63
50% NPK+ 100% plant compost (PC)	1.80	0.25	1.60
50% NPK+ 100% animal compost (AC)	1.93	0.28	1.70
100% plant compost (PC)	1.60	0.21	1.30
100% animal compost (PC)	1.70	0.22	1.50
50% plant compost +50% animal compost	1.60	0.20	1.40
50% (NPK+ plant compost+ animal compost)	2.50	0.34	2.05
100% (NPK+ plant compost+ animal compost)	2.88	0.39	2.50

Source: [92].

Table 9.
Effect of organic and mineral fertilization on NPK contents on fruits of Ewaise mango.

Treatment	Total sugar (%)	Reducing sugar (%)	Nonreducing sugar (%)	TSS (°Brix)	Titratable acidity (%)
FYM plus soil microbes	23.30	11.23	12.11	25.72	0.23
Improved compost	19.83	9.12	10.68	24.46	0.25
Vermicompost	21.25	10.36	10.92	25.15	0.22
Soil microbes	22.50	10.74	11.57	25.02	0.19

Source: [100].

Table 10.
Effect of organic fertilizers on fruit biochemical properties of banana.

7.4 Fruit organoleptic attributes

Organic production of fruit improves fruit quality, viz. fruit taste and color, keeping the quality of the fruits than conventionally produced fruits. Crops grown organically typically have higher sensory and long-term storage properties, according to Rembialkowska [101]. Several studies have conclusively shown that produce from organic farms tastes and smells better. More total sugars were present in organic fruits, which likely contributed to customers' perceptions of a better flavor [101]. According to studies, strawberries that are grown organically have superior overall acceptance, flavor, sweetness, and appearance than those that are grown conventionally (**Table 11**) [103].

7.5 Soil health and microbial population

Organic manuring and composting have a significant and positive influence on microbial community and activity in the soil. Increased population of bacteria, fungi, actinomycetes, and total diazotrophs were recorded by Reddy et al. [104] in organic fields compared to the conventional soils. By comparing the organic treatment to the synthetic fertilizer and control treatment, a significantly increased level of soil respiration and mineralizable nitrogen content were also observed. Significantly extended

Treatment	Appearance	Flavor	Taste	Acceptability
FYM (10 kg) + NC (1.25 kg) + VC (5 kg) + Ash (6.6 kg)	3.17	2.83	3.27	3.07
FYM (10 kg) + NC (1.25 kg) + VC (5 kg) + Ash (14.20 kg)	3.25	3.18	3.28	2.67
FYM (15 kg) + NC (1.875 kg) + VC (7.5 kg) + Ash (2.36 kg)	3.22	3.16	3.11	3.07
FYM (15 kg) + NC (1.875 kg) + VC (7.5 kg) + Ash (9.94 kg)	3.55	3.20	3.80	3.80
N_0 + P_0 + K_0 + Triple green manuring (TGM)	2.67	2.54	3.22	2.80
AMF(25 g) + *Azospirillum*(50 g) + PSB(50 g) + *T. harzianum* (50 g)	3.00	2.97	3.16	3.05
FYM (10 kg) + NC (1.25 kg) + VC (5 kg) + Ash (6.6 kg) + N_0 + P_0 + K_0 + TGM	3.22	2.83	3.28	3.05
FYM (10 kg) + NC (1.25 kg) + VC (5 kg) + Ash (6.6 kg) + AMF(25 g) + *Azospirillum*(50 g) + PSB(50 g) + *T. harzianum*(50 g)	3.07	2.72	3.13	3.10
FYM (10 kg) + NC (1.25 kg) + VC (5 kg) + Ash (6.6 kg) + N_0 + P_0 + K_0 + TGM	3.24	2.61	2.77	3.07
Poultry manure (10 kg) + Neem cake (1.875 kg)	3.33	3.63	3.33	3.13
200 g N + 50 g P + 200 g K	3.00	2.33	3.00	2.67
N_0 + P_0 + K_0 or control	2.00	2.16	2.00	2.05

Source: [102].

Table 11.
Effect of different organic manures and biofertilizers on fruit organoleptic quality of banana cv. Grand Naine (AAA) (1–4 scale basis).

Treatment	Bacteria (cfu/g of soil)
Azotobacter chorococcum (AC) + Panchagavya 3%	2.6×10^6
Azospirillum brasilense (AB) + Panchagavya 3%	2.7×10^6
Glomus musseae (AM) + Panchagavya 3%	2.3×10^6
AC + AB + Panchagavya 3%	2.9×10^6
AC + AM + Panchagavya 3%	2.9×10^6
AB + AM + Panchagavya 3%	3.0×10^6
AB+ AM + Panchagavya 3%	3.1×10^6
Panchagavya 3%	2.6×10^6
N:P:K (1000:500:1000 g plant^{-1} year^{-1})	2.0×10^6

Source: [87].

Table 12.
Effect of biofertilizers on soil bacteria of mango orchard (cv. Himsagar).

amount of soil respiration and mineralizable nitrogen content were also noted in organic treatment compared to synthetic fertilizer and control treatment. Similarly, Sau et al. [87] noticed that the application of organic manures along with biofertilizers substantially increased soil microbial population, which improved soil health as well as availability of other essential nutrient elements and thereby the growth and productivity of the tree (**Table 12**).

8. Organic farming in spice crop improvement

8.1 Spices growth and productivity

Organic manures contain nutrients and small quantities of growth boosters. It corresponds to the fundamental factor that revitalizes the growth cycle by reducing the physical, chemical, and physiological imbalances. Application of soil, mine spoil, and coir pith vermicompost at the ratio of (1:1:1) and RDF as an integrated approach enhances the height of the plant, yield, and number of leaves compared to mine, spoil integrated with RDF. Integrated use of 50% N through vermicompost, 50% N, and 100% P and K through chemical fertilizers along with *Azospirillum* is found effective in increasing the bulb yield of onion. When *Azotobacter* is added to the various combinations, a marked rise in plant height and no. of leaves is observed as compared to the treatments with organic manures alone [105]. According to Somasundaram et al. [106], application of panchgavya at 3% improves yield in comparison with the sole application of RDF. According to Somasundaram et al. [106], panchgavya application at 3% increases yield compared to RDF's single use. It was recorded that higher plant height, number of leaves, leaf area, flowers per plant, and flower weight were observed compared to cow dung extract, cow urine, and vermicast extract when vermiwash was used as a spray [107]. This indicates a major increase in the panchgavya spray on yield components @ 3 percent [108]. Vermicompost, when applied with vermiwash at 1:1, improves yield and plant height of chili [109]. Howlader and Gomasta [110] noted that chili yield is largely governed by applying cowdung and poultry manure as organic amendments besides chemical inputs. Kamal and Yousuf [111] measured taller turmeric plants (79.30 cm) with a

Treatment	Mother rhizomes plant^{-1}	Primary rhizomes plant^{-1}	Fresh rhizome yield (t/ha)	Cured rhizome yield (t/ha)
Cowdung (15 t/ha)	1.46	3.87	21.17	4.636
Poultry manure (7.0 t/ha)	1.81	4.80	27.30	5.18
Mustard cake (2.0 t/ha)	1.55	4.03	22.80	4.59
Neem cake (2.0 t/ha)	1.75	5.19	29.48	5.59
Control	0.43	2.27	14.84	2.38

Source: [111].

Table 13.
Yield and quality of turmeric as influenced by different organic manures.

Treatment	Essential oil content (%)		Methyl chavicol (%)	
	Main crop	Ratoon	Main crop	Ratoon
FYM (10 t ha^{-1}) + 100% recommended N through FYM	0.42	0.32	60.07	50.23
FYM (10 t ha^{-1}) + 100% N through FYM + biofertilizer	0.45	0.41	63.78	59.67
FYM (10 t ha^{-1}) + 75% N through FYM	0.35	0.23	57.57	50.15
FYM (10 t ha^{-1}) + 75% N through FYM + biofertilizer	0.40	0.25	62.92	55.38
FYM (10 t ha^{-1}) + 50% N through FYM	0.35	0.22	56.08	43.28
FYM (10 t ha^{-1}) + 50% N through FYM + biofertilizer	0.34	0.21	59.67	51.22
NPK (160:80:80 kg ha^{-1})	0.46	0.43	49.52	53.90
FYM 10 t ha^{-1} + NPK (160:80:80 kg ha^{-1})	0.48	0.45	52.62	44.17

Source: [114].

Table 14.
Effect of different levels FYM, inorganic fertilizer, and biofertilizers on the oil content of sweet basil.

maximum number of leaves (5.40), and leaf area (44.09) produced rhizomes which had the highest fresh and dry weight (256.21 and 40.35 g, respectively) after neem cake manuring in the field. Total yield (6.85 t ha^{-1}) was notably higher than control or organic fertilization (**Table 13**).

8.2 Spices quality characters

Spices are largely used for their aroma and pungency present in them. These quality parameters are the reflection of the growing environment of the crop. Poultry manure followed by goat manure was significantly superior with regard to yield, nutrient uptake, and enhanced piperine and oleoresin content of black pepper [112]. Soeparjono [113] reported that bokashi: charcoal husk: coco pea media composition having the organic fertilizer concentration (4.5 cc/l) produced ginger with zingerone level (1.88%) and oleoresin level (1.57%). In sweet basil, recommended FYM (10 tha^{-1}) along with recommended NPK (160:80:80 kg ha^{-1}) recorded the highest essential oil content (0.48 and 0.45%) and essential oil yield (199.7 and 107.58 kg ha^{-1}) in the main crop and ratoon, respectively [114]. The addition of recommended FYM (10 t ha^{-1}) and N through FYM together with biofertilizers in the main crop and in the ratoon resulted in the highest proportion of methyl chavicol (63.78 and 59.67%, respectively) in sweet basil (**Table 14**).

9. Organic farming and agricultural sustainability

A growing human population's desire for food, increased environmental dangers brought on by agriculture, and rising risks of food chain contamination and related health issues due to excessive use of agrochemicals are just a few of the factors drawing attention to modern agriculture worldwide. The fertility stability of the majority of soils is decreased by ongoing, intensive cropping without an equivalent addition of nutrients. Because of this, arable areas need increasingly more nutrients to produce the same amount of crop. Sustainable agriculture is a system that can generate plenty of food without diminishing the earth's definite assets or contaminating its environment. Sustainable agriculture maintains long-term ecological efficiency without depleting its natural resource base or harming the health of its consumers. It includes management techniques for raising crops and animals. Thus, sustainable agricultural management includes preserving soil organic matter; choosing ecologically and locally suited crops; increasing agricultural and biological diversity; preventing land degradation; strengthening biogeochemical cycles; and safeguarding environmental health. The objective of sustainable development of agriculture is "to increase food and enhance food security

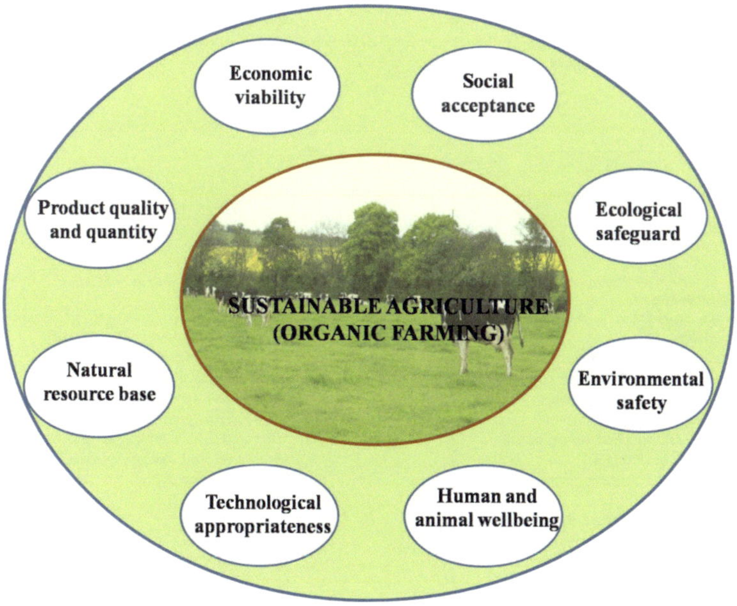

Figure 8.
Organic farming approach meeting the aspects of sustainable agriculture system.

in an environmentally sound way so as to contribute to sustainable natural resource management." Therefore, sustainable agriculture has few key indicators: (a) economic viability, (b) social acceptance, (c) ecological safeguard, (d) environmental safety, (e) human and animal well-being, (f) technological appropriateness, (g) natural resource base, (h) product quality and quantity. Organic farming is a kind of composite culture system that fulfills all aspects of a sustainable agriculture system (**Figure 8**).

10. Pathways and opportunities for future organic farming

The mounting environmental, economic, and social impacts of conventional agriculture call for a transformation of agriculture to more innovative farming systems. Transitioning to organics from conventional can be economically challenging and more information intensive. But once after transition, the net return per acre for organic compared to conventional farms is generally higher because of good yields and price premiums. Organic agriculture develops coinnovation among farmers, farm advisors, and scientists [115, 116] and enhances collaboration and communication between farmers as well as between farmers and consumers. The greatest way to manage waterways and create buffer zones between agriculture and nature conservation areas is through organic agricultural systems [8, 117]. Organic farming has room for growth: From 1% of the cropland today being organic to 10 to 20% by

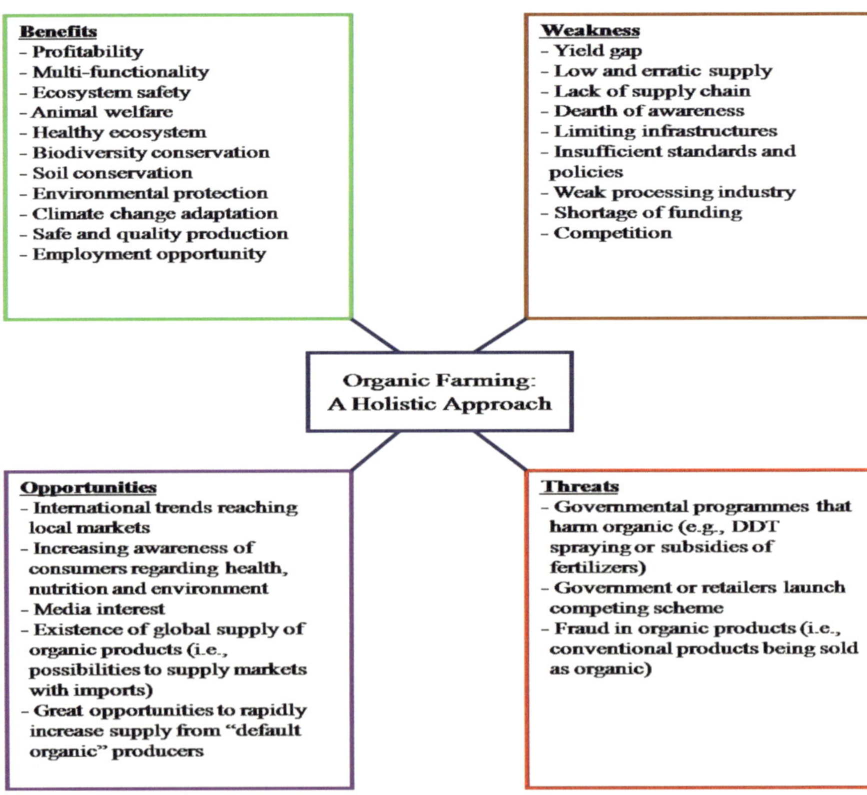

Figure 9.
Benefits, limitations, future opportunities, and threats of organic farming as a holistic approach to crop production.

DOI: http://dx.doi.org/10.5772/intechopen.1001589

2050 [9]. According to research, organic farming can produce an abundance of food at reasonable prices while also preserving the environment, boosting farm finances, and improving the health of farmers and farm workers. Consumers are seeking out organic and alternatively grown foods at grocery stores and farmers' markets. Active participation of farmers, along with sound organic regulation and more multi-actor cooperation, will definitely help in achieving future organic sustainability.

Therefore, the holistic organic farming is summarized in **Figure 9**.

11. Conclusion

Organic farming is a holistic production management system that promotes and enhances agroecosystem health, including biodiversity, biological cycles, and soil biological activity, and consequently, it is an efficient and promising approach for sustainable agriculture within a circular and green economy. Organic farming responds positively to all sustainable agriculture and rural development objectives, helps maintain soil fertility, and improves crop production and socioeconomic conditions of the farmers. Global agriculture must minimize its negative impacts and achieve productivity gains if it is to be sustainable, foster rural development, and support peoples' livelihoods. Besides the various benefits and strengths, organic food and farming systems can contribute to solving these challenges.

Author details

Emrul Kayesh[1*], Joydeb Gomasta[1], Nadira Bilkish[2], Khadiza Akter Koly[1] and Sharmila Rani Mallick[1]

1 Department of Horticulture, Bangabandhu Sheikh Mujibur Rahman Agricultural University, Gazipur, Bangladesh

2 Biotechnology Division, Bangladesh Agricultural Research Institute, Gazipur, Bangladesh

*Address all correspondence to: ekayeshhrt@bsmrau.edu.bd

References

[1] SRUC. Taking a holistic approach to farming. 2022. Available from: https://www.sruc.ac.uk/all-news/taking-a-holistic-approach-to-farming/. [Accessed: December 21, 2022].

[2] Savory RM. Is there an Ultimate Use for Historians? Reflections on Safavid History and Historiography. The Annual Noruz Lecture Series: 16 March 1995. Washington, D.C.: Foundation for Iranian Studies; 1995

[3] Garnett T, Godfray C. Sustainable intensification in agriculture. Navigating a course through competing food system priorities. In: Food Climate Research Network and the Oxford Martin Programme on the Future of Food. UK: University of Oxford; 2012

[4] Smith OM, Cohen AL, Rieser CJ, Davis AJ, Taylor JM, Adekunle AW, et al. Organic farming provides reliable environmental benefits but increases variability in crop yields: A global meta-analysis. Frontiers in Sustainable Food Systems. 2019;**3**:1-10. DOI: 10.3389/fsufs.2019.00082

[5] Leifeld J. How sustainable is organic farming? Agriculture. Ecosystems & Environment. 2012;**150**:121-122

[6] Niggli U. Sustainability of organic food production: Challenges and innovations. Proceedings of the Nutrition Society. 2015;**760**(1):83-88. DOI: 10.1017/S0029665114001438

[7] Morgera E, Caro CB, Durán-Marín G. Organic Agriculture and the Law. FAO Legisl. Rome: Food and Agriculture Organization of the United Nations; 2012

[8] Gomiero T, Pimentel D, Paoletti MG. Environmental impact of different agricultural management practices: Conventional vs. organic agriculture. Critical Reviews in Plant Sciences. 2011;**30**(1-2):95-124. DOI: 10.1080/07352689.2011.554355

[9] Reganold JP, Wachter JM. Organic agriculture in the twenty-first century. Nature Plants. 2016;**2**(February):15221. DOI: 10.1038/NPLANTS.2015.221

[10] Sandhu HS, Wratten SD, Cullen R. Organic agriculture and ecosys-tem services. Environmental Science and Policy. 2010;**13**(1):1-7. DOI: 10.1016/j.envsci.2009.11.002

[11] Lampkin N. From conversion payments to integrated action plans in the European Union. In: Organic Agriculture: Sustainability, Markets and Policies. Washington, D.C., USA: OECD workshop on organic agriculture; 2003, 2002. pp. 313-328. CABI, 2003. DOI: 10.1079/9780851997407.0313

[12] Nandwani D, Nwosisi S. Global trends in organic agriculture. In: Nandwani D, editor. Organic Farming for Sustainable Agriculture. Switzerland: Springer International Publishing; 2016. pp. 1-35. DOI: 10.1007/978-3-319-26803-3_1

[13] Atreya K, Subedi BP, Ghimire PL, Khanal SC, Pandit S. A review on history of organic farming in the current changing context in Nepal. Archives of Agriculture and Environmental Science. 2020;**5**(3):406-418. DOI: 10.26832/24566632.2020.0503024

[14] Conford P. The Origins of the Organic Movement. Edinburgh, UK: Floris Books; 2001. p. 287

[15] Vogt G. Geschichte des ökologischen Landbaus im deutschsprachigen Raum.

2001. Available from: www.orgprints.org/1110/. [Accessed: December 21, 2022]

[16] Heckman J. A history of organic farming: Transitions from sir Albert Howard's war in the soil to USDA national organic program. Renewable Agriculture and Food Systems. 2006;**21**(3):143-150. DOI: 10.1079/RAF2005126

[17] Paull J. Biodynamic agriculture: The journey from Koberwitz to the world, 1924-1938. Journal of Organic Systems. 2011;**6**(1):27-41

[18] Taverne D. The march of the unreason. In: Science, Democracy and the New Fundamentalism. Oxford, U.K: Oxford University Press; 2005. p. 310

[19] Trewavas A. A critical assessment of organic farming-and-food assertions with particular respect to the UK and the potential environmental benefits of no-till agriculture. Crop Protection. 2004;**23**:757-781

[20] FAO (Food and Agriculture Organization of the United Nations). Online Database. 2015. Available from: http://faostat.fao.org/site/377/default.aspx#ancor. [Accessed: August, 2015]

[21] Willer H, Trávníček J, Meier C, Schlatter B, editors. The World of Organic Agriculture. Statistics and Emerging Trends 2022. Research Institute of Organic Agriculture FiBL, Frick, and IFOAM – Organics International, Bonn. Hachenburg, Germany: Druckerei Hachenburg; 2022

[22] Reddy BS. Organic farming: status, issues and prospects–a review. Agricultural Economics Research Review. 2010;**23**(2):343-358. DOI: 10.22004/ag.econ.97015

[23] Chandrashekar HM. Changing scenario of organic farming in India: An overview. International NGO Journal. 2010;**5**:34-39

[24] Gonciarov M, Tapaloaga D, Neagu I. Principles and standards of organic agriculture. Journal of Biotechnology. 2014;**185**:S76. DOI: 10.1016/j.jbiotec.2014.07.259

[25] IFOAM. Principles of Organic Agriculture. Bonn: International Federation of Organic Agriculture Movements; 2005

[26] IFOAM. In: Kölling A, editor. Organic Food and Farming: A System Approach to Meet the Sustainability Challenge. Belgium: IFOAM; 2010. p. 5

[27] Eyhorn F, Muller A, Reganold JP, Frison E, Herren HR, Luttikholt L, et al. Sustainability in global agriculture driven by organic farming. Nature. 2019;**2**:253-255. DOI: 10.1038/s41893-019-0266-6

[28] Ponisio LC, M'Gonigle LK, Mace KC, Palomino J, de Valpine P, Kremen C. Diversification practices reduce organic to conventional yield gap. Proceedings of the Royal Society B: Biological Sciences. 2015;**282**:20141396. DOI: 10.1098/rspb.2014.1396

[29] Seufert V, Ramankutty N, Foley JA. Comparing the yields of organic and conventional agriculture. Nature. 2012;**485**(7397):229-232. DOI: 10.1038/nature11069

[30] Baret P, Marcq P, Mayer C. 2015. Research and organic farming in Europe. Available from: https://orgprints.org/id/eprint/29412/7/Full%20report_R4ORG.pdf. [Accessed: December 21, 2022]

[31] Kovács-Hostyánszki A, Espíndola A, Vanbergen AJ, Settele J, Kremen C, Dicks LV. Ecological intensification

to mitigate impacts of conventional intensive land use on pollinators and pollination. Ecology Letters. 2017;**20**:673-689. DOI: 10.1111/ele.12762

[32] Seufert V, Ramankutty N. Many shades of gray—The context dependent performance of organic agriculture. Science Advances. 2017;**3**:e1602638. DOI: 10.1126/sciadv.1602638

[33] Crowder DW, Reganold JP. Financial competitiveness of organic agriculture on a global scale. Proceedings of the National Academy of Sciences of the United States of America. 2015;**112**:7611-7616. DOI: 10.1073/pnas.1423674112

[34] Ashely R, Bishop A, Dennis J, French J, Gardam P, Butler L, et al. Intensive Organic Vegetable Production: Integrated Development. Hobert, Australia: Rural Industries Research and Development Corporation; 2007. p. 46

[35] Smukler SM, Jackson LE, Murphree L, Yokota R, Koike ST, Smith RF. Transition to large-scale organic vegetable production in the Salinas Valley, California. Agriculture, Ecosystems & Environment. 2008;**126**(3-4):168-188

[36] Rahmann G, Oppermann R, Paulsen HM, Weißmann F. Good, but not good enough? Research and development needs in organic farming. Applied Agriculture and Forestry Research. 2009;**59**(1):29-40

[37] Brook L. Organic Agriculture Helps Solve Climate Change. 2022. Available from: https://www.nrdc.org/experts/lena-brook/organic-agriculture-helps-solve-climate-change. [Accessed: December 21, 2022]

[38] Bengtsson J, Ahnström J, Weibull AC. The effects of organic agriculture on biodiversity and abundance: A meta-analysis. Journal of Applied Ecology. 2005;**42**:261-269

[39] Hole DG, Perkins AJ, Wilson JD, Alexander IH, Grice PV, Evans AD. Does organic farming benefit biodiversity? Biological Conservation. 2005;**122**:113-130

[40] Rundlöf M, Nilsson H, Smith HG. Interacting effects of farming practice and landscape context on bumble bees. Biological Conservation. 2008;**141**:417-426

[41] Stolze M, Piorr A, Häring AM, Dabbert S. The environmental impacts of organic farming in Europe. Organic Farming in Europe: Economics and Policy. 2000;**6**:143

[42] Skinner C, Gattinger A, Muller A, Mäder P, Flieβbach A, Stolze M, et al. Greenhouse gas fluxes from agricultural soils under organic and non-organic management—A global meta-analysis. Science of the Total Environment. 2014;**468**:553-563

[43] Borron S. Building Resilience for an Unpredictable Future: How Organic Agriculture Can Help Farmers Adapt to Climate Change. Rome: FAO; 2006

[44] Holt-Giménez E. Measuring farmers' agroecological resistance after hurricane Mitch in Nicaragua: A case study in participatory, sustainable land management impact monitoring. Agriculture, Ecosystems & Environment. 2002;**93**(1-3):87-105

[45] Ditlevsen K, Sandøe P, Lassen J. Healthy food is nutritious, but organic food is healthy because it is pure: The negotiation of healthy food choices by danish consumers of organic food. Food Quality and Preference. 2019;**71**:46-53

[46] Tscharntke T, Clough Y, Wanger TC, Jackson L, Motzke I, Perfecto I, et

al. Global food security, biodiversity conservation and the future of agricultural intensification. Biological Conservation. 2012;**151**(1):53-59

[47] Baranski M, Srednicka-Tober D, Volakakis N, Seal C, Sanderson R, et al. Higher antioxidant and lower cadmium concentrations and lower incidence of pesticide residues in organically grown crops: A systematic literature review and meta-analyses. British Journal of Nutrition. 2014;**112**(5):794-811. DOI: 10.1017/S0007114514001366

[48] Nair A, Ngouajio M. Integrating rowcovers and soil amendments for organic cucumber production: Implications on crop growth, yield, and microclimate. Hort Science. 2010;**45**(4):566-574

[49] Chatterjee R, Bandyopadhyay S, Jana JC. Evaluation of vegetable wastes recycled for vermicomposting and its response on yield and quality of carrot (*Daucus carota* L.). International Journal of Recycling of Organic Waste in Agriculture. 2014;**3**:1-7

[50] Chauhan A, Kumar S, Singh AP, Gupta M. Vermicomposting of vegetable wastes with cowdung using three earthworm species *Eisenia foetida*, *Eudrilus eugeniae* and *Perionyx excavatus*. Nature and Science. 2010;**8**(1):34-42

[51] Niggli U. Organic agriculture: a productive means of low-carbon and high biodiversity food production. In: Trade and Environment Review-Promoting poles of clean growth to foster the transition to a more sustainable economy. United Nations Conference on Trade and Development (UNCTAD), CH-Geneva, chapter 3.II; 2010. pp. 112-118

[52] Simmons AM, Kousik CS, Levi A. Combining reflective mulch and host plant resistance for sweetpotato whitefly (Hemiptera: Aleyrodidae) management in watermelon. Crop Protection. 2010;**29**(8):898-902

[53] Alimi T, Ajewole OC, Awosola O, Idowu EO. Organic and inorganic fertilizer for vegetable production under tropical conditions. Journal of Agricultural and Rural Development. 2007;**1**:120-136

[54] Bruno RDLA, Viana JS, Silva VFD, Bruno GB, Moura MFD. Production and quality of seeds and roots of carrot cultivated under organic and mineral fertilization. Horticultura Brasileira. 2007;**25**:170-174

[55] Pôrto ML, Alves JDC, de Souza AP, Araújo RDC, de Arruda JA. Nitrate production and accumulation in lettuce as affected by mineral nitrogen supply and organic fertilization. Horticultura Brasileira. 2008;**26**:227-230

[56] Ma CH, Chen JH, Yang RY, Palada MC, Ou SC, Lin YH, et al. Monitoring soil and vegetable quality under six fertilization strategies in organic and conventional farming systems. In: The 9th International Conference of the East and Southeast Asia Federation of Soil Science Societies, Seoul, Republic of Korea. 2009. pp. 373-374

[57] Bhattacharyya P, Chakraborty G. Current status of organic farming in India and other countries. Indian Journal of Fertilisers. 2005;**1**(9):111

[58] Singh SK, Yadava RB, Chaurasia SNS, Prasad RN, Singh R, Chaukhande P, et al. Producing organic vegetables for better health. Indian Horticulture. 2016;**61**(1):5-8

[59] Singh SK, Yadav RB, Singh J, Singh B. Organic Farming in Vegetables. IIVR Technical Bulletin no. 77. Varanasi: ICAR-IIVR; 2017. p. 47

[60] Rembialkowska E. Organic farming as a system to provide better vegetable quality. In: International Conference on Quality in Chains. An Integrated View on Fruit and Vegetable Quality. Acta Horticulturae. 2003;**604**:473-479. DOI: 10.17660/ActaHortic.2003.604.52

[61] Maggio A, De Pascale S, Paradiso R, Barbieri G. Quality and nutritional value of vegetables from organic and conventional farming. Scientia Horticulturae. 2013;**164**:532-539

[62] Sultana N, Mannan MA, Khan SAKU, Gomasta J, Roy T. Effect of different manures on growth, yield and profitability of small scale Brinjal (egg-Plant) cultivation in Gunny bag. Asian Journal of Agricultural and Horticultural Research. 2022;**9**(1):52-60. DOI: 10.9734/ajahr/2022/v9i130136

[63] Ramesh P, Panwar NR, Singh AB, Ramana S. Effect of organic manures on productivity, soil fertility and economics of soybean (Glycine max)-durum wheat (Titicum durum) cropping system under organic farming in Vertisols. The Indian Journal of Agricultural Sciences. 2008;**78**(12):1033-1037

[64] Rajendran TR, Venugopalan MV, Tarhalkar PP. Organic Cotton Farming. Nagpur: Technical Bulletin No. 1/2000, CICR; 1999. p. 37

[65] Rauton SRW. Soil Fertility Management in Organic Vegetable Production [Thesis]. USA: Clemson University; 2007

[66] Thamburaj S. Tomato responds to organic gardening. Kissan World. 1994;**21**:10-49

[67] Jose D, Shanmugavelu KG, Thamburaj S. Studies on the efficiency of organics vs. inorganic forms of N in brinjal. Indian Journal of Horticulture. 1988;**45**:100-103

[68] Yadav P, Singh P, Yadav RL. Effect of organic manures and nitrogen levels on growth, yield and quality of okra. Indian Journal of Horticulture. 2006;**63**(2):215-217

[69] Howlader MIA, Gomasta J, Rahman MM. Integrated nutrient Management for Tomato in the southern region of Bangladesh. International Journal of Innovative Research. 2019;**4**(3):55-58

[70] Singh B, Upadhaya NC. Organic potato production in plains. In: Souvenir of the International Conference on Organic Agriculture, 22-24 June, 2011, Patna. 2011

[71] Upadhayay NC, Rawal S, Kumar P. Effect of organic and inorganic sources of nutrients on potato production and soil fertility. In: Pathak RK, Kishun R, Khan RM, Ram RA, editors. Organic Farming in Horticulture. Lucknow: CISH; 2004. pp. 290-293

[72] Manjunath BL, Mahajan GR, Ramesh R, Singh NP. Effect of improved nutrient management on grain yield of rice (Oryza sativa) and soil health under organic management. Indian Journal of Agronomy. 2016;**61**(1):25-32

[73] Amanullah M. Response of lowland rice varaties to reclamation practices in costal saline soils. Journal of Applied Science and Research. 2008;**4**(7):871-874

[74] Pulleman M, Jongmans A, Marinissen J, Bouma J. Effects of organic versus conventional arable farming on soil structure and organic matter dynamics in a marine loam in the Netherlands. Soil Use and Management. 2003;**19**(2):157-165

[75] Powlson DS, Prookes PC, Christensen BT. Measurement of soil microbial biomass provides an early

indication of changes in total soil organic matter due to straw incorporation. Soil Biology and Biochemistry. 1987;**19**(2):159-164

[76] Kumari N, Singh CS, Prasad J, Singh MK, Kumar R. Influence of organic nutrient sources on productivity of rice (Oryza sativa)-based cropping systems in Jharkhand. Indian Journal of Agronomy. 2013;**58**(3):277-281

[77] Singh YV, Dhar DW, Agarwal B. Influence of organic nutrient management on basmati rice (Oryza sativa)–wheat (Triticum aestivum)–greengrsam (Vigna). Indian Journal of Agronomy. 2011;**56**:169-175

[78] Bhadoria PBS. Importance of organic manures in improving quality of rice and okra. Environment and Ecology. 2002;**20**(3):628-633

[79] Bahadur A, Singh J, Singh KP, Upadhyay AK, Rai M. Morphophysiological, yield and quality traits in lettuce (*Lactuca sativa*) as influenced by use of organic manures and biofertilizers. Indian Journal of Agricultural Science. 2009;**79**(4):282-285

[80] Lairon D. Nutritional quality and safety of organic food. A review. Agronomy for Sustainable Development. 2010;**30**:33-41

[81] Olle M, Williams IH. Organic cultivation of vegetables. Sustainable Agriculture Reviews. 2021;**52**:1-19

[82] Barański M, Rempelos L, Iversen PO, Leifert C. Effects of organic food consumption on human health; the jury is still out! Food & Nutrition Research. 2017;**61**(1):1287333

[83] Masarirambi MT, Hlawe MM, Oseni OT, Sibiya TE. Effects of organic fertilizers on growth, yield, quality and sensory evaluation of red lettuce (Lactuca sativa L.)'Veneza Roxa'. Agriculture and Biology Journal of North America. 2010;**1**(6):1319-1324

[84] Hadayat N, De Oliveira LM, Da Silva E, Han L, Hussain M, Liu X, et al. Assessment of trace metals in five most-consumed vegetables in the US: Conventional vs. organic. Environmental Pollution. 2018;**243**:292-300

[85] De Ponti T, Rijk B, Van Ittersum MK. The crop yield gap between organic and conventional agriculture. Agricultural Systems. 2012;**108**:1-9. DOI: 10.1016/j.agsy.2011.12.004

[86] Dheware RM, Nalage NA, Sawant BN, Haldavanekar PC, Raut RA, Munj AY, et al. Effect of different organic sources and biofertilizers on yield and quality production in mango cv. Alphonso. Journal of Pharmacognosy and Phytochemistry. 2020;**9**(2):97-99

[87] Sau S, Mandal P, Sarkar T, Das K, Datta P. Influence of bio-fertilizer and liquid organic manures on growth, fruit quality and leaf mineral content of mango cv. Himsagar. Journal of Crop and Weed. 2017;**13**(1):132-136

[88] Manju PR, Pushpalatha PB. Effect of organic nutrients on the yield and quality of Banana cv. Nendran (Musa spp., AAB). International Journal of Bio-resource and Stress Management. 2022;**13**(2):179-186

[89] Jaiprakash, Prasad VM, Bahadur V, Dinesh. Studies on impact of organic and inorganic fertilizers on growth, yield and quality of guava (*Psidium guajava*) Allahabad Safeda under sub tropical climatic conditions. Biological Forum – An International Journal. 2021;**13**(3a):285-288

[90] Raghavan M, Hazarika BN, Das S, Ramjan M, Langstieh LB. Integrated

nutrient management in litchi (Litchi chinensis Sonn.) cv. Muzaffarpur for yield and fruit quality at foothills of Arunachal Pradesh. IJCS. 2018;**6**(3):2809-2812

[91] Jadczak D, Grzeszuczuk M, Kosecka D. Quality characteristics and content of mineral compounds in fruit of some cultivars of sweet pepper (*Capsicum annum* L.). The Elemental Journal. 2010;**15**(3):509-515

[92] Harhash M, Ahmed M. Effect of organic and mineral fertilization on yield and fruit quality of Ewaise mango cultivar. Alexandria Journal of Soil and Water Sciences. 2018;**2**(2):102-115

[93] Easmin S, Hoque M, Saikat M, Kayesh E. Influence of organic and inorganic fertilizers on growth, yield and physio-chemical properties of papaya. Annals of Bangladesh Agriculture. 2021;**24**(2):69-83. DOI: 10.3329/aba.v24i2.55785

[94] Turemis N. The effects of different organic deposits on yield and quality of strawberry cultivar Dorit (216). Acta Horticulturae. 2002;**567**:507-510

[95] Singh B, Upadhyay NC. Organic potato production in plains. In: Souvenir of the International Conference on Organic Agriculture, Patna. 22-24 Jun 2011

[96] Leskinen M, Vaisanen HM, Vestergaard J. Chemical and sensory quality of strawberry cultivars used in organic cultivation. Acta Horticulturae. 2002;**567**:523-526

[97] Cayuela JA, Vidueira JM, Albi MA, Gutierrez F. Influence of the ecological cultivation of strawberries (*Fragaria X Ananassa* cv. Chandler) on the quality of the fruit and on their capacity for conservation. Journal of Agricultural and Food Chemistry. 1997;**45**:1736-1740

[98] Abu-Zahra TR, Al-Ismail K, Shatat F. Effect of organic and conventional systems on fruit quality of strawberry (*Fragaria* X *Ananassa* Duch) grown under plastic house conditions in the Jordan Valley. Acta Horticulturae. 2007;**741**:159-172

[99] Sharma K, Negi M. Effect of organic manures and inorganic fertilizers on plant growth of strawberry (Fragaria x ananassa) cv. Shimla delicious under mid-hill conditions of Uttarakhand. Journal of Pharmacognosy and Phytochemistry. 2019;**8**(2):1440-1444

[100] Rahman JI, Hazarika DN, Borah B, Bhattacharjee D. Effect of organic manures and inorganic fertilizer on the fruit quality of Banana. Biological Forum–An International Journal. 2021;**13**(4):908-912

[101] Rembialkowska E. Quality of plant products from organic agriculture. Journal of the Science of Food and Agriculture. 2007;**87**(15):2757-2762. DOI: 10.1002/jsfa.3000

[102] Hema R, Bhagavan BVK, Sudhavani V, Umakrishna K. Effect of Sorganic manures and bio-fertilizers on yield and fruit quality of banana cv. Grand Naine (AAA). International Journal of Bio-resource and Stress Management. 2016;7(4):832-836

[103] Reganold JP, Andrews PK, Reeve JR, Carpenter-Boggs L, Schadt CW, Alldredge JR, et al. Fruit and soil quality of organic and conventional strawberry agroecosystems. PLoS One. 2010;**5**(9):e12346

[104] Reddy YT, Kurian RM, Ganeshamurthy AN, Pannerselvam P.

Effect of organic nutrition practices on papaya (cv. Surya) fruit yield, quality and soil health. Journal of Horticultural Sciences. 2010;**5**(2):124-127

[105] Jayathilake PKS, Reddy IP, Srihari D, Neeraja G, Reddy R. Effect of nutrient management on growth, yield and yield attributes of rabi onion (*Allium cepa* L.). Vegetable Science. 2002;**29**:184-185

[106] Somasundaram E, Sankaran N, Meena S, Thiyagarajan TM, Chandragiri KK, Panneerselvam S. Response of greengram to varied concentrations of Panchakavya (organic nutrition) foliar application. Madras Agricultural Journal. 2003;**90**(1/3):169-172

[107] Shivsubramanian K, Ganeshkumar M. Influence of vermiwash on biological productivity of Marigold. Madras Agricultural Journal. 2004;**91**:221-225

[108] Yadav BK, Christopher L. Effect of organic manures and Panchagavya spray on yield attributes, yield and economics of rice. Crop Research. 2006;**31**(1):6-10

[109] George S, Giraddi RS, Patil RH. Utility of vermiwash for the management of Thrips and mites on chilli (*Capsicum annuum* L.) amended with soil organics. Karnataka Journal of Agricultural Science. 2007;**20**(3):657-659

[110] Howlader MIA, Gomasta J. Integrated nutrient Management for Chilli in the southern region of Bangladesh. International Journal of Innovative Research. 2019;**4**(1):18-21

[111] Kamal MZU, Yousuf MN. Effect of organic manures on growth, rhizome yield and quality attributes of turmeric (Curcuma longa L.). The Agriculturists. 2012;**10**(1):16-22

[112] Hamza S. Effect of organic farming on soil quality, nutrient uptake, yield and quality of Indian spice. In: The 18th World Congress of Soil Science. Philadelphia, Pennsylvania, USA. 2006

[113] Soeparjono S. The effect of media composition and organic fertilizer concentration on the growth and yield of red ginger rhizome (Zingiber officinale Rosc.). Agriculture and Agricultural Science Procedia. 2016;**9**:450-455

[114] Al-Mansour B, Kalaivanan D, Suryanarayana MA, Umesha K, Nair AK. Influence of organic and inorganic fertilizers on yield and quality of sweet basil (*Ocimum basilicum* L.). Journal of Spices and Aromatic Crops. 2018;**27**(1):38-44

[115] Cornell S, Berkhout F, Tuinstra W. Opening up knowledge systems for better responses to global environmental change. Environmental Science & Policy. 2013;**2**:60-70

[116] Koutsouris A. Facilitating agricultural innovation systems: A critical realist approach. Studies in Agricultural Economics. 2012;**114**:64-70

[117] Tuomisto HL, Hodge ID, Riordran P, Macdonald DW. Does organic farming reduce environmental impacts? – A meta-analysis of European research. Journal of Environmental Management. 2012;**112**:309-320

Chapter 7

Perspective Chapter: Conservation and Enhancement of Soil Health for Sustainable Agriculture

Pratik Ramteke, Vijay Gabhane, Prakash Kadu, Vilas Kharche and Samrat Ghosh

Abstract

Despite increasing crop yields, the indiscriminate use of chemical fertilizers in conventional agriculture damages soil health, reduces crop productivity, and negatively impacts agricultural sustainability. Therefore, restoring soil health and the environment is imperative. Higher crop productivity can be achieved with natural fertilizers such as biofertilizers, vermicompost, green manures, farmyard manures, and crop residues, which are a sustainable approach to nourishing the soil and the environment. This chapter addresses the importance of healthy soils, how they can be influenced by agricultural inputs and practices, and strategies for enhancing soil health.

Keywords: soil health, management interventions, fertilizers, organics, sustainable agriculture

1. Introduction

Ensuring sustainability in agriculture and meeting the increasing demand for food depend on maintaining water and soil quality, soil organic matter (SOM), recycling and storing nutrients, efficient use of natural resources, and controlling soil degradation [1]. In addition to air and water, the soil is also critical to life on Earth [2]. As a materially and morphologically diverse ecosystem, the soil is critical for its ecosystem services, including nutrient supply, long-term soil and crop productivity maintenance, and preservation of environmental quality, requiring appropriate characterization of its physical, chemical, and biological properties. Soil health plays a pivotal role in the sustainability of agriculture. Soil health is defined as 'The continuing potential of a given type of soil to function as a vital living system within managed or natural ecosystem limits, to sustain animal and plant productivity, to maintain or improve the quality of the air and water, and to support human health and habitation' [3]. The term soil health is not limited to increasing crop productivity. It also includes other functions, such as maintaining an appropriate balance between different soil functions, plant and animal health, and environmental interaction and regulation [4].

World literature shows that the productivity of soils per hectare is very low and constantly decreasing due to faulty agricultural practices. Current agricultural practices rely heavily on inorganic fertilizers and pesticides; over the years, they have impaired biological activities in the soil and processes such as nutrient transformation, thus affecting nutrient availability in the soil [5, 6]. The introduction of improved crop varieties with high yield potential, chemical fertilizers, pesticides, weedicides, and farm equipment such as tractors, drills, harvesters etc. increased crop productivity. However, their long-term effects on soil health are a serious concern. Recent research reports indicate that the overuse of fertilizers and pesticides has seriously affected soil health in many parts of the world, leading to a decline in crop productivity [7]. This decline in crop productivity has been attributed to imbalanced nutrient ratios, micronutrient deficiencies, reduced microbial activity, and disturbed soil physical structure due to the heavy use of agricultural equipment [8].

2. Impact of current agricultural practices on soil health

A wide range of inefficient or faulty agricultural practices can be classified into physical, chemical, and biological categories (**Figure 1**) that affect the productive capacity of the soil, as briefly illustrated below.

2.1 Physical soil degradation

Physical soil degradation occurs when agricultural practices degrade the physical properties of the soil to such an extent that its critical functions are impaired. For example, intensive tillage often results in soil compaction and hardening at the plowing depth when heavy farm machinery is used repeatedly during tillage. Burning crop residues, repetitive tillage, little or no recycling of agricultural wastes, and other agricultural practices that reduce SOM content make soils more susceptible to physical degradation.

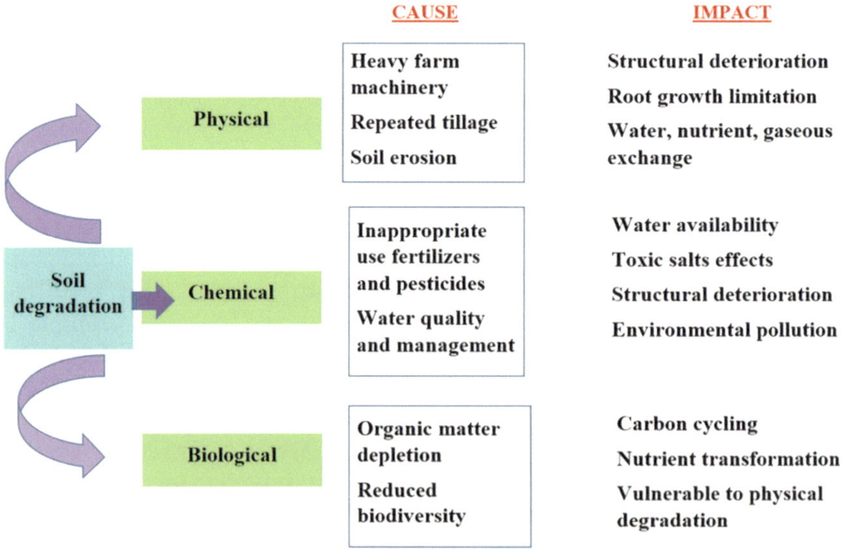

Figure 1.
Flowchart depicting the types, causes, and impacts of soil degrading processes on soil functioning.

2.2 Chemical soil degradation

It refers to the adverse change in the chemical environment of soils and affects their yield potential. Of the common agricultural practices, inappropriate chemical fertilizers and pesticides contribute significantly. Evidence shows that heavy fertilizer use over long periods can lead to increased soil acidification, which negatively impacts soil productivity [9, 10]. The chemicals used in controlling pests (weeds, insects, etc.) also harm soil organisms, which are crucial to soil health and productivity.

2.3 Biological soil degradation

A significant part of our biodiversity is found in the soil. Soil organic matter is the basis of life, and living organisms are critical to maintaining soil productivity. The range and diversity of microorganisms in soil have decreased due to the decline in organic matter recycling. Agricultural practices that do not emphasize integrated nutrient management (INM), involving the best use of crop residues in agriculture, burning crop residues (CR), etc., deplete SOM, which in turn leads to a loss of soil biology and thus contributes to soil degradation. Monocropping, i.e., continuous cultivation of the same crop without crop rotation, also reduces soil biodiversity.

Therefore, researchers worldwide have implemented several research options that could improve soil health in one or more ways. The proven agricultural practices to improve soil health are described in the following sections.

3. Strategies to enhance soil health and sustain agricultural food production

3.1 Increase inputs of organic matter

Soil organic matter has several useful functions: It improves soil structure, is a nutrient storehouse, increases the water holding capacity (WHC) and cation exchange capacity (CEC) of the soil, chelates micronutrients and dissolves phosphates, etc. Therefore, its availability in the soil is paramount in improving soil health and crop productivity. However, soils in tropical and subtropical climates are naturally deficient in organic carbon; therefore, management practices to increase soil organic carbon (SOC) are strongly emphasized. These practices include the addition of CR, reduced tillage, compost and vermicompost (VC), biochar, and green manures (GM). Apart from increasing SOM, these practices add significant amounts of plant-available nutrients and reduce nutrient requirements from fertilizers [11].

3.2 Tillage practices

A brief comparison of conservation (CA) and conventional tillage and their effects on soil is shown in **Table 1**. The summary of studies showing the effects of CA or tillage with or without CR is given in **Tables 2** and **3**. Baker et al. [27] defined CA as an umbrella term that generally refers to minimum tillage (MT), direct drilling, no-tillage (NT), and/or ridge-tillage aimed at conserving some resources.

The core idea of classifying agricultural practice for CA is that at least 30% CR should be left on the soil surface, with subsequent conservation of time, power, fuel, soil, and soil properties. Thus leaving CR on the soil alone does not adequately

Parameter	Conventional tillage	Conservation tillage
Soil physical health	Very poor	Comparatively good
Soil biological health	Lowest due to frequent disturbance	More diverse and healthy
Soil fertility	Low, heavily rely on inorganic fertilizers	Medium, organics supplement mineral fertilizers
Soil disturbance	Very high	Minimal
Soil erosion	Maximum	Comparatively reduced
Soil Aggregates	Breakdown	Protection and enhancement
Soil surface	Bare	Well protected with crop residues
Soil organic matter	losses	Build-up
Water infiltration	Lowest after clogging of soil pores	Best
Overall soil health	Poor	Good

Table 1.
A comparison of the impact of conventional vs. conservation tillage practices on soil.

describe all CA practices. The overall goals of CA include improving and sustaining agricultural production and conserving the environment, soil, and human health. FAO states that CA practices should be resource efficient and include three essential ideas: Crop rotation, minimal soil disturbance, and maintaining soil cover through residues. Controlled traffic was recently added to this list by FAO.

Agricultural practices and modern tillage cause SOM to decrease over time due to increased oxidation, leading to soil degradation, loss of biological fertility, and long-term soil resilience [28]. Mineralization of SOM can improve yields in the short term by releasing nitrogen (N), but there is always some leaching of nutrients into the subsoil. This is particularly important in soils under tropical climates, where SOM is rapidly degraded so that SOC levels are low after only one or two decades of intense soil tillage.

In contrast, cropping under zero-tillage (ZT) has resulted in a build-up of SOC in the surface layers [29] since permanent soil cover is maintained with crop residue. By using NT, it is possible to minimize SOM losses and build soil C and N stocks [30]. Though tillage is sometimes effective in reducing compaction, it is also one of the significant causes of compaction, especially tillage breaks the soil aggregates, disturbs the surface soil, and accentuates soil runoff and erosion. Repeated passes with a tractor are used for seedbed preparation or to maintain a fallow pasture. Tillage exposes the soil to the air, increasing evaporative loss of soil moisture and thus affecting soil biotic activity, which is essential for healthy soil. The biological properties of the soil are altered by this form of disturbance, which has numerous negative implications on soil productivity. For instance, some microorganisms can be severely damaged, negatively impacting soil biodiversity. Repeated plowing can disrupt the extraradical hyphal network of fungi, resulting in poor nutrient supply to plants. To address the problems described above, farmers in agricultural systems that rely on tillage for good crop yields need to increase organic matter input through compost, VC, or GM.

Reference	Location	Soil type	Treatments studied	Results reported
[12]	Delhi, India	Sandy loam	Zero tillage (ZT), Permanent beds (PB), Conventional tillage (CT)	Adoption of PB/ZT resulted in ~22.5% higher SQI than CT. ZT had 22, 18, 25, 28, 29, and 15% higher water-stable aggregates (WSA), hydraulic conductivity (HC), dehydrogenase enzyme (DHA), B-glucosidase activity, available phosphorus (Av. P), and potassium (Av. K), respectively, over CT.
[13]	Pakistan	Sandy clay loam	Minimum tillage (MT), CT, Deep tillage (DT)	With MT, nitrate is less likely to leach into the soil, and bulk density (BD) was about 10% lower than with DT.
[14]	Germany	Silt loam	CT and Reduced tillage (RT)	RT had greater air capacity, HC, macro porosity, and pore connectivity but lower BD.
[15]	China	Clay loam	CT, and NT	Microbial biomass was 21% higher after NT treatment than after CT treatment. Additionally, a higher fungus to bacterial ratio was seen after NT treatment compared to CT treatment.
[16]	Ohio, USA	Loam	CT and NT	The NT practice had 68, 18, 60, and 53% higher mineralizable carbon, basal soil respiration, total carbon (TC), and total nitrogen (TN) than the CT practices.
[17]	Romania	Clay loam	NT, chisel tillage, and CT	NT systems were found to have higher TC, non-labile fraction, very labile fraction than CT treatments.
[18]	Denmark	Sandy loam	Residue retention + tillage (NT and plowing)	In plowed land, the residue retention had a significant effect soil C while there was no discernible effect in NT soils.
[19]	South Africa	Alluvial origin, (Haplic Cambisol)	CT, NT with crop rotation and residue removal, retention, and biochar.	NT treatment had 23% higher SOC, over CT. NT increased PR, MWD and BD of soil compared to CT.
[20]	Northern Ethiopia	Vertisol	Permanent bed (PB), conventional tillage (TRAD), and terwah (a traditional water conservation technique (TERW))	SOM and aggregate stability followed order: PB > TERW>TRAD, while reverse was true for runoff and soil loss amounts.
[21]	USA	Silt loam	NT, disk (DP), chisel (CP), moldboard plow (MP), and NT with winter wheat cover crop (NTW)	At 0–15 cm depth, MP exhibited comparable BD to NT, NTW, and CP; however, the geometric mean diameter (GMD) of aggregates was much greater under NT and NTW.

Table 2.
Summary of studies revealing the effect of tillage practices on soil health.

Reference	Location	Soil type	Treatments studied	Results reported
[22]	Punjab India	Sandy clay loam	Rice straw incorporation at varying doses @ 0, 5, 7.5, and 10 t ha^{-1}	Incorporating rice straw (7.5 and 10 t ha^{-1}) increased soil properties such as aggregation, porosity, water retention, and nutrient availability.
[23]	USA	Silt loam	NT and chisel tillage practices with no, partial, and complete residue removal	When all crop residue was left in the field, the SOC stock under chisel tillage was 13% lower than in NT plots. In comparison to tilled plots, NT plots showed 5% and 39% greater BD and PR, respectively, while residue removal considerably enhanced PR under NT.
[24]	Punjab India	Sandy loam	CT, NT with and without residue (R), and Deep tillage (DT)	NTR and DT had approximately 30 and 37% lower HC, respectively, and 7 and 22% lower PR than CT.
[25]	China	Silty clay	CT with crop residues incorporation (CRI), rotary tillage with (RT + CRI), NT with CR retention; rotary tillage with CR removed (RTO)	NT treatment had higher SOC while humic compounds were higher under RT and CT treatments.
[26]	Central Mexico	Sandy loam	CT, NT with 0, 33, 66, 100% residue cover and planting of vetch (*Vicia* sp.) and ayocote bean (*Phaseolus vulgaris* L.) in different combinations	Over CT, NT with crop residue increased aggregate stability by around 22%, TOC by 48%, Av. P by 80%, and Av. K by 11%.

Table 3.
Summary of studies revealing the effect of tillage practices with crop residue (incorporation, retention, removal) on soil health.

3.3 Organic source of nutrients

The mechanism of how long-term organic management improves soil health and agricultural production is illustrated in **Figure 2**. The studies that demonstrate the effect of organics (cover crops, crop rotations, manures, green manures, and vermicompost) on soil health are listed in **Table 4**. Cover crops (CC) (legumes or non-legumes) are used in cropping systems as a nutrient management tool [43] that protects the soil from raindrops and soil erosion and alters the temperature regime, thus improving soils [44]. The type of CC to be included in the cropping system is goal-oriented. For example, the legume CC is a source of nutrients, especially N, for the following crop [45], while grasses are mainly used to reduce soil erosion and NO3-N leaching [46]. Initially, CC has a protective effect on the soil, and later, when they decompose, they add significant amounts of SOM, which improves soil aggregate stability and microbial activity. Incorporating legumes CC has several advantages, namely the supply of essential nutrients when they decompose, biological N fixation, reducing the need for N fertilizer [47], and the extensive root system of legumes, which explores nutrients in the subsoil [48]. However, soil nutrient enrichment with legumes varies depending on the type of CC, the amount of dry matter produced, the ability to assimilate nutrients,

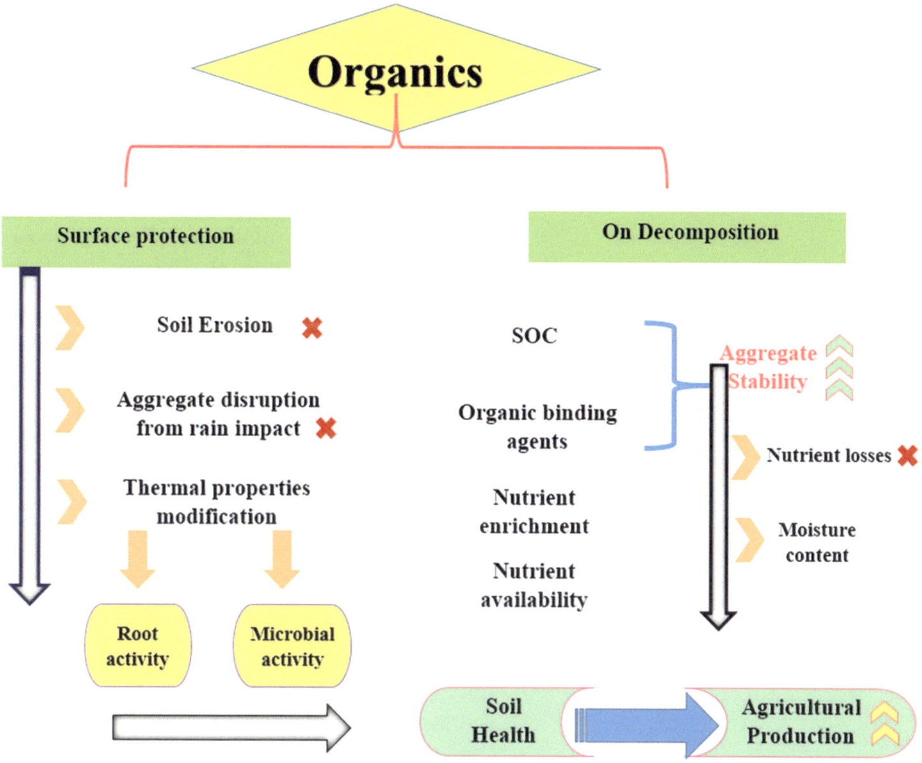

Figure 2.
Flowchart depicting the impact of long-term agricultural use of organics on soil health and crop production.

and the root system. Therefore, including CC between the rows of the main crops can lead to improved soil health [49].

An adequate and balanced supply of plant nutrients is essential for improving soil health. Improving the availability of less available nutrients in the soil is becoming increasingly important. In this regard, crop rotation involving the cultivation of different crops with different rooting habits, nutrient requirements, and leaf litter deposition can be efficient, which helps to regulate soil nutrient supply [50]. For example, the water requirements of rice and sugarcane are very high, which lowers the water table. If a crop with lower water demand (millet) is grown instead, this could help conserve water and nutrients in the soil. A meta-analysis conducted by Venter et al. [51] found that crop rotation increased the diversity and richness of microbial communities. Similarly, microbial diversity due to crop rotation resulted in beneficial changes in soil physicochemical properties [52], improved water use efficiency, and controlled temperature fluctuations [53]. Thus, crop rotation can manage soil and soil fertility, improve soil workability, increase nutrient and water availability, reduce soil losses due to erosion and crusting, and recycle nutrients in the soil, ultimately improving soil health [54]. However, increasing crop diversity in crop rotation is an effective strategy for long-term resilience [55].

SOC can be maintained or restored by the application of organic manure. In addition, manure increases total N content, which is essential for plant growth. Manure provides better soil structure by increasing SOC and keeping soil pH in an optimum range for

Reference	Location	Soil type	Treatments studied	Results reported
[31]	Indo-Gangetic plains (India)	Typic Ustocrept	Farmyard manure (FYM), RDF and their combination (IPNS), Inclusion of forage cowpea or forage berseem	Continuous rice-wheat cultivation with RDF raised soil BD, while incorporating fodder berseem or cowpea every third year helped reduce it.
[32]	Meghalaya (India)		Control, RDN through FYM, RDN through poultry manure (PM), RDN through vermicompost (VC) as well as their varying doses	In comparison to control, organics (FYM, PM, vermicompost) enhanced TOC and microbial biomass carbon by 57 and 62%, respectively. There was an increase in available N, P, and K inorganic treatments over control.
[33]	Central Chile	Coarse-textured soil	cover crops + and N fertilization	The addition of *L. multiflorum* improved soil organic pools and microbial activity.
[34]	Italy	loam soil	Two species of legume cover crop and non-legume cover crop	Cover cropping with legumes increased the SOC content of soil.
[35]	western Illinois	Silty clay loam	Continuous corn (CCC), corn-soybean (CS) rotation for 2 years, corn–soybean–wheat rotation for 3 years, and continuous soybean (SSS).	The 3-year CSW rotation and CCC had higher total N and WSA than the CS or SSS rotations. In comparison to SSS, Av. K levels were higher in CCC and CSW. In terms of storing nitrogen and preserving soil aggregates, the two crop rotations, CCC and CS, were comparable.
[36]	New York	Silt loam soil	Crop rotation with Soybean, corn, wheat, and clover.	Soybean-wheat/clover-corn rotation showed the increased infiltration rates and highest earthworm populations.
[37]	Akola (India)	Vertisol	RDF, FYM, Wheat Straw, Gliricidia green leaf manuring and their combinations.	Adopting minimum tillage with 50% N through *gliricidia* GLM and compensation of RDF through chemical fertilizers helps improve soil quality with higher yield and maximum net returns.
[38]	North China plain	Silty loam	control, fertilizers, and FYM	Continuous FYM application for 15 years significantly increased soil N and organic matter contents, and enzyme activities.
[39]	Brazil	Fluvisol	Effect of four legume species Cajanus, Crotalaria, Canavalia, and Mucuna used as green manure	Crotalaria plots had the highest soil P and K concentrations, while Mucuna plots had the highest soil Ca content. When compared to other green manure species, the plot containing Mucuna had a higher level of soil microbial biomass. Mucuna was more effective in enhancing the biological properties of the soil, but Crotalaria appeared to be more effective at enhancing the chemical characteristics.

Reference	Location	Soil type	Treatments studied	Results reported
[40]	Himalayan region of Kashmir, Pakistan	Loam	Repeated application of wheat straw residue and poultry manure alone or in combination with urea	Organic amendments, either alone or in combination with UN, significantly enhanced soil physical properties by lowering BD and PR, while increased aggregate stability and HC.
[41]	Turkey	i. sandy loam, ii. loam and iii. clay	Vermicompost application	Vermicomposting considerably enhanced organic matter content and wet aggregate stability in all three textural soils.
[42]	Bangladesh	Silt clay loam	Control, cowdung, green manure, compost, and rice straw and three levels of N.	Different organic materials that are applied over a long period of time enhance soil SOC, total N, P, and S, and lowered pH. Increased in SOC maximum under green manure.

Table 4.
Summary of studies revealing the effect of organics on soil health.

crop growth [56]. Organic manure differs significantly in its influence on soil properties. The nature and composition of manure primarily affect its residence time in the soil and thus influence soil properties. Therefore, it is always advantageous to use a variety of organic sources. For example, well-decomposed compost may not improve soil aggregation but can rapidly increase the soil's biological activity and nutrient content, persisting for a short period, while dairy cow manure may stimulate soil aggregation [57].

Vermicompost (VC) is a nutrient-rich organic amendment produced by earthworms [58]. A VC consists of earthworm casts, humic substances, seeds or cocoons, and partially decomposed bedding materials. VC is usually added to soil as a source of essential plant nutrients and to reduce fertilizer dose, but simultaneously, it improves soil microbial composition, diversity, and thus nutrient transformation, structural stability [59], and soil health. Adding VC to soil alone or with chemical fertilizers can improve soil structural stability and WHC [60]. In VC-rich soils, earthworm populations proliferate, resulting in porous soil with good aeration, water absorption, and drainage. The application of VC enhances soil microbial diversity. Numerous N-fixing bacteria have been reported to reside in earthworm burrows [61].

Green manuring is the addition of undecomposed green plants to improve soil nutrient content and supply to subsequent crops. It can also be considered a system for incorporating green plants into the soil at a green stage before flowering in the same or another field. GM technology improves nutrient supply and soil fertility, structure, and WHC, curbs soil erosion, and increases soil microbial populations. This practice is environmentally friendly and poses no threat to soil, water, and air [62]. Like chemical fertilizers, they do not negatively affect soil properties and food production. GM plants with high nutrient concentrations and low C/N ratio are more valuable as organic fertilizers in crop production [63]. Using GM crops in conjunction with proper residue management and crop rotation can help maintain soil health, promote soil fertility, support nutrient cycling in deeper soil profiles [64], limit weed growth, and eliminate the need for external fertilizers [65, 66]. GM Plants are often referred to as soil-building plants because they are grown primarily for the benefit of the soil. In addition to improving the soil's physical, chemical, and biological properties, GM plants enrich the soil with organic matter and nutrients [67–70]. They also facilitate soil and nutrient conservation, promote biological activity, reduce soil compaction, increase soil porosity and water permeability, and ultimately improve soil health [71–73] and crop productivity [74, 75]. GM Intercropping plants pull nutrients and prevent leaching from the soil. Nutrients contained in GM plants become plant-available upon decomposition and can feed the following crop. This cycle of nutrient recycling contributes to a healthier soil environment. GM Plants also increase phosphorus utilization by crops [76, 77], reducing nitrate leaching and the need for nitrogen fertilizer for subsequent crops [78]. By growing GM crops between main crops, soils are protected from erosion, degraded soils become productive again, and chemical fertilizers can be replaced to some extent [79, 80].

3.4 Biochar

The studies showing the effect of biochar on soil health are listed in **Table 5**. Biochar is the carbon-rich byproduct of biomass pyrolysis formulated under oxygen-limited conditions, intentionally applied to soils, and optimized for agronomic and environmental benefits [89]. Biochar has many of the same characteristics as charcoal, such as the presence of stable recalcitrant organic carbon [90], but differs from similar materials in its intended use as a soil amendment [91] and as a long-term carbon storage material [92]. It is possible to produce biochar from various materials, including agricultural crop residues, municipal waste, forestry waste, and animal manure [93]. There are several fundamental properties of biochar, including pH,

Reference	Location	Soil type	Char type	Results reported
[81]	China	Clay loam	Maize straw	Increase in SOC by 28%, TN, Available P, and K
[82]	Korea	clay	Rice hull-derived biochar	Increase in WSA Reduced ESP
[83]	Pakistan	Aridisol	Corn cob	Increased C seq Enhanced microbial C
[84]	USA	Entisol	Switchgrass	Increased soil moisture by 38% Reduced BD by 15%
[85]	Punjab India	Sandy loam	Rice straw-derived biochar	Higher SOC, available P, and K Increased Nutrient use efficiency
[86]	Australia	Sandy loam	Acacia green waste	About 23% increment in SOC was observed for biochar.
[87]	A meta-analysis Kenya			In general, adding biochar to soils alone or in combination with fertilizer increases the soil's availability of P.
[88]	Kenya	Variables texture sites		Biochar addition slightly increased pH, Av. porosity, and N mineralization from native SOM

Table 5.
Summary of studies revealing the effect of biochar type on soil health.

specific surface area, CEC, and porosity, which are affected by the feedstock and the production process [94]. These properties affect how biochar interacts with soil constituents, especially physically, chemically, and biologically, and its fate within an ecosystem [95]. As a soil amendment, biochar can maintain crop productivity by improving the soil's physical environment, nutrient availability [96], and nutrient supply to plants, reducing leaching losses and the need for fertilizer [97, 98].

Biochar also stimulates microbial activity, diversity, and nutrient transformation [99]. Biochar can improve soil WHC and reduce greenhouse gas emissions [100]. In addition, biochar can control the contaminants in soil: toxicity, mobility, and bio-availability [101]. It is well known that biochar has tremendous potential to mitigate global climate change by sequestering carbon in the soil. For example, biochar application has increased SOC by 4.9–6.3 g kg^{-1} [102], 4.3 kg^{-1} [103], and 0.52% [104]. However, after pyrolysis of organic plant material, the long-term potential of biochar to sequester C depends on the composition of stable and resistant forms of organic C.

3.5 Inorganic fertilizers

Although inorganic or chemical fertilizers can improve crop growth and yield relatively quickly, chemical fertilizers are associated with certain disadvantages. Studies on long-term fertilization show that much of the increase in crop production yields in recent decades can be attributed to mineral fertilizers [105]. Higher productivity also results in more plant residues returning to the soil after harvest, which boosts SOM in the long run. According to Körschens et al. [106], NPK fertilization increased SOC by 10% compared to the control. One notable exception is when the pH of a crop is subsided by ammonium or urea fertilizer applications. In such cases, yield is significantly reduced and may fall below that of an unfertilized crop [107]. The acidic condition of the soil may decrease the soil aggregates, making the soil susceptible to erosion.

It has been found that soil biotic activity is significantly affected by soil pH [107]. Since chemical fertilizers are highly soluble, their continued use leads to groundwater contamination. These chemicals subsequently combine with clay to form hard pans in the soil that impede the growth of plant roots in the soil [108]. Chemical fertilizers also destroy soil structure, leaving highly compacted soils with a restricted pore network and air circulation [99]. The life of beneficial soil microorganisms, such as bacteria that fix N, is threatened by the repeated use of inorganic fertilizers [109, 110]. Mineral fertilizers differ in their ability to alter specific microbial communities over time. In general, fungi have demonstrated the benefits of mineral fertilization even when soil pH had little effect on their growth [23, 111–113].

Nevertheless, Kirchmann et al. [109] found lower fungal biomass in fertilized soils in two long-term field studies in Sweden. They attributed this to N-rich residues with a reduced C/N ratio, which may have favored bacterial activity over that of fungi. According to Melkamu and Alemayehu [10], applying N or sulfate fertilizers led to a decrease in microbial C, accompanied by a decrease in soil pH.

3.6 Need-based input application

To prevent the inefficient use of external inputs, the golden rule is to use them as needed and control their waste, such as tillage, water, fertilizers, etc. Fertilizers are crucial in crop production, as nutrients determine more than 50% efficiency.

Nitrogenous fertilizers need special treatment. In addition, research on the site-specific application of N and P, the timing of the application, and placement is essential for determining the exact amount of nutrients and calibrating the amount of fertilizer needed to increase the efficiency of their use. For example, in soils with a high P-fixing capacity, application of P near the root zone is generally recommended, especially for water-soluble P fertilizers (SSP), while citric acid-soluble P fertilizers (dicalcium phosphate) can be applied as such. Split application is recommended in the case of N and K since a significant amount of these nutrients is lost through leaching. Depending on soil conditions, rainfall intensity and duration, and crop requirements, two to three splits are recommended for most crops. This synchronizes nutrient supply with crop needs at critical growth stages.

In addition, using coated technologies such as sulfur-coated urea, neem-coated urea, urea granules, nitrification inhibitors, and urease inhibitors is also very promising to reduce N losses from the agroecosystem. The coated materials provide a controlled release of N in the soil, increasing its availability and accessibility to plant roots. Nitrification inhibitors can reduce nitrate leaching and denitrification, especially in flooded soils.

3.7 Role of Nano fertilizers

The waste of nutrients in conventional fertilizers is alarming, leading to calls for environmentally friendly fertilizers with high use efficiency. In this context, nanotechnology has emerged as a potential substitute for conventional fertilizers. As nano-fertilizers, they contribute to nutritional management by improving nutrient use efficiency. In response to various environmental factors, such as heat, moisture, and other unfavorable circumstances, the nano-fertilizer distributes nutrients in a controlled manner. Using nanoscale fertilizers, it may be possible to control the release of nutrients and properly deliver the right amount of nutrients to plants.

3.8 Precision farming and agricultural sustainability

Precision farming, also known as site-specific agriculture, aims to identify how external and controllable factors such as fertilizer, seeding rate, herbicide, pesticide use, and available water affect crop response to personalize soil and crop management practices to the unique circumstances of each field. The basic idea is to provide nutrients precisely for a particular site. Because there is high soil variability, the concept of precision agriculture measures these variations in soil properties to make the proper adjustments in fertilizer inputs, rates, application method and timing, seed rates, and pesticide doses in near "real-time."

4. Conclusions

Since the advent of agriculture, people have used various chemicals to enrich the soil to increase crop yields. This overuse of chemicals damages the environment and the health of the soil. Although fertilizers and pesticides are used to meet the world's food needs, they are now a cause for concern; as a result, we have lost a large area of fertile land worldwide. In this context, maintaining soil fertility and restoring the health of stressed or degraded soils are paramount. The use of cost-effective natural practices such as FYM, vermicompost, green manures, biofertilizers, and integrated nutrient management, as well as conservation farming practices such as reduced tillage and no-till, have proven to be the best methods to improve and sustain agricultural production, soil health, and the environment. Recent advances in the use of nutrients, i.e. nano fertilizers, need-based nutrient management, and the concept of precision agriculture, can also be explored to improve soil health and agricultural sustainability.

Acknowledgements

No financial support is received in connection with publishing this document. I thank Mr. Kartik Madankar for his assistance throughout all aspects of our study.

Conflict of interest

The authors declare no conflict of interest.

Author details

Pratik Ramteke[1*], Vijay Gabhane[1], Prakash Kadu[1], Vilas Kharche[1] and Samrat Ghosh[2]

1 Dr. Panjabrao Deshmukh Krishi Vidyapeeth, Akola, Maharashtra, India

2 Bidhan Chandra Krishi Vishwavidyalaya, Mohanpur, West Bengal, India

*Address all correspondence to: pratik-soil@pau.edu

References

[1] Srivastava P, Balhara M, Bhoopander G. Soil health in India: Past history and future perspective. In: Giri B, Varma A, editors. Soil Health Soil Biology. India: Springer; 2020. pp. 1-21

[2] Biswas B, Nirola R, Biswas JK, Pereg L, Willett IR, Naidu R. Environmental microbial health under changing climates: State, implication and initiatives for high-performance soils. In: Lal R, Francaviglia R, editors. Sustainable Agriculture Reviews. Cham: Springer; 2019. DOI: 10.1007/978-3-030-26265-5_1

[3] Doran JW, Zeiss MR. Soil health and sustainability: Managing the biotic component of soil quality. Applied Soil Ecology. 2000;**15**:3-11

[4] Doran JW. Soil health and global sustainability: Translating science into practice. Agriculture Ecosystem Environment. 2002;**88**(2):119-127

[5] Li X, Rui J, Xiong J, Li J, He Z, Zhou J, et al. Functional potential of soil microbial communities in the maize rhizosphere. PLoS One. 2014;**9**(11):e112609. DOI: 10.1371/journal.pone.0112609

[6] Damodaran T, Sah V, Rai RB, Sharma DK, Mishra VK, Jha SK, et al. Isolation of salt tolerant endophytic and rhizospheric bacteria by natural selection and screening for promising plant growth-promoting rhizobacteria (PGPR) and growth vigour in tomato under sodic environment. African Journal of Microbiology Research. 2013;7(44):5082-5089

[7] Singh MV. Micronutrient deficiencies in crops and soils in India. In: Alloway BJ, editor. Micronutrient Deficiencies in Global Crop Production. Dordrecht: Springer; 2008. pp. 93-125

[8] Lal R. Enhancing crop yields in the developing countries through restoration of the soil organic carbon pool in agricultural lands. Land Degradation Development. 2005;**17**:197-209. DOI: 10.1002/ldr.696

[9] Ozlu E, Kumar S. Response of soil organic carbon, pH, electrical conductivity, and water stable aggregates to long-term annual manure and inorganic fertilizer. Soil Science Society of America Journal. 2018;**82**(5):1243-1251

[10] Melkamu YB, Alemayehu M. Impact of crop production inputs on soil health: A review. Asian Journal of Plant Science. 2017;**16**(3):109-131

[11] Alori ET, Adekiya AO, Adegbite KA. Impact of agricultural practices on soil health. In: Giri B, Varma A, editors. Soil Health Soil Biology. India: Springer; 2020. pp. 89-99

[12] Parihar CM, Singh AK, Jat SL, Dey A, Nayak HS, Mandal BN, et al. Soil quality and carbon sequestration under conservation agriculture with balanced nutrition in intensive cereal-based system. Soil and Tillage Research. 2020;**202**:1046-1053

[13] Khan S, Shah A, Nawaz M, Khan M. Impact of different tillage practices on soil physical properties, nitrate leaching and yield attributes of maize (Zea mays L.). Journal of Soil Science and Plant Nutrition. 2017;**17**(1):240-252

[14] Schlüter S, Großmann C, Diel J, Wu GM, Tischer S, Deubel A, et al. Long-term effects of conventional and reduced tillage on soil structure, soil ecological and soil hydraulic properties. Geoderma. 2018;**332**:10-19

[15] Zhang Z, Cai W, Yang T, Zhu Y, Yu YX. Long-term field fertilization

affects soil nitrogen transformations in a rice-wheat-rotation cropping system. Journal of Plant Nutrition and Soil Sciences. 2012;**175**:939-946

[16] Aziz I, Mahmood T, Islam KR. Effect of long term no-till and conventional tillage practices on soil quality. Soil and Tillage Research. 2013;**131**:28-35

[17] Topa D, Cara IG, Jităreanu G. Long term impact of different tillage systems on carbon pools and stocks, soil bulk density, aggregation and nutrients: A field meta-analysis. Catena. 2021;**199**:105102

[18] Gómez-Muñoz B, Jensen LS, Munkholm L, Olesen JE, Møller Hansen E, Bruun S. Long-term effect of tillage and straw retention in conservation agriculture systems on soil carbon storage. Soil Science Society of America Journal. 2021;**85**(5):1465-1478

[19] Nyambo P, Chiduza C, Araya T. Effect of conservation agriculture on selected soil physical properties on a haplic cambisol in Alice, eastern cape, South Africa. Archives of Agronomy and Soil Science. 2022;**68**(2):195-208

[20] Oicha T, Cornelis WM, Verplancke H, Nyssen J, Govaerts B, Behailu M, et al. Short-term effects of conservation agriculture on Vertisols under tef (Eragrostis tef (Zucc.) trotter) in the northern Ethiopian highlands. Soil and Tillage Research. 2010;**106**(2):294-302

[21] Nouri A, Lee J, Yoder DC, Jagadamma S, Walker FR, Yin X, et al. Management duration controls the synergistic effect of tillage, cover crop, and nitrogen rate on cotton yield and yield stability. Agriculture, Ecosystems & Environment. 2020;**301**:107007

[22] Ramteke PR, Vashisht BB, Sharma S, Jalota SK. Assessing and ranking influence of rates of Rice (Oryza sativa L.) straw incorporation and N fertilizer on soil physicochemical properties and wheat (Triticum aestivum L.) yield in Rice-wheat system. Journal of Soil Science and Plant Nutrition. 2022;**22**:515-526. DOI: 10.1007/s42729-021-00665-z

[23] Villamil MB, Little J, Nafziger ED. Corn residue, tillage, and nitrogen rate effects on soil properties. Soil and Tillage Research. 2015;**151**:61-66

[24] Kahlon MS, Chawla K. Effect of tillage practices on least limiting water range in Northwest India. International Agrophysics. 2017;**31**(2):183-194. DOI: 10.1515/intag-2016-0051

[25] Tang H, Xiao X, Li C. Impact of tillage practices on soil aggregation and humic substances under double-cropping paddy field. Agronomy Journal. 2020;**112**(1):624-632. DOI: 10.1002/agj2.20051

[26] Roldán A, Caravaca F, Hernández MT, Garcıa C, Sánchez-Brito C, Velásquez M, et al. No-tillage, crop residue additions, and legume cover cropping effects on soil quality characteristics under maize in Patzcuaro watershed (Mexico). Soil and Tillage Research. 2003;**72**(1):65-73

[27] Baker CJ, Saxton KE, Ritchie WR. No-Tillage Seeding: Science and Practice. 2nd ed. Oxford, UK: CAB International; 2002

[28] Lal R. Sustainable land use systems and resilience. Soil resilience and sustainable land use. In: Greenland DJ, Szabolcs I, editors. Proc. Symp. Held in Budapest, Including the Second Workshop on the Ecological Foundations of Sustainable Agriculture (WEFSA II), 28 September-2 October 1992. Oxford, UK: CAB International; 1994. pp. 99-118

[29] Campbell CA, McConkey BG, Zentner RP, Selles F, Curtin D.

Long-term effects of tillage and crop rotations on soil organic C and N in a clay soil in southwestern Saskatchewan. Canadian Journal of Soil Science. 1996;**76**:395-401

[30] Bayer C, Mielniczuk J, Amado TJC, Martin-Neto L, Fernandes SV. Organic matter storage in a sandy loam Acrisol affected by tillage and cropping systems in southern Brazil. Soil and Tillage Research. 2002;**54**:101-109

[31] Singh VK, Dwivedi BS, Mishra RP, Shukla AK, Timsina J, Upadhyay PK, et al. Yields, soil health and farm profits under a rice-wheat system: Long-term effect of fertilizers and organic manures applied alone and in combination. Agronomy. 2019;**9**(1):1. DOI: 10.3390/agronomy9010001

[32] Kumar M, Baishaya LK, Ghosh DC, Gupta VK, Dubey SK, Das A, et al. Productivity and soil health of potato (Solanum tuberosum L.) field as influenced by organic manures, inorganic fertilizers and biofertilizers under high altitudes of eastern Himalayas. Journal of Agricultural Science. 2012;**4**(5):223-234. DOI: 10.5539/jas.v4n5p223

[33] Salazar O, Balboa L, Peralta K, Rossi M, Casanova M, Tapia Y, et al. Effect of cover crops on leaching of dissolved organic nitrogen and carbon in a maize-cover crop rotation in Mediterranean Central Chile. Agriculture Water Management. 2019;**212**:399-406

[34] Mazzoncini M, Sapkota TB, Barberi P, Antichi D, Risaliti R. Long-term effect of tillage, nitrogen fertilization and cover crops on soil organic carbon and total nitrogen content. Soil and Tillage Research. 2011;**114**(2):165-174

[35] Zuber SM, Behnke GD, Nafziger ED, Villamil MB. Crop rotation and tillage effects on soil physical and chemical properties in Illinois. Agronomy Journal. 2015;**107**(3):971-978

[36] Katsvairo T, Cox WJ, Van EH. Tillage and rotation effects on soil physical characteristics. Agronomy Journal. 2002;**94**(2):299-304

[37] Sonune BA, Kharche VK, Gabhane VV, Jadhao SD, Mali DV, Katkar RN, et al. Sustaining soil health and cotton productivity with tillage and integrated nutrient management in Vertisols of Central India. Indian Journal of Soil Conservation. 2021;**49**(1):1-11

[38] Liang Q, Chen H, Gong Y, Yang H, Fan M, Kuzyakov Y. Effects of 15 years of manure and mineral fertilizers on enzyme activities in particle-size fractions in a North China plain soil. European Journal of Soil Biology. 2014;**60**:112-119

[39] Nilza SC, Antnio BBO, Maristella MCP, Vicente PCN, Ricardo SDS, Joo RDC, et al. Short-term effect of different green manure on soil chemical and biological properties. African Journal of Agricultural Research. 2015;**10**(43):4076-4081

[40] Khaliq A, Abbasi MK. Improvements in the physical and chemical characteristics of degraded soils supplemented with organic–inorganic amendments in the Himalayan region of Kashmir, Pakistan. Catena. 2015;**126**:209-219

[41] Aksakal EL, Sari S, Angin I. Effects of vermicompost application on soil aggregation and certain physical properties. Land degradation & development. 2016;**27**(4):983-995

[42] Akter F, Rahman MM, Alam MA. Soil chemical properties as influenced by long term manuring and nitrogen fertilization in Bangladesh. Asian Journal of Soil Science and Plant Nutrition. 2019;**4**(4):1-9

[43] SSSA. Glossary of soil science term. In: Soil Science Society. Madison, Wisconsin: Soil Science Society of America, Inc.; 2008. p. 134. DOI: 10.2136/2008.glossary soilscienceterms

[44] Ruffo ML, Bollero GA. Modelling rye and hairy vetch residue decomposition as a function of degree days and decomposition days. Agronomy Journal. 2003;**95**:900-907

[45] Smith MS, Frye WW, Varco JJ. Legume winter cover crops. Advances in Soil Science. 1987;**7**:95-139

[46] Meisinger JJ, Hargrove WL, Mikkelsen RL, Williams JR, Benson VW. Effects of cover crops on groundwater quality. In: Hargrove WL, editor. Cover Crop for Clean Water. Ankeny, Iowa: Soil and Water Conservation Society; 1991. pp. 9-11

[47] Singh Y, Singh B, Ladha JK, Khind CS, Gupta RK, Meelu OP, et al. Long-term effects of organic inputs on yield and soil fertility in the rice-wheat rotation. Soil Science Society of America Journal. 2004;**68**:845-853

[48] Gathumbi SM, Cadisch G, Buresh RJ, Giller KE. Subsoil nitrogen capture in mixed legume stands as assessed by deep nitrogen-15 placement. Soil Science Society of America Journal. 2003;**67**:573-582

[49] Fagaria NK, Baligar VC, Bailey BA. Role of cover crops in improving soil and row crop productivity. Commun. Soil Science and Plant Nutrition. 2005;**36**(19-20):2733-2757

[50] Stockdale EA, Lampkin NH, Hovi M, Keatinge R, Lennartsson EKM, MacDonald DW, et al. Agronomic and environmental implications of organic farming systems. Advances in Agronomy. 2001;**70**:261-327

[51] Venter ZS, Jacobs K, Hawkins HJ. The impact of crop rotation on soil microbial diversity: A meta-analysis. Pedobiologia. 2016;**59**(4):215-223

[52] Dias T, Dukes A, Antunes PM. Accounting for soil biotic effects on soil health and crop productivity in the design of crop rotations. Journal of the Science of Food and Agriculture. 2015;**95**(3):447-454

[53] Kennedy AC. Bacterial diversity in agroecosystems. In: Invertebrate Biodiversity as Bioindicators of Sustainable Landscapes. Agriculture, Ecosystems & Environment. 1999;**74**:65-76

[54] NRSC (Natural Resources Conservation Service). Rotations for Soil Fertility: Small Scale Solutions for your Farm [Internet]. 2009. Available from: https://www.nrcs.usda.gov/Internet/FSE_DOCUMENTS/stelprdb1167375

[55] Woodyard J, Kladivko E. Four Strategies to Improve Your Field's Soil Health. Purdue Agronomy, West Lafayette: Purdue Extension; 2017

[56] Ozlu E, Sandhu SS, Kumar S, Arriaga FJ. Soil health indicators impacted by long-term cattle manure and inorganic fertilizer application in a corn-soybean rotation of South Dakota. Scientific Reports. 2019;**9**(1):1-11. DOI: 10.1038/s41598-019-48207-z

[57] SARE (Sustainable Agriculture Research & Education). Enhancing Biota and Improving Soil Health [Internet]. 2012. Available from: https://www.sare.org/Learning-Center/Books/Manage-Insects-on-Your-Farm/Text-Version/Putting-it-All-Together/Enhancing-Biota-and-Improving-Soil-Health

[58] Dominguez J. State of the art and new perspectives on vermicomposting research. In: Edwards CA, editor.

Earthworm Ecology. Boca Raton: CRC Press; 2004. pp. 401-424

[59] Arancon NQ, Edwards CA, Babenko A, Cannon J, Galvis P, Metzger JD. Influences of vermicomposts, produced by earthworms and microorganisms from cattle manure, food waste and paper waste, on the germination, growth and flowering of petunias in the greenhouse. Applied Soil Ecology. 2008;**39**:91-99. DOI: 10.1016/j.apsoil.2007.11.010

[60] Manivannan S, Balamurugan M, Parthasarathi K, Gunasekaran G, Ranganathan LS. Effect of vermicompost on soil fertility and crop productivity of beans (Phaseolus vulgaris). Journal of Environmental Biology. 2009;**3**:275-281

[61] Lazcano C, Dominguez J. The use of vermicompost in sustainable agriculture: Impact on plant growth and soil fertility. In: Miransari M, editor. Soil nutrients. New York: Nova Science; 2011. pp. 1-23

[62] Yang L, Bai J, Liu J, Zeng N, Cao W. Green manuring effect on changes of soil nitrogen fractions, maize growth, and nutrient uptake. Agronomy. 2018;**8**:261. DOI: 10.3390/agronomy8110261

[63] Talgre L, Lauringson E, Roostalu H, Astover A, Makke A. Green manure as a nutrient source for succeeding crops. Plant Soil Environment. 2012;**58**(6):275-281

[64] Yadav GS, Lal R, Meena R, Babu S, Das A, Bhoumik SN, et al. Conservation tillage and nutrient management effects on productivity and soil carbon sequestration under double cropping of rice in north eastern region of India. Ecological Indicators. 2019;**105**:303-331

[65] Melero S, Riuz JC, Porrai Herencia JF, Madejon E. Chemical and biochemical properties in a silty loam soil under conventional and organic management. Soil and Tillage Research. 2006;**90**:162-170

[66] Drinkwater LE, Wagoner P, Sarrantonio M. Legume based cropping systems have reduced carbon and nitrogen losses. Nature. 1998;**396**:262-265

[67] Golec AFC, Pérez PG, Lokare C. Effective microorganisms: Myth or reality? Revista Peruana de Biología. 2007;**14**(2):315-319

[68] Pandey AK, Singh MK. Importance and uses of green manuring in field crops. Rashtriyakrishi. 2016;**11**(2):35-35

[69] Nayak JJ, Vaidya AK. Green manure in crop production and soil health. International Journal of Innovative Research, Science, Engineering and Technolgy. 2018;**7**(6):7378-7381

[70] Zaccheo PVC, Neves CSVJ, de Cinque MD, Zorzenoni TO, Higashibara LR, Piccinin GG, et al. Green manure in fruit culture: Aspects on soil quality and use in agriculture. African Journal of Agricultural Research. 2016;**11**(17):1469-1474. DOI: 10.5897/AJAR2015.10416

[71] Selvi RV, Kalpana R. Potentials of green manure in integrated nutrient management for rice: A review. Agricultural Reviews. 2009;**30**(1):40-47

[72] Doran JW, Fraser DG, Culik MN, Liebhardt WC. Influence of alternative and conventional agricultural management on soil microbial process and nitrogen availability. American Journal of Alternative Agriculture. 1988;**2**:99-106

[73] Schutter M, Dick R. Shifts in substrate utilization potential and structure of soil microbial communities in response to carbon substrates. Soil Biology and Biochemistry. 2001;**33**(11):1481-1491

[74] Bhattarai N, Vaidya GS, Baral B. Effect of mycorrhizal soil and green manures on growth of Ipil Ipil (Leucaena diversifolia L.). Scientific World. 2012;**10**(10):66-69

[75] Cavigelli MA, Thien SJ. Phosphorus bioavailability following incorporation of green manure crops. Soil Science Society of America Journal. 2003;**67**(4):1186-1194

[76] Bah AR, Zaharah AR, Hussin A. Phosphorus uptake from green manures and phosphate fertilizers applied in an acid tropical soil. Communications in Soil Science and Plant Analysis. 2006;**37**:2077-2093

[77] Fagaria NK. Green manuring in crop production. Journal of Plant Nutrition. 2007;**30**(5):691-719

[78] Xie Z, Tu S, Shah F, Xu C, Chen J, Han D, et al. Substitution of fertilizer-N by green manure improves the sustainability of yield in double-rice cropping system in South China. Field Crop Research. 2016;**188**:142-149

[79] Biederman LA, Harpole WS. Biochar and its effects on plant productivity and nutrient cycling: A meta-analysis. Global Change Biology Bioenergy. 2013;**5**:202-214. DOI: 10.1111/gcbb.12037

[80] Thomas SC, Frye S, Gale N, Garmon M, Launchbury R, Machado N, et al. Biochar mitigates negative effects of salt additions on two herbaceous plant species. Journal of Environment Management. 2013;**129**:62-68

[81] Xiao Q, Zhu LX, Zhang HP, Shen LXY, Y F, Li S Q. Soil amendment with biochar increases maize yields in a semi-arid region by improving soil quality and root growth. Crop and Pasture Science. 2016;**67**:495-507

[82] Kim HS, Kim KR, Yang JE, Ok YS, Owens G, Nehls T, et al. Effect of biochar on reclaimed tidal land soil properties and maize (Zea mays L.) response. Chemosphere. 2016;**142**:153-159

[83] Arif M, Ilyas M, Riaz M, Ali K, Shah K, Haq IU, et al. Biochar improves phosphorus use efficiency of organic-inorganic fertilizers, maize-wheat productivity and soil quality in a low fertility alkaline soil. Field Crop Research. 2017;**214**:25-37

[84] Novak JM, Busscher WJ, Watts DW, Amonette JE, Ippolito JA, Lima IM, et al. Biochar impact on soil-moisture storage in an ultisol and two aridisols. Soil Science. 2012;**177**:310-320

[85] Gupta RK, Hussain A, Singh Y, Sooch SS, Kang JS, Sharma S, et al. Rice straw biochar improves soil fertility, growth and yield of rice–wheat system on a sandy loam soil. Experimental Agriculture. 2019;**56**(1):1-14. DOI: 10.1017/S0014479719000218

[86] Abujabhah IS, Bound SA, Doyle R, Bowman JP. Effects of biochar and compost amendments on soil physico-chemical properties and the total community within a temperate agricultural soil. Applied Soil Ecology. 2016;**98**:243-253

[87] Gao S, DeLuca TH, Cleveland CC. Biochar additions alter phosphorus and nitrogen availability in agricultural ecosystems: A meta-analysis. Science of The Total Environment. 2019;**654**:463-472

[88] Kätterer T, Roobroeck D, Andrén O, Kimutai G, Karltun E, Kirchmann H, et al. Biochar addition persistently increased soil fertility and yields in maize-soybean rotations over 10 years in sub-humid regions of Kenya. Field Crops Research. 2019;**235**:18-26

[89] Wang JZ, Xiong KY. Biochar stability in soil: Meta-analysis of decomposition

and priming effects. Global Change Biology Bioenergy. 2015;**8**:512-523. DOI: 10.1111/gcbb.12266

[90] Diatta AA, Fike JH, Battaglia ML, Galbraith JM, Baig MB. Effects of biochar on soil fertility and crop productivity in arid regions: A review. Arabian Journal of Geoscience. 2020;**13**(14):1-17

[91] Duku MH, Gu S, Hagan EB. Biochar production potential in Ghana—A review. Renewable and Sustainable Energy Reviews. 2011;**15**:3539-3551

[92] Sohi S, Lopez-Capel E, Krull E, Bol R. Biochar, climate change and soil: A review to guide future research. CSIRO Land and Water Science Report. 2009;**5**(09):17-31

[93] Joseph S, Taylor P. The production and application of biochar in soils. Advances in Biorefineries. 2014:525-555. DOI: 10.1533/9780857097385.2.525

[94] Joseph S, Graber E, Chia C, Munroe P, Donne S, Thomas T, et al. Shifting paradigms: Development of high-efficiency biochar fertilizers based on nano-structures and soluble components. Carbon Management. 2013;**4**:323-343

[95] Liang B, Lehmann J, Solomon D, Kinyangi J, Grossman J, O'Neill B, et al. Black carbon increases cation exchange capacity in soils. Soil Science Society of America Journal. 2006;**70**:1719-1730. DOI: 10.2136/sssaj2005.0383

[96] Glaser B, Lehmann J, Zech W. Ameliorating physical and chemical properties of highly weathered soils in the tropics with charcoal—A review. Biology and Fertility of Soils. 2002;**35**:219-230

[97] Kloss S, Zehetner F, Wimmer B, Buecker J, Rempt F, Soja G. Biochar application to temperate soils: Effects on soil fertility and crop growth under greenhouse conditions. Journal of Plant Nutrition and Soil Science. 2014;**177**:3-15

[98] Lehmann J, Silva Júnior JPD, Steiner C, Nehls T, Zech W, Glaser B. Nutrient availability and leaching in an archaeological Anthrosol and a Ferralsol of the Central Amazon basin: Fertilizer, manure and charcoal amendments. Plant and Soil. 2003;**249**:343-357. DOI: 10.1023/A:1022833116184

[99] Karhu K, Mattila T, Bergström I, Regina K. Biochar addition to agricultural soil increased CH4 uptake and water holding capacity—Results from a short-term pilot field study. Agriculture Ecosystem Environment. 2011;**140**:309-313

[100] Singh B, Singh BP, Cowie AL. Characterisation and evaluation of biochars for their application as a soil amendment. Soil Research. 2010;**48**:516-525

[101] Ahmad M, Rajapaksha AU, Lim JE, Zhang M, Bolan N, Mohan D, et al. Biochar as a sorbent for contaminant management in soil and water: A review. Chemosphere. 2014;**99**:19-33. DOI: 10.1016/j.chemosphere.2013.10

[102] Zhang A, Liu Y, Pan G, Hussain Q, Li L, Zheng J, et al. Effect of biochar amendment on maize yield and greenhouse gas emissions from a soil organic carbon poor calcareous loamy soil from Central China plain. Plant and Soil. 2011;**351**(1-2):263-275

[103] Chan K, Van Zwieten L, Meszaros I, Downie A, Joseph S. Agronomic values of greenwaste biochar as a soil amendment. Soil Research. 2008;**45**(8):629-634

[104] Shanthi P, Renuka R, Sreekanth N, Babu P, Thomas A. A study of the fertility and carbon sequestration potential of rice soil with respect to application of biochar and selected amendments. Annals of Environmental Science. 2013;**7**:17-30

[105] Rothamsted Research. Guide to the Classical and Other Long-Term Experiments, Datasets and Sample Archive. Bury St. Edmunds, Suffolk, UK: Premier Printers Ltd; 2006

[106] Körschens M, Albert E, Armbruster M, Barkusky D, Baumecker M, Schalk BL, et al. Effect of mineral and organic fertilization on crop yield, nitrogen uptake, carbon and nitrogen balances, as well as soil organic carbon content and dynamics: Results from 20 European long-term field experiments of the twenty-first century. Archives of Agronomy and Soil Science. 2013;**59**:1017-1040

[107] Zhang H, Wang B, Xu M. Effects of inorganic fertilizer inputs on grain yields and soil properties in a long-term wheat-corn cropping system in South China. Communications in Soil Science and Plant Analysis. 2008;**39**:1583-1599

[108] Schroder L, Zhang H, Girma K, Raun WR, Penn CJ, Payton ME. Soil acidification from long-term use of nitrogen fertilizers on winter wheat. Soil Science Society of America Journal. 2011;**75**:957-964

[109] Fierer N, Ackson RB. The diversity and biogeography of soil bacterial communities. Proceedings of the National Academy of Sciences of the USA. 2006;**103**:626-631

[110] Sarfaraz I. The effects of chemical fertilizers on soil. Hunker [Internet]. 2019. Available from: https://www.hunker.com/13427782/the-effects-of-chemical-fertilizers-on-soil

[111] Böhme L, Langer U, Böhme F. Microbial biomass, enzyme activities and microbial community structure in two European long-term field experiments. Agriculture Ecosystem, Environment. 2005;**109**:141-152

[112] Zhong W, Gu T, Wang W, Zhang B, Lin X, Huang Q, et al. The effects of mineral fertilizer and organic manure on soil microbial community and diversity. Plant and Soil. 2010;**326**:511-522

[113] Kirchmann H, Schön M, Börjesson G, Hamnér K, Kätterer T. Properties of soils in the Swedish long-term fertility experiments: VII. Changes in topsoil and upper subsoil at Örja and Fors after 50 years of nitrogen fertilization and manure application. Acta Agriculturae Scandinavica, Section B Soil and Plant Sciences. 2013;**63**:25-36

Chapter 8

Effect of Passive and Forced Aeration on Composting of Market Solid Waste

Mohammad Atauzzaman and Quazi Hamidul Bari

Abstract

The book chapter discusses the influence of aeration on the decomposition of solid wastes available on the market. The vegetable waste, paper waste, and sawdust as a filler were mixed intensively in a ratio of 75:10:15. The temperature of composting mass inside the reactors was recorded intermittently daily. The weights of total sample and volatile solids were measured for both passive and forced aeration composting tests before and after composting the mixed waste. The temperature rose to a maximum of 52°C with passive aeration and 54°C with forced aeration. The percent decrease in the total sample was higher with forced aeration than with passive aeration. The volatile solids reduced over time at the end of these tests. The degree of volatile solids degradation of the mixed waste over time by the forced aeration was determined for a series of the composting processes. The amount of volatile solids and total sample were determined at 2- to 4-day intervals. The percent reduction in volatile solids and total sample was found to be 4 to 55% and 3 to 68%, respectively. The percent reduction in volatile solids increased with time. The chapter helps to understand the recycling possibility of the mixed waste in the form of compost.

Keywords: composting, forced aeration, organic fertilizer, passive aeration, recycling, volatile solids degradation

1. Introduction

The idea of integrated solid waste management has been developed due to the increase in municipal waste, the decrease in landfill capacity, the increase in waste management costs, communal disagreement to waste management conveniences, and anxieties about the threats related to the solid waste management [1]. The huge amount of organic solid wastes, both putrescible and non-putrescible, is one of the major socio-environmental problems in many developing countries. A number of health and environmental problems may arise due to the solid waste mismanagement. The common problems associated with the improper disposal of solid waste include esthetic nuisance, odor nuisance, water and air pollution, fire hazards, financial losses, disease transmission, etc. [2]. Solid waste management is now becoming an immense task owing to urbanization, explosion of population, fund deficiency, poverty, etc. Landfilling, gasification, incineration, pyrolysis, etc., are some of the

IntechOpen

waste disposal methods that are effective and, however, have adverse influences on the public health and environment as well. When managed properly, composting is a workable way with numerous benefits, namely the bio-fertilizer production, comparatively low water and air pollution, income generation, low operating costs, etc. [3]. Ecological imbalance and environmental degradation happen constantly because of the improper solid waste planning and management. Composting is an environmentally friendly and sustainable technique to manage the putrescible content of organic solid waste [4].

Composting is a biological conversion of the putrescible organic content of municipal solid wastes to decrease the material weight and volume and produce a humus-like material. Both aerobic and anaerobic practices have taken places in the waste management [5]. Putrescible parts of the organic materials are biodegraded to a stable end product, which can be used as an organic fertilizer [6]. The end product residual after the microbial action in the composting process of organic waste is known as humus or compost [7]. The most common practice in treating the solid waste is aerobic composting due to its easiness and operative treatment and requires the air diffusion through the waste [8]. Huge bacterial action quickens the breakdown of the organic content in the thermophilic phase [9, 10]. A minimum oxygen level is constantly continued to confirm high organic quality [11]. Moisture, pH, temperature, carbon to nitrogen ratio (C: N), etc., are the major factors, which have an effect on the composting process and contribute to the effectiveness of this process [12]. High organic materials and macronutrients in the waste have a high potential in the organic fertilizer production [13]. Organic fertilizers improve soil fertility, reduce soil salinity, require less irrigation water, resist pests and insects, accelerate rapid plant growth, and reduce dependency on expensive chemical fertilizers [14]. The application of compost in agricultural soils significantly increases the water holding capacity and thereby reduces the irrigation water demand of the land as described in several literatures [15–17]. Compost is often stigmatized when it is made from organic waste. Using compost is a great way to add nutrients to the soil and restore a healthy ecosystem [18]. In developing countries, the high organic part in the municipal solid waste is perfect for the composting process. Composting is well-suited with other forms of recycling [19].

Composting, one of the simplest methods of organic waste stabilization, is the most efficient method of treating organic waste and making a good organic fertilizer. The market solid waste contains numerous nutrients, namely potassium, phosphorus, nitrogen, etc., that contribute to the growth of various plants. In view of this, the study was investigated with the market solid waste for determining the variation of temperature in passive and forced aeration composting tests, for determining the degree of the volatile solids degradation of the mixed waste, and for investigating the recycling possibility of the mixed waste in the form of compost [20].

2. Materials and methods

The study was conducted to determine the influence of aeration, both passive and forced, on the composting of market solid waste. The details of waste materials, reactor and aerator type, measurement of temperature, determination of moisture content and volatile solids, determination of variation of temperature, and determination of rate of volatile solids degradation are described in this section.

2.1 Waste materials

The vegetable waste, paper waste, and sawdust were selected as the mixed waste and collected at Khulna's local market. The vegetable waste and paper waste were then cut into small pieces less than 10 mm in size. The vegetable waste, paper waste, and sawdust as a filler were intensively mixed in a ratio of 75:10:15. The different types of waste materials, namely vegetable waste, paper waste, sawdust, and mixed waste, are shown in **Figure 1**.

2.2 Reactor and aerator type

Twenty thermo-fluxes, each with a capacity of one liter, were used as reactors for the composting processes. The height and diameter of the reactors were 270 and 100 mm, respectively. The reactors are heat insulated to retain the self-heat inside the reactor and to ensure the increase in temperature up to the thermophilic range of 50 to 60°C [21]. The wet mixed waste of 400 to 450 g is put in the reactor. Small pieces of polyurethane sheeting were placed over the mixed waste to protect the whole system from the leakage of self-generated heat of the mixed waste during composting and to keep the system thermodynamically open [22]. Five aerators (Super Pump SP-780) with an air flow of 500 ml/min were used for the purpose of aeration. The types of reactors and aerators used for the composting processes of the mixed waste are shown in **Figure 2**. Air tubes of 5 mm in diameter and 1000 mm in length were connected with the aerator to the reactors. Four air tubes were used to connect each aerator to four reactors.

Figure 1.
Different types of waste materials (a) vegetable waste, (b) paper waste, (c) sawdust, (d) mixed waste.

Figure 2.
Types of reactors and aerators used for composting process of mixed waste: (a) reactor, (b) aerator.

2.3 Measurement of temperature

Two types of thermometers were used to record the temperature in the room and the temperature generated in the mixed waste within the reactors during the composting processes. The different types of thermometers used for recording the temperature during composting processes are shown in **Figure 3**. Thermometers were inserted up to midway into the reactors and temperature readings were recorded several times a day.

2.4 Determination of moisture content and volatile solids

Using a precise digital balance, the weight of the pan (w_1) was measured. A small amount of the mixed waste was placed in the pan. The weight of the pan plus wet sample (w_2) was measured. The pan with wet sample was then placed in an oven for 24 hours at 105 ± 5°C. The weight of the pan plus dry sample (w_3) was measured. A desiccator was used for controlling the moisture of the mixed waste. The measurement of weight of the mixed waste for determining the moisture content and volatile solids is shown in **Figure 4**. The quantity of moisture content was determined using Eq. (1).

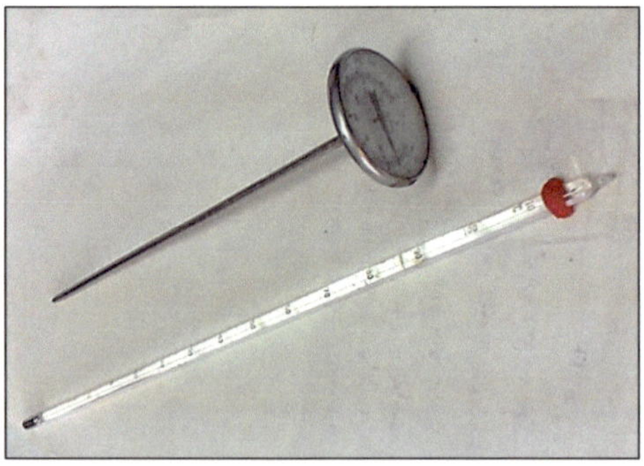

Figure 3.
Different types of thermometers used for recording temperature during composting process.

Figure 4.
Measurement of weight of mixed waste for determining moisture content and volatile solids.

$$\text{Moisture Content}(\%) = (w_2 - w_3)/(w_2 - w_1) \quad (1)$$

The pan with oven-dried sample was placed in a muffle furnace for 5 hours at 550 ± 15°C. The weight of the pan plus fixed sample (w4) was measured. The quantity of volatile solids was determined using Eq. (2).

$$\text{Volatile Solids}(\%) = (w_3 - w_4)/(w_3 - w_1) \quad (2)$$

2.5 Determination of variation of temperature

Three runs of composting tests were conducted according to the previous study [22]. The first and second runs were conducted with 6 reactors (3 reactors for passive aeration and 3 reactors for forced aeration) to determine the temperature variation in the composting processes. The details of the first and second runs for the determination of variation of temperature during the composting processes are discussed below.

2.5.1 Variation of temperature in composting process with passive aeration

The first run was conducted with three reactors for the replication of the composting process with passive aeration. After cutting into small pieces, the vegetable waste (75% of wet weight) and paper waste (10% of wet weight) were intensively mixed with 15% sawdust. The mixed waste of approximately 450 g was placed in three reactors. These reactors were filled with the mixed waste and shaken gently. The reactor openings were sealed with small pieces of polyurethane sheeting. The thermometers were inserted midway into the reactors to record the temperature generated inside the reactors. The test arrangement with temperature measurement for passive aeration composting process is shown in **Figure 5**. The moisture content, volatile solids, and total sample of the mixed waste were calculated before and after the composting process. The temperature readings were recorded intermittently daily for 30 days.

Figure 5.
Test arrangement with temperature measurement for passive aeration composting process.

2.5.2 Variation of temperature in composting with forced aeration

The second run was conducted with three reactors for the replication of the composting process with forced aeration. After cutting into small pieces, the vegetable waste (75% of wet weight) and paper waste (10% of wet weight) were intensively mixed with 15% sawdust. Air tubes of 5 mm in diameter and 1000 mm in length were inserted midway into each reactor from the air pump before filling the reactors with the mixed waste. These reactors were filled with the mixed waste and shaken gently. The reactor openings were sealed with small pieces of polyurethane sheeting. The thermometers were inserted midway into the reactors to record the temperature generated inside the reactors. The test arrangement with temperature measurement for forced aeration composting process is shown in **Figure 6**. Air was passed through the mixed waste inside the reactor at the rate of 500 ml/min for 8 hours a day during the day. The moisture content, volatile solids, and total sample of the mixed waste were calculated before and after the composting process. The temperature readings were recorded intermittently daily for 30 days.

Figure 6.
Test arrangement with temperature measurement for forced aeration composting process.

2.6 Determination of rate of volatile solids degradation

The third run was conducted with 20 reactors for composting the mixed waste with forced aeration following the process as described in Section 2.5.2. The test arrangement for determining the rate of volatile solids degradation is shown in **Figure 7**. The temperature readings were recorded intermittently daily for 50 days. The dry solids, volatile solids, moisture content, and total sample of the mixed waste were calculated at 2- to 4-day intervals.

3. Results and discussion

The study was conducted for determining the variation of temperature over time during the composting process of market solid waste with passive and forced aeration conditions and for determining the rate of volatile solids degradation over time for a series of composting processes with forced aeration conditions. The results obtained from this study are discussed below.

3.1 Passive aeration composting

The variation of temperature during composting process with passive aeration is shown in **Figure 8**. The temperature of composting mass inside reactor-1 increased from 26 to 42°C within the first 7 days. The maximum temperature was 52°C. The temperature slowly dropped to 32°C after 30 days of composting process. The temperature of composting mass inside reactor-2 increased from 26 to 44°C within the first 7 days. The maximum temperature was 49°C. The temperature dropped to 28°C after 30 days of composting process, distant from the ambient temperature of 17°C. A similar pattern was also observed for the composting mass inside reactor-3. The temperature of composting mass increased from 26 to 40°C within the first 7 days. The maximum temperature was 51°C. The temperature dropped to 30°C after 30 days of composting process. Initially, the temperature of composting mass inside the reactors rose rapidly for a few days and then fell slowly in all the reactors. The degradation of the putrescible part of the mixed waste starts within a short period of the passive aeration composting process, which leads to rise in temperature of composting mass

Figure 7.
Test arrangement for determining rate of volatile solids degradation.

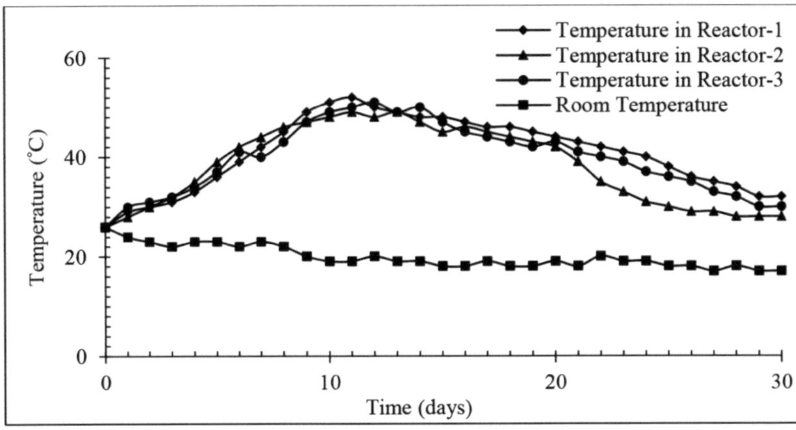

Figure 8.
Variation of temperature during composting with passive aeration.

inside reactors for a few days. After that the amount of the putrescible part of the mixed waste is reduced significantly, resulting a slight drop in temperature over time. Since no air was blown into the reactors and small pieces of polyurethane sheeting served as insulating materials, there was a slow cooling effect in the composting mass, consequently a significant difference from the ambient temperature of 17°C within 30-day period. The temperature follows a similar pattern for all the reactors as indicated in **Figure 8** confirming the replicability of the experiment as described in the literature [23].

The initial weights of the mixed waste in reactor-1, reactor-2, and reactor-3 were 450, 455, and 435 g, respectively, and the final weights after 30 days of composting process with passive aeration were 370, 380, and 355 g, respectively. The moisture content of the mixed waste was initially 69.4% and increased to 77.9%. The volatile solids of the mixed waste were initially 94.4% and reduced to 90.3%. The percent reductions in dry solids, volatile solids, moisture, and total sample in reactor-1 were found to be 40.60, 43.15, 7.72, and 17.78%, respectively. The percent reductions in dry solids, volatile solids, moisture, and total sample in reactor-2 were found to be 35.06, 37.52, 8.30, and 16.48%, respectively. The percent reductions in dry solids, volatile solids, moisture, and total sample in reactor-3 were found to be 37.87, 40.57, 9.80, and 18.39%, respectively. The average percent reductions in dry solids, volatile solids, moisture, and total sample in these three reactors after 30 days of composting process with passive aeration were found to be 37.84, 40.41, 8.61, and 17.55%, respectively.

3.2 Forced aeration composting

The variation of temperature during composting process with forced aeration is shown in **Figure 9**. The temperature of composting mass inside reactor-4 increased from 26 to 42°C within the first 7 days. The maximum temperature was 50°C as comparable with the literature [24]. The temperature dropped to 19°C after 30 days of composting process, near to the ambient temperature of 17°C. The temperature of composting mass inside reactor-5 increased from 26 to 39°C within the first 7 days. The maximum temperature was 53°C. The temperature dropped to 20°C after 30 days of composting process. Similarly, the temperature of composting mass inside reactor-6 increased from 26 to 43°C within the first 7 days. The maximum temperature

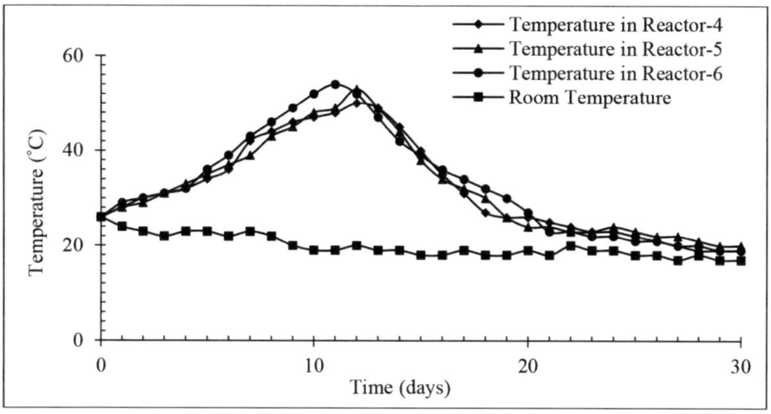

Figure 9.
Variation of temperature during composting with forced aeration.

was 54°C. The temperature dropped to 19°C after 30 days of composting process. Initially, the temperature of composting mass inside the reactors rose rapidly for a few days and fell rapidly over 20 days and then slowly decreased in all these three reactors. The degradation of the putrescible part of the mixed waste starts within a short period of the forced aeration composting process, which leads to a rapid rise in temperature of composting mass inside the reactors for a few days. After that, the amount of the putrescible part of the mixed waste is reduced significantly, resulting a sharp drop in temperature within rest of the days. This is due to the air flow into the reactors having ambient temperature and after that become hotter as in the reactor temperature. Finally, it carries more moisture and heat from the reactor, since the moisture carrying capacity of air increases exponentially with the temperature.

The initial weights of the mixed waste in reactor-4, reactor-5, and reactor-6 were 450, 455 and 465 g, respectively, and the final weights after 30 days of composting process with forced aeration were 235, 265, and 270 g, respectively. The moisture content of the mixed waste was initially 69.4% and reduced to 66.1%. The volatile solids of the mixed waste were initially 94.4% and reduced to 88.2%. The percent reductions in dry solids, volatile solids, moisture, and total sample in reactor-4 were found to be 42.12, 45.15, 50.27, and 47.78%, respectively. The percent reductions in dry solids, volatile solids, moisture, and total sample in reactor-5 were found to be 36.21, 39.12, 44.21, and 41.76%, respectively. The percent reductions in dry solids, volatile solids, moisture, and total sample in reactor-6 were found to be 37.60, 41.70, 43.85, and 41.94%, respectively. The average percent reductions in dry solids, volatile solids, moisture, and total sample in these three reactors after 30 days of composting process with forced aeration were found to be 38.64, 41.99, 46.11, and 43.83%, respectively.

3.3 Volatile solids degradation

Composting is a self-heating biological transformation of organic waste, which produces a stable end product, for example, organic fertilizer. Thermophilic phase is a very dynamic phase where high bacterial action leads to enhance the organic matter degradation [9]. The maximum temperature of composting mass in both passive and forced aeration tests was within the range of composting process (below 60°C). The percent reduction in total sample was higher with forced aeration than with passive aeration. The composting process with forced aeration was selected for determining the rate of volatile solids degradation over time. The temperature increased from 30 to 52°C. The moisture content of the mixed waste was initially 68.9% and reduced to 41.7%. The volatile solids of the mixed waste were initially 92.4% and reduced to 87.4%. The volatile solids reduced over time at the end of the composting. The percent reductions in dry solids, volatile solids, moisture, and total sample in reactor-1 after 3 days of the composting were found to be 6.99, 7.23, 2.43, and 3.85%, respectively. The percent reductions in dry solids, volatile solids, moisture, and total sample over time are shown in **Figure 10**. The percent reductions in dry solids, volatile solids, moisture, and total sample increased over time. The temperature of composting mass inside the reactors rose significantly after a few days under the forced aeration composting condition. The temperature was higher in composting mass inside the reactors than the ambient, resulting in a rapid reduction in volatile solids and moisture content of the mixed waste. The rate of volatile solid reduction increased with temperature and was usually double in every 10°C temperature increase. The moisture reduced in composting mass as the moisture carrying capacity of exhaust air increased exponentially with the temperature. For this combined reduction effect, that is volatile solids degradation and dryness, the amount of dry solids and total mass also reduced significantly over time. The percent reduction rates in dry solids, volatile solids, moisture, and total sample were not uniform in the biochemical reaction of the mixed waste. The gradient or slope of the trend line of the reduction in volatile solids

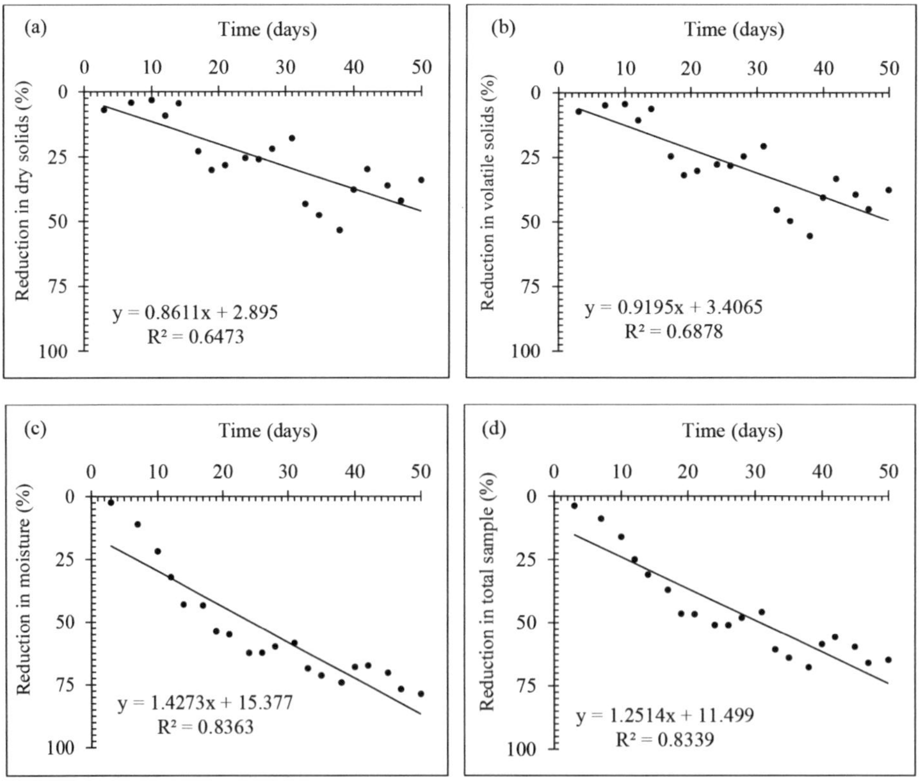

Figure 10.
Reductions of different parameters during forced aeration composting (a) dry solids, (b) volatile solids, (c) moisture, and (d) total sample.

with time curve was used to approximate the percent rate of reduction in biochemical reaction of the mixed waste during the forced aeration composting process. The rate of reduction or degradation of volatile solids was found to be 0.92% per day.

4. Conclusions

The first run with passive aeration and the second run with forced aeration were conducted to determine the variation of temperature over time during the composting processes. The third run with forced aeration was conducted to determine the rate of volatile solids degradation over time for a series of composting processes. For the composting process with passive aeration, the temperature of composting mass inside the reactors rose rapidly for a few days from 26 to 52°C and then fell slowly to 28°C, distant from the ambient temperature of 17°C. For the composting process with forced aeration, the temperature of composting mass inside the reactors rose rapidly for a few days from 26 to 54°C and fell rapidly to 26°C for a few days and then dropped slowly to 19°C, near to the ambient temperature of 17°C. During the composting process, the biochemical reaction in the mixed waste produced heat, rising the temperature inside the reactors. The percent reductions in volatile solids were found to range from 4 to 55%. The percent reductions in total sample were found to range from 3 to 68%. The percent reductions in dry solids, volatile solids, moisture, and total sample increased with time. The reduction or degradation rate of volatile solids was found to be 0.92% per day. The mixture of vegetable waste and paper waste was well degraded during composting processes in both aeration conditions. There is a possibility of recycling the putrescible part of organic solid waste in the form of compost for its further beneficial usages.

Acknowledgements

The authors would like to express their sincere gratitude to the staff of the Environmental Engineering Laboratory of the Department of Civil Engineering, Khulna University of Engineering and Technology, Khulna, Bangladesh, for assisting the research work.

Conflict of interest

The authors declare no conflict of interest.

Author details

Mohammad Atauzzaman[1*] and Quazi Hamidul Bari[2]

1 Department of Civil Engineering, Pabna University of Science and Technology, Pabna, Bangladesh

2 Department of Civil Engineering, Khulna University of Engineering and Technology, Khulna, Bangladesh

*Address all correspondence to: atauzzaman@gmail.com; atauzzaman@pust.ac.bd

References

[1] United States Environmental Protection Agency. The Solid Waste Dilemma: An Agenda for Action. Draft Report. Washington, DC: United States Environmental Protection Agency. EPA/530-SW-88-052; 1988

[2] Ahmed MF, Rahman MM. Water Supply & Sanitation: Rural and Low Income Urban Communities. Dhaka: ITN-Bangladesh; 2000

[3] Taiwo AM. Composting as a sustainable waste management technique in developing countries. Journal of Environmental Science and Technology. 2011;**4**(2):93-102. DOI: 10.3923/jest.2011.93.102

[4] Lohani SR. Composting: A better solution for managing Nepal's increasing solid waste. Journal of the Institute of Engineering. 2017;**13**(1):215-220

[5] Tchobanoglous G, Theisen H, Vigil S. Integrated Solid Waste Management: Engineering Principles and Management Issues. New York: McGraw-Hill, Inc.; 1993. DOI: 10.3126/jie.v13i1.20369

[6] Tchobanoglous G, Burton FL. Wastewater Engineering: Treatment, Disposal and Reuse. New Delhi: Tata McGraw-Hill Publishing Company Limited; 1995

[7] Peavy HS, Rowe DR, Tchobanoglous G. Environmental Engineering. New York: McGraw-Hill Book Company; 1985

[8] Kulkarni SJ. Aerobic composting - a short review. International Journal of Research & Review. 2017;**4**(2):73-75

[9] Sarkar S, Pal S, Chanda S. Optimization of a vegetable waste composting process with a significant thermophilic phase. Procedia Environmental Sciences. 2016;**35**:435-440. DOI: 10.1016/j.proenv.2016.07.026

[10] Meena AL, Karwal M, Dutta D, Mishra RP. Composting of night soil and horse manure with leaves as organic substrate. Agriculture & Food: E-Newsletter. 2021;**3**(1):85-90. DOI: 10.13140/RG.2.2.13546.95689

[11] Neugebauer M, Sołowiej P, Piechocki J, Czekała W, Janczak D. The influence of the C: N ratio on the composting rate. International Journal of Smart Grid and Clean Energy. 2017;**6**(1):54-60. DOI: 10.12720/sgce.6.1.54-60

[12] Gonawala SS, Jardosh H. Organic waste in composting: A brief review. International Journal of Current Engineering and Technology. 2018;**8**(1):36-38. DOI: 10.14741/ijcet.v8i01.10884

[13] Alamgir M, Ahsan A. Characterization of MSW and nutrient contents of organic component in Bangladesh. Electronic Journal of Environmental, Agricultural and Food Chemistry. 2007;**6**(4):1945-1956

[14] Roy TK, Rahman S, Dev PK. Compost fertilizer from municipal solid wastes and its application in urban agro-forestry nurseries: A case study on Khulna city. Journal of Bangladesh Institute of Planners. 2013;**6**:191-199

[15] Nayeema N, Bari QH, Shafique N, Hossain F. Effects of using organic compost on water holding capacity of Dhaka soil. In: Proceedings of the 3rd International Conference on Civil Engineering for Sustainable Development (ICCESD'16); 12-14 February, 2016;

Khulna, Bangladesh. Khulna, Bangladesh: Khulna University of Engineering and Technology; 2016. pp. 252-257

[16] Faysal SW, Bari QH, Hossain M, Ratul SJ. Effect of using compost on water holding capacity of soil. In: Proceedings of the Waste Safe 2015- 4th International Conference on Solid Waste Management in Developing Countries; 15-17 February, 2015; Khulna, Bangladesh. Khulna, Bangladesh: Khulna University of Engineering and Technology; 2015. pp. 176-180

[17] Reza MS, Bari QH, Saha GC, Hasan MA. Increase water holding capacity of soil using compost in southern region of Bangladesh. In: Proceedings of the International Conference on Industrial Waste Management and Process Efficiency; 12-14 November, 2012; Gazipur, Bangladesh: Dhaka University of Engineering and Technology; 2012. pp. 579-586

[18] Ishola TM, Ishola ET. Composting and sustainable development. In: Encyclopedia of Sustainability in Higher Education. Cham, Switzerland: Springer International Publishing; 2019. pp. 1-8. DOI: 10.1007/978-3-319-63951-2_122-1

[19] Hoornweg D, Thomas L, Otten L. Composting and its applicability in developing countries. In: Urban Waste Management Working Paper Series, No. 8. Washington, DC: Urban Development Division, World Bank; 2000

[20] Atauzzaman M, Bari QH. Effect of passive and forced aeration on composting of market solid waste. International Journal of Engineering & Technology. 2020;**9**(1):182-186. DOI: 10.14419/ijet.v9i1.30301

[21] Alamin M, Bari QH. Extent of degradation in three stage co-composting of fecal sludge and solid waste. Journal of the Air & Waste Management Association. 2022;**72**(8):914-924. DOI: 10.1080/10962247.2022.2064936

[22] Koenig A, Bari QH. Application of self-heating test for indirect estimation of respirometric activity of compost: Theory and practice. Compost Science and Utilization. 2000;**8**(2):99-107. DOI: 10.1080/1065657X.2000.10701755

[23] Bari QH, Koenig A. Measuring solid waste compost maturity: A review and practice. In: Proceedings of the International Conference on Bangladesh Environment; 19-21 December, 2002; Dhaka, Bangladesh: Bangladesh Poribesh Andolon; 2002. pp. 712-722

[24] Zhang H, Sun X, Shan H, Xue C, Wang J. Composting: Phases and factors responsible for efficient and improved composting. Compost Science & Utilization. 2020;**28**(3-4):158-168. DOI: 10.1080/1065657X.2021.1949409

Chapter 9

Role of Entomopathogenic Nematodes in Organic Farming and Sustainable Development

Kajol Yadav, Lovely Bharti and Ashok Kumar Chaubey

Abstract

Chemical fertilizers and pesticides are presently accumulating in the environment harming the ecosystem, causing pollution, and spreading some of the diseases. Nematodes can be considered as entomopathogenic (EPN) if they fulfill criteria for entomo-pathogenicity when they bearing a pathogenic bacterium within a dauer juveniles juvenile, releasing the bacterium inside the host, actively seeking out and penetrating the host, rapid insect death, nematode and bacterial reproduction, reassociation of the pathogenic bacteria with new generations of dauer juveniles, and emergence of IJs from the cadaver so that the cycle can be repeated. Synthetic chemical pesticides have various disadvantages which include crop and soil contamination; killing of beneficial fauna and flora; resistance development in insects and adverse effects due to contamination in food chain; and other environment-related issues. To minimize pesticides contamination, EPN were identified as biological control agents and most suitable natural enemies of problematic insects because they reduce risk to humans and other related vertebrates.

Keywords: chemical fertilizers, entomopathogenic, pesticides, pathogenicity, biological control

1. Introduction

Around the world, growing vegetables has become a significant source of revenue for farmers. Vegetable fields make up about 7% of all croplands worldwide, and this number is typically greater in richer nations [1]. Vegetable fields differ from crop fields in that they apply more nitrogen, produce their crops more intensively, use more tillage and irrigation, and have more planting-harvest cycles throughout the year [2]. For instance, fertilizer inputs in vegetable cultivation reached 600 kg of nitrogen per hectare per year [3], compared to 300 kg of nitrogen per hectare per year in cereal cropping systems [4]. It has been demonstrated that intensive farming decreases soil biodiversity, which is crucial for the health of ecosystems [5, 6]. Therefore, it is critical to look into potential impacts of intensive agricultural methods on soil nematode communities. According to [7, 8], organically managed farmlands have been increasing to about 107 ha globally and are anticipated to continue growing. Organic farming systems are generally thought to be more sustainable than conventional farming systems [9]. In India, organic vegetable growing is likewise showing an upward trend. According to two global meta-analyses, organic farming benefits the soil biota [10]; however, a thorough understanding of whether and how organic management affects the community structure of soil nematodes and related functions is currently lacking. Although the effects of organic farming on soil nematodes have been evaluated in grasslands [11], arable fields [12]; orchards [13]; and vegetable fields [14]. Organic farming may have a negative impact on soil nematodes. Change depends on the crop type, soil texture, and past land use [15]. Because nematode abundance and community composition can be related to edaphic and climatic fluctuations at different scales, the impacts of organic farming on soil nematode community may depend on spatial scale [16]. Finally, the comparative impact of organic farming on soil nematode communities across various vegetable types is largely unknown because previous studies assessing the effect of organic farming on soil nematodes frequently focused on single vegetable types such as tomato [14], green peppers [17], and asparagus [18].

Insect-parasitic nematodes, known as entomopathogenic nematodes (EPNs), have been described from 23 nematode families [19]. Among all nematodes that have been studied for insect biocontrol, Steinernematidae, and Heterohabdichidae have received the most attention because they possess many properties of effective biocontrol agents [19]. It has been used as a classical antiseptic and as an adjunctive biological control agent. Most of the applied research has focused on its potential as an adjunctive biological control agent applied to floods [20]. Extensive research over the past three decades has demonstrated both successes and failures in pest control of crops, ornamental plants, lawns, and peat [21, 22].

Extensive studies have demonstrated both success and failure in controlling pests on crops, trees, ornamental plants, lawns, and peat [23]. EPNs occur naturally in soil and are divided into the Steinernematidae and Heterorhabditidae families [22]. EPNs form a persistent or stress-tolerant stage known as infectious juvenile (IJ) [24]. This developmentally retarded stage also plays an important role in nematode dispersal in soil, as nematodes actively seek out and infect suitable insect hosts [25]. In addition, IJs play a role in transferring entomopathogenic bacteria from one host to another. After localizing and invading the insect host, IJs migrate to the hemolymph, where they recover from their arrested developmental state and release their bacterial symbionts. The bacteria multiply, release toxins, and kill the insect within 24–72 hours. EPNs traverse the soil by following chemical signals (chemotaxis). Through chemotaxis, they recognize hosts in their environment or areas where hosts are likely [26]. Several studies have shown that migration of EPNs in soil is also affected by other factors such as CO_2, plant exudates, pH, temperature, electrical potential, and VOCs [27]. The use of EPNs in biological control has traditionally been associated with the control of soil-dwelling pests [23]. Studies over the past two decades indicate that they may also control airborne pests, but only under certain circumstances [28]. The reduced effectiveness of EPNs in controlling aerial pests is primarily due to exposure to ultraviolet light [29], temperature extremes [30], and inadequate humidity [31]. These factors are important for EPN survival. For this reason, outdoor EPNs are less efficient against airborne pests, although previous laboratory tests have shown much higher efficiencies [32]. The most common EPN formulation is an aqueous suspension [33]. Equipment intended for the application of pesticides, fertilizers, or irrigation can be used for EPN applications. Backpack, hand or tractor sprayers, and sprinklers are suitable for this. IJs can pass through spray tubes of at least 500 μm diameters and withstand pressures up to 2000 kPa [34]. In addition, IJ withstands short-term exposure (2–24 hours) to many chemical and biological pesticides, fungicides, herbicides, fertilizers, and growth regulators and can be mixed in a tank and treated as such. It can be applied together with the product [35]. The combination of nematodes and insecticides in tank mixes can provide a cost-effective alternative to foliar-integrated pest management (IPM) systems. Because nematodes are sensitive to UV light, nematodes should be applied to plants in the evening, early morning, or during cloudy weather when the radiation is less intense [36]. Nematode survival and foliar efficacy are improved to varying degrees by adding various adjuvant with anti-drying (e.g., glycerol and various polymers) or UV-protective (brightener) effects to the spray mixture [37], but much more needs to be done to improve survival after application. Arguably, the greatest potential for the use of EPNs against foliar pests is in combination with other biological control agents [38] or selective chemicals [39]. In the IPM program, EPNs are considered particularly safe biological agents [40]. Due to their specific effects, they pose no environmental risks other than chemical pesticides [41]. Since EPN was first used in the United States in his 1841 to control the *Popyria japonica* Newman beetle, no cases of environmental damage by EPN have been documented [42]. Using nematodes is safe for users. EPN and its associated bacteria do not harm animals or plants [43].

1.1 Impact of chemical pesticides

1.1.1 Impact on environment

Chemical pesticides can infect soil, water, lawn, and other undergrowth. Insecticides not only kill insects and weeds, but can also be toxic to a wide variety of organisms, including birds, fish, beneficial insects, and unwanted plants. Insecticides are generally the most toxic class of pesticides, but herbicides can also pose risks to non-target organisms.

1.1.2 Impact on humans

Application pesticideswill kill the disease causing insects thereby increasing the food and fiber production. There is now irresistible evidence that some of these chemicals pose potential risks to humans and other living organisms and have adverse effects on the environment [44]. No population is completely resistant from exposure to pesticides, and potentially severe health effects are disproportionately borne by people in developing countries and high-risk groups in each country [45]. Worldwide, pesticide poisoning causes approximately 1 million deaths and chronic illnesses annually [46]. Groups at high risk of exposure to pesticides include production workers, formulators, sprayers, mixers, shippers, and farm workers. The processes involved are not without risk, and the potential for hazards during manufacturing and formulation can increase. Industrial environments pose increased risks to workers as they work with a wide variety of toxic chemicals, including pesticides, raw materials, toxic solvents, and inert carriers.

Organochlorine pesticides can infect the tissues of virtually all life forms on Earth, the air, lakes and seas, the fishes that live in them, and the birds that feed on them [47]. The National Academy of Sciences in United States found that the DDT metabolite DDE caused eggshell thinning and that the bald eagle population declined primarily due to exposure to DDT and its metabolites [48]. Certain environmental chemicals, including pesticides, called endocrine disruptors, are known to produce adverse effects by mimicking or antagonizing natural hormones in the human body, and their long-term low-dose exposure may have implications for human health, including immune suppression, hormone disruption, decreased intelligence, abnormalities in reproduction, and cancer etc. [49].

1.1.3 Impact on food materials

A program entitled "Monitoring of Pesticide Residues in Products of Plant Origin in the European Union" has been set up in the European Union since 1996 to determine the extent of pesticide residues in food. In 1996, seven pesticides (acephate, clopyrifos, clopyrifos-methyl, methamidophos, iprodione, procymidone, and chlorothalonil) and two groups (benomyl group and maneb group, *i.e.,* dithiocarbamates) were tested to apples, tomatoes, lettuce, strawberries, and grapes [50].

In India, the first report of pesticide poisoning was in Kerala in 1958, where the pesticide caused mortality more than 100 people from eating wheat flour contaminated with parathion [51]. This led to the dedication of the special committee set up by ICAR on the Harmful Effects of Pesticides (ICAR Task Force Report, 1972). In an interdisciplinary study evaluating pesticide residues in selected foods collected in different states of the country (Surveillance of Food Contaminants in India, 1993), DDT residue was detected approximately 82% of 2205 samples of milk collected from 12 states. Approximately 37% of the samples contained DDT residues above the acceptable limit of 0.05 mg/kg (whole milk basis). The maximum amount of DDT residue detected was 2.2 mg/kg. The proportion of samples containing residues above acceptable limits was highest in Maharashtra (74%), followed by Gujarat (70%), Andhra Pradesh (57%), Himachal Pradesh (56%), and Punjab (51%). In other federal states, this percentage was less than 10%. Data from 186 samples from 20 brands of infant formula showed that approximately 70 and 94% of the samples had DDT and HCH isomers at maximum levels of 4.3 and 5.7 mg/kg (fat basis), respectively. Measuring chemicals in the entire diet provides the best estimate of human exposure and potential risk. Risk to consumers can be assessed relative to toxicologically acceptable intake levels. The average total intake of DDT and BHC by adults was 19.24 and 77.15 mg/day, respectively [52]. Fatty foods were the main source of these contaminants. Another study reported that her average daily intake of HCH and DDT by an Indian was 115 and 48 mg per person, respectively, which is the amount observed in most developed countries [53].

1.1.4 Impact on soil

Numerous transformation products (TPs) from a wide range of pesticides have been documented [54]. Pesticides and TPs can be classified as follows: (i) Pesticides that exhibit such behavior include organochlorine DDT, endosulfan, endrin, heptachlor, lindane, and their TPs. Most of them are now banned from agriculture, but their residues still exist. (ii) Polar insecticides are represented mainly by herbicides, but also carbamates, fungicides, Phosphorus insecticide TP are also included. They can be moved from soil by runoff and leaching, thereby constituting a problem for the supply of drinking water to the population. Herbicide pesticide TPs are unquestionably the most studied pesticide TPs in soil. Numerous metabolic pathways that involve transformation through hydrolysis, methylation, and ring cleavage and result in the production of several toxic phenolic compounds. Pesticides and their TPs are retained in soil to varying degrees, depending on the interaction of soil and pesticide properties. The most influential soil property is organic matter content. The higher the organic matter content, the greater the adsorption of pesticides and TPs. The soil's ability to retain positively charged ions in exchangeable form is important for paraquat and other positively charged pesticides. Strong mineral acids are required to extract these chemicals, but no analytical improvements or studies have been reported in recent years. Soil pH is also important. Adsorption increases with decreasing soil pH for ionizable pesticides (2,4-D, 2,4,5-T, picloram, atrazine, etc.) [55].

1.1.5 *Impact on soil fertility*

Severe treatment of soil with pesticides can reduce the number of beneficial soil microorganisms. Soil scientist Dr. Elaine Ingham said "Overuse of chemical fertilizers and pesticides has similar effects on soil organisms as overuse of antibiotics by humans. The indiscriminate use of chemicals may be effective for a few years, but after a while there will not be enough beneficial soil organisms to hold the nutrients" [56]. Mycorrhizal fungi grow with the roots of many plants and aid in nutrient absorption. These fungi can also be damaged by herbicides in the soil. One study found that both oryzalin and trifluralin inhibited the growth of certain types of mycorrhizal fungi [57]. Roundup has been shown to be toxic to mycorrhizal fungi in laboratory studies, with some adverse effects observed at concentrations lower than those found in soil after normal application [58]. Triclopyr was also found to be toxic to some mycorrhizal fungi [58], and oxadiazon decreased mycorrhizal fungal spores numbers [59].

1.1.6 *Impact on surface and groundwater*

Pesticides can enter surface waters through runoff from treated crops and soil. Pesticide infection of water is widespread. Results of a comprehensive series of studies conducted by the United States Geological surveys (USGS) conducted in major river basins across the country have returned surprising results. More than 90% of water and fish samples from all rivers contained one or more pesticides [60]. Pesticides were detected in all major river samples with mixed agricultural and urban land use impacts and in 99% of urban stream samples [61]. According to the USGS, more pesticides were found in municipal than agricultural waterways [62]. The herbicides 2,4-D, diuron, and prometon, and the insecticides chlorpyrifos and diazinon, all commonly used by urban homeowners and school districts, are among the most common in surface and groundwater sources nationwide. Was among the 21 pesticides found in the United States [63]. Trifluralin and 2,4-D were detected in water samples from 19 of the 20 river basins studied [64]. The USGS also found that pesticide levels in urban waterways often exceed guidelines for protecting aquatic life [65].

Groundwater contamination by pesticides is a global problem. According to the USGS, at least 143 different pesticides and 21 transformation products have been detected in groundwater, including pesticides from all major chemical classes. Evidence has been found in groundwater in more than 43 countries in the last 20 years [66]. A study in India found that 58% of drinking water samples collected from various hand pumps and wells around Bhopal were contaminated with organochlorine pesticides exceeding EPA standards [67]. When groundwater becomes contaminated with toxic chemicals, it can take years for the contaminants to dissipate or be cleaned up. Cleaning can be very expensive and complicated, if not impossible [66].

1.1.7 *Impact on non-target fauna and flora*

Pesticides are common pollutants found in non-target organisms in soil, air, water, and urban landscapes. Once there, it can cause plant and animal damage, ranging from beneficial soil microbes and insects to non-target plants, fish, birds and other wildlife. Drift occurs, even from terrestrial instruments [68]. Drift can represent a loss of 2–25% of the applied chemical and can extend over distances of a few meters to hundreds of kilometers. Up to 80–90% of applied pesticides can volatilize within days of application [69]. Research on this topic is limited, but research continues to find pesticide residues in the air. According to the USGS, airborne pesticides were detected in all US sample areas [56]. Herbicides are designed to kill plants, so it is not surprising that direct application to such plants, or drifting or volatilizing, can harm or kill desirable species. Formulated herbicides have been shown to volatilize untreated plants with enough vapors to cause severe damage to other plants [70]. In addition to outright killing non-target plants, exposure to insecticides can have sublethal effects on plants [71]. Exposure to the herbicide glyphosate can severely affect seed quality [72]. It may also increase susceptibility to certain plant disease. This poses a particular threat to endangered plant species.

Entomopathogenic nematodes (EPNs) are members of the soil biota and provide biological control of arthropod pests, an important ecosystem service in agriculture. Their infective juvenile (IJ) stage occurs naturally in soils where arthropods can coexist and serve as hosts, ranging from marine areas to alpine areas, natural to agroecosystems, and even heavily polluted. It is distributed in soils that have been extensively treated [73]. Research on EPNs in the context of agroecology and applied soil ecology has increased in recent decades [74]. Several studies have demonstrated how changes in soil properties affect EPN communities and related organisms such as nematophagous fungi (NFs) and free-living nematodes (FLNs), their natural enemies, and potential competitor [75]. EPNs are considered particularly safe biological agents [40]. Due to their specific effects, they pose no environmental risks other than chemical pesticides [41]. Since EPN was first used to control the *Popylia japonica* Newman beetle in the United States in 1841, no cases of environmental damage from EPNs have been documented [42]. Using nematodes is safe for users. EPN and its symbiont bacteria are harmless to mammals and plants [43].

1.2 Identification of entomopathogenic nematodes

The first species of entomopathogenic nematodes was described morphologically in 1923. Adults of first and second generations and third stage IJs of *Steinernema* and *Heterorhabditis* possess some distinctive morphological features which are very important from the taxonomic point of view. However, it was become a monotonous task to categorize the increasing number of species with these taxonomic characteristics. Therefore, certain ratios and De Man Indices were created in order to delineate the species more appropriately. These ratios are based on the following characteristics, *viz*. tail length; position of excretory pore, nerve ring, and pharynx length. Besides these, males acquire some prominent characters such that spicule and gubernaculums. Analysis and measurement of these traits are playing a key role for identifying species. The vulva, which is a well-known feature of the females of entomopathogenic nematodes, its position, and associated structure are also important traits, provided by taxonomists with a clear method for identifying the species. The SEM investigations of first-generation males revealed the complete structure of gubernaculums and spicules, the presence or absence of caudal mucron, the position of the copulatory papillae, the morphology of spermatozoons [76], and the presence or absence of small cuticular projections, or epiptygmata, which protect the opening of the female vagina [77]. Head contour, cephalic horns, the lateral field, tail length and shape, and so forth are some of the crucial traits of taxonomic significance for IJs [78].

Due to the increase in the number of species, morphological characterization no longer provides accurate results and makes molecular characterization essential for species identification. Morphology is completely determined by the external characteristics of the specimen; however, some genes tend not to reveal themselves in phenotypes despite it having conserved portions that are crucial from a taxonomic perspective. In addition, morphology is a laborious task that needs for qualified taxonomists with the necessary knowledge. In order to confirm the taxonomic position of a certain species and its validity, this creates a necessity for molecular identification and characterization. Advancements in the molecular techniques help in the precise identification and placement of the species in its appropriate position in the classification. Moreover, the phylogenetic relationships of the species with the other species of a genus and with other orders are also established by utilizing the modern and advanced tools of molecular characterization. A number of molecular techniques are being used for more precise identification of EPNs as immunological techniques [79]; isoenzyme patterns [80]; total protein patterns [81]; and RFLP detection within total genomic DNA [82].

1.3 Life cycle of entomopathogenic nematodes

Steinernema and *Heterorhabditis* have comparable life cycles. Between a free-living stage and a parasitic stage, both genera maintain stability. An outermost cuticle acts like a barrier between the environment and the free-living form of EPNs. The invasive EPN stage, also known as the infective juvenile (IJ) or J3 stage, is encapsulated and unable to feed since its mouth and anus are sealed [83]. To be capable of surviving without a host for several months, they actually have enormous lipid storage [84]. It has been observed that IJs of *Steinernema* live much longer in the environment than IJs of *Heterorhabditis*, while having similar lipid reserves, it may be explained by the IJs' motile behavior. According to findings, Heterorhabditis IJs nictate between 70 and 90% of their lives, compared to *Steinernema* IJs' 50 to 80% [85], and as a result, lipid reserves are depleted more quickly in *Heterorhabditis* IJs. Infective juveniles wait for insect larvae up to 20 cm deep in soil [77]. In case of *Steinernema*, IJs invade the insect larvae through natural openings such as the mouth, anus, spiracles, and wounds [86]. However, in case of *Heterorhabditis,* the IJs are also able to penetrate the insect body by directly scratching their cuticle because of the presence of a large anterior tooth [87]. IJs lose their cuticle and release the entomopathogenic bacteria (EPB) after fettling entering in the insect body. Together, they eventually kill the insect. On usually, 3 days after an insect infestation, IJs begin feeding on the insect cadaver and eventually reach the fourth stage juvenile (J4), which divides into males and females. First generation (G1) females lay eggs after mating, either in the external medium or still inside the female's body and these eggs hatch into first-stage juveniles (J1). Depending on the quantity of food is left in the insect cadaver at that point, types of situations are possible. In the case of insufficient food, J1 immediately transformed into the second-stage juvenile (J2) in 2 or 3 days. Before becoming an infective juvenile, J2 stopped feeding and had a molt while still in the pre-infective stage. Then the newly generated IJ emerge from the depleted insect cadaver to actively look for another susceptible insect prey. On the other hand, if there is an abundance of food in the cadaver, both males and females can reproduce numerous generations in the same cadaver. After hatching from the eggs of the G1 females, J1 turns into J2, non-infectious J3, and J4 before becoming the second generation (G2) of adults. After mating, G2 females release eggs that develop into J1s, starting a new cycle. EPNs typically reproduced for two to three generations before the food sources in the insect cadaver are completely depleted, [86]. After insect invasion by IJs, the entire reproductive cycle lasts 7 to 14 days and is mostly dependent on temperature. After mating with male, both *Steinernema* and *Heterorhabditis* females lay eggs in the dead insects cadaver. Eggs that have been hatched usually develop juveniles that grow and develop to be amphimictic adults [88]. The reproductive life cycle of the majority of *Steinernema* has sexually distinct partners, G1 males and females, while all *Heterorhabditis* IJs after insect infection become self-fertilizing hermaphrodite females. However, amphimictic *Heterorhabditis* adults are developed by the second generation. It is interesting to note that IJs from the *S. hermaphroditum* species can mature into self-fertilizing hermaphrodite females, rather like IJs from *Heterorhabditis* do. The unusual characteristic of this *Steinernema* species has been proposed to support the independent but converging evolution with *Heterorhabditis* described by Poinar and previously described [89]. *Heterorhabditis* EPNs reproduce hermaphroditically, which greatly reduces or impairs the genetic diversity of the offspring. *Heterorhabditis's* hermaphrodite behavior makes it possible for a single IJ to infect a host and molt into a hermaphrodite female [90], whereas at least two *Steinernema* IJs must penetrate an insect larva and reach maturity into male and female to cause infection. This undoubtedly represents *Heterorhabditis* species a survival edge over *Steinernema* species. The process of fertilizing the female's eggs with sperm occurs during male and female mating. Male releases spermatozoids into the female's vulva along with its spicule, which it uses to develop spermatozoids. In the uterus, the female's eggs are fertilized by the male's sperm. For hermaphrodites female who are self-fertile, sperm is formed and stored in spermatic vesicles, which are described as a distal enlargement of the uterus. When the female initiates to lay eggs, the sperm in the spermatic vesicles automatically fertilizes them. Since females are longer and bigger than males, males need to find a strategy of scanning the full female body in order to find the vulva. The two ways that a male identified the vulva on a female body. These two reproductive strategies highlight still another difference between *Steinernema* and *Heterorhabditis*, namely that males stick to females and slither all over their bodies until it finds the vulva, but both the female and male heads of *Heterorhabditis* point in the opposite directions [91]. According to *Steinernema*, the males behave like a ring around the female body. Until reaching the vulva, the male coils entirely around and along the female body [26]. There are some safeguards that have been adopted, to prevent multiple males from mating with the same female. In *Heterorhabditis* species, after mating, a male releases a mating plug that closes the vulva, preventing other males from mating with the same female [92]. In *Steinernema* species, it has been proved that virgin females generate various chemicals that attract males and that their production reduces off after mating [26]. Moreover, male of *S. longicaudum* required virgin conspecific females to mature in their immediate surroundings [93].

1.4 Mode of action

After mating, most of the eggs are preserved inside the maternal body of the EPN. Following that, the offspring grow and feed inside the maternal body. This process is called endotokia matricida which is derived from the Greek word ενδο ("endo", inside) and τοκοσ ("tocos", birth), and from the Latin ("mater", mother and "caedere", kill). This term was coined by Maupas, (2015) when he first characterized the *Caenorhabditis elegans*. This phenomenon helps the progeny by protecting it and, in the case of EPNs, by giving it a high-lipid food source, particularly whenever the infected insect cadaver is about to be exhausted. When endotokia matricida is encouraged in cases where there is a lack of food, this phenomenon takes place to the first generation of juveniles even when there is still a plenty of food. It follows that it is clear that the size of the vulnerable insect will have an impact on the growth and survival of EPNs. According to some authors, *Steinernema* IJs are ineffective at controlling micro-insect pests [94]. The infectivity of four different *Steinernema* species in insects smaller than 5 mm in length was recently demonstrated by Bastidas and coworkers [95], who also came to the conclusion that *Steinernema* and *Heterorhabditis* nematodes cannot survive in the environment for very long if there are no larger insects available for them to complete their life cycle.

2. Biological control

Biological control is an eco-friendly and effective means of reducing or mitigating pests through the use of natural enemies [96]. It relies on predation, parasitism, herbivores, or other natural mechanisms, but involves an active human management role [97]. According to Dreistadt (2007) "*Biological control is the beneficial action of predators, parasites, pathogens, and competitors in controlling pests and their damage*". Biological control now becomes an interdisciplinary science combining entomology, microbiology, plant pathology, weed science, and virology with the goal to reduce and control pathogens, microorganisms, insects, and plants alike, which can cause damage to crop plants [98]. The different biological control agents have been used time to time, and their success and failures have been extensively reviewed. Their use in bio-control has increased over recent decades.

Classical biological control involves usage of an exotic, usually co-evolved, biological control agent for permanent establishment and long-term pest control [99]. Classical biological control focuses on finding natural enemies, introduces them into the area of the target pest, and permanently establishes them so that they will provide continuing pest control with little or no additional human intervention. Augmentation on the other hand involves deliberate discharge of natural agent that does not occur in good numbers and thus are incapable of reducing pest below damaging level [100]. There are two general approaches to augmentation: inundative releases and inoculative releases. The former involves usage of large number of natural enemies for immediate pest control by disseminating them on the crops fields multiple times, while later method involves release of small natural enemies at given intervals in order to keep pest populations below economic injury level. The last type, conservation biological control involves measures of modifying existing practices to further enhance specific natural enemies of other organisms to reduce the effect of pests.

Biological control agents include bacteria, fungi, viruses, nematodes, or protozoa that can infect and kill the host. Some of these agents can kill and infect insects and are referred to as entomopathogens. Among entomopathogens, there is class of nematodes which parasitize insects only and are referred to as entomopathogenic nematodes (EPN) which are associated with entomopathogenic bacteria. The entomopathogenic nematodes, belonging to the families, Steinernematidae [101] and Heterorhabditidae [102], are associated with bacteria which belong to the family Enterobacteriaceae [103], *Xenorhabdus* [104] in case of steinernematids and *Photorhabdus* [105] in case of *Heterorhabditids* which reside in their alimentary canal. At present two well-known genera, *Steinernema* and *Heterorhabditis* of EPNs, are globally described and consisting more than 100 species and 21 species, respectively [106]. They all are lethal duo, capable of killing the host within short duration and hence are considered as good bio-control agents [107]. The bacteria mostly resides in the alimentary canal of third stage juveniles, called infective juveniles (IJ) which is only free living stage in the life cycle of EPN. They live freely in moist soils and move in search of their insects hosts. Once, they come in contact with the insect host, they enter in their body either through natural openings or by abrading the skin. After their entry in the insect host, they release their endosymbiont bacteria and kill the host within 24 to 48 hours. They feed on the insect cadaver and produce their adult generations, which mate inside the cadaver and released the eggs which hatched into juvenile stages. Once the survival resources are depleted, they move out the cadaver and search for the next host.

2.1 Entomopathogenic nematodes in insect pest management

Researchers focused on this area when *Steinernema glaseri*, the first EPN species introduced as a biocontrol agent, was used in the United States against the Japanese beetle, *Popillia japonica* [108]. These organisms reemerged as effective biocontrol agents in the 1960s and 1970s, with *Steinernema carpocapsae* (also known as *Neoaplectana carpocapsae*) serving as the primary biocontrol agent [109]. With the advancement of fermentation technology, several species of EPN (including *S. carpocapsae, S. scapterisci, S. feltiae, S. glaseri,* and *Heterorhabditis megidis*) have been mass produced commercially and are sold in market for the use by growers in formulations suitable for short term storage [110]. The mass production of IJs of EPNs is easy and cost effective. The preferred method of application is inundative release [111].

S. carpocapsae, S. feltiae, S. kraussei, S. glaseri, S. riobrave, Heterorhabditis bacteriophora, and *H. megidis* are some of the well-known most frequently used and successfully deployed nematodes as biopesticides. This characteristic is attributed to their simple and easy mass production in liquid culture [112]. Cultivation using live insect hosts (*in vivo*) requires cheap start-up costs, low levels of technology, and high nematode quality but cost-effective efficiency. *In vitro* solid or liquid culture is a cost-effective method, with liquid culture requiring the largest start-up capital.

2.2 Bio-formulations using entomopathogenic nematodes

One of the traditional ways to prevent losses from insect pests has been to use chemicals; although, nowadays days, due to several unjustified side effects, pest control relies on many other solutions in addition to pesticides. The term "Integrated Pest Management" refers to the combination of all these options (IPM). IPM is a pest management technique based on a systems approach that considers the entire ecosystem of the orchard. Continuous use of hazardous chemicals, at high doses against agricultural key pests, has led to major problems such as pest resurgence resulting from development of resistance and destruction of natural enemies. Also, the enormous use of pesticides is not only costly affair but also due to its residual effects, is directly or indirectly harmful to animals, other non-targeted soil fauna and human beings too. In search of new avenues in biological control, the importance of EPN has been highlighted as an environment-friendly pest control method. The successful market introduction of an EPN-based product requires a reliable species-specific isolate and stable formulation having more than 6 months of its shelf-life period when stored at room temperature (20–25°C) such formulations are on high demand. Unfortunately, no species-specific nematode formulation has been developed so far which could achieve the goal.

EPN formulation is a process of the transformation of living entities into a product that can be applied by practical methods. Few factors affect their application part in field conditions which include market value; crop and target insects; formulation type and shelf life; usage directions; technical support; cost and others [113]. Generally, EPN formulation contains an active ingredient, a carrier and additives. There are different types of EPN formulations present in market which are synthetic sponges [114], gels [115], clay and powder [116], or infected cadavers form [117]. Few factors like soil moisture; soil texture; water content; temperature; and UV radiation may impact drastically affect the infectivity of EPNs, storage and formulation and development of industrial product. To increase the strategies for optimization of effectiveness, timing of application and type of formulations gives best results [118]. Several formulations which have been used before include clay; polyether polyurethane sponge; anhydrobiotic nematodes; bait and activated charcoal [119].

Good storage and formulation strategies can be developed in order to ensure nematode survival and maintenance of increased infectivity [119]. This is why it is crucial to have an understanding of nematode ecology and also analyze which environmental factors affect their activity and infectivity [120]. EPN application rates vary widely, ranging between 7400 and 1,500,000 IJs/m^2 [121], but 250,000 IJs/m^2 being a common recommended rate for commercial applications [122]. Currently, S. carpocapsae, S. feltiae, S. kraussei, S. glaseri, S. riobrave, H. bacteriophora (CAB Reviews 2018 13, No. 058) http://www.cabi.org/cabreviews and Heterorhabditis megidis are the most commonly used and successfully applied nematodes due to their easily production in liquid culture [112] (**Table 1**).

In India, various studies have been conducted to improve formulations in terms of storage, shelf life, application techniques, virulence control, etc. *Heterorhabditis indica, Steinernema abbasi, S. bicornutum, S. Carpocapsae,* and *S. riobrave* are contained in a variety of carrier materials such as talc, alginate capsules, wheat bran pellets, sodium alginate beads, vermiculite, spray vehicles or hydrogels (Hussaini et al. [123, 124]; Gupta [125]; Vyas et al. [126]). However, it is currently only in the comprehensive research stage. National Institute of Plant Health Management (Hyderabad), Indian Agricultural Research Institute (New Delhi), National Agricultural and Insect Resources Board (Bangalore), Multiplex Biotech Pvt. Ltd., Ajay Biotech (India) Ltd., and Pest Control (India) Pvt. Ltd. conduct research to provide EPN formulated products with improved shelf life (**Tables 1** and **2**).

S. No.	EPN species	Product Name	Country
1.	*S. carpocapsae*	ORTHO Biosafe	USA
2.	*S. carpocapsae*	Biovector	USA
3.	*S. carpocapsae*	Exhibit	USA
4.	*S. carpocapsae*	XGNAT	USA
5.	*S. carpocapsae*	Helix	Germany
6.	*S. carpocapsae*	Boden Niitzlinge	Switzerland
7.	*S. carpocapsae*	Sanoplant	USA
8.	*S. carpocapsae*	Proactant	USA
9.	*S. carpocapsae*	Green Commandos-Ecomax	India*
10.	*S. feltiae*	Manget	USA

S. No.	EPN species	Product Name	Country
11.	*S. feltiae*	Entonem	USA
12.	*S. feltiae*	Nemasys	USA
13.	*S. feltiae*	Stealth	UK
14.	*S. riobrave*	Vector MG	USA
15.	*S. kushidai*	SDS Biotech	Japan
16.	*H. bacteriophora*	Otinem	USA
17.	*H. bacteriophora*	E-nema	Germany
18.	*H. bacteriophora*	Soil Commandos-Ecomax	India*

**Product launched in 2011 but withdrawn from the market due to inconsistency of results in the field.*

Table 1.
Worldwide used entomopathogenic nematodes-based formulations.

S. No.	EPN species	Production/ formulation	Reference
1.	*S. bicornutum*	Bait as alginate capsule	[123]
2.	*S. carpocapsae*	Alginate capsule	[123]
3.	*S. carpocapsae*	Wheat bran pellets	[123]
4.	*S. riobrave*	Spray-adjuvants	[126]
5.	*Heterorhabditis indica*	Talc	[124]
6.	*S. abbasi*	Talc	[124]
7.	*S. carpocapsae*	Talc	[124]
8.	*S. carpocapsae*	Pearl (sod. alginate)	[125]
9.	*S. carpocapsae*	Vermiculite	[124]
10.	*Steinernema .f abbasi = (Symbiobacterium thermophilum)*	Hydrogel	[127, 128]

Table 2.
Entomopathogenic nematodes-based product used in India.

3. Various formulations and their applications

In comparison with foliar pests, the performance of EPN has shown more success in controlling soil-born insect pests. Juveniles' intolerance to fluctuations of desiccation [31], temperature [30], and UV radiation [129] is a crucial factor in the failure of foliar applications of EPN Schroer and Ehlers (2005) recently used *S. carpocapsae* on cabbage to attack the foliar insect *Plutella xylostella*, which was mixed with the surfactant Rimulgan and the polymer xanthan to create the best circumstances for nematode infection on the plant surface. As an integrated pest control tool, EPN effectiveness has also been proven to be compatible with chemical insecticides, fungicides, and acaricides [130]. As a result, these chemicals can frequently be tank mixed and applied with other pesticides. Some insecticides, including Imidacloprid [131], tefluthrin [132], neonicotinoid [133] and *Bacillus thuringiensis* [134] were found to be synergistic with EPN. Hence, before EPN is released into the field, its formulation and integration with pesticides and surfactants should be carefully evaluated. Nematode control is regarded as more crucial due to the lack of possible nematicides, the high cost of field application, and the current trend toward eco-friendly pest and disease control methods.

Nematicides should not be used in the current era of globalization. Nematicides will probably no longer be available because they are expressly prohibited in organic farming. Root-knot nematode, which causes serious harm to various crops, affects a huge range of vegetables, fruits, and pulse crops in India. Since EPN is already well-established for the control of insect pests, applying it to the management of plant parasitic nematodes would be extremely cost-effective (PPN). There are various reports that indicate EPN was used to reduce PPN [135]. The impacts of the microorganisms *Xenorhabdus* spp. and *Photorhabdus* spp. that are linked to *Steinernema* and *Heterorhabiditis*, respectively, have suppressed selected species of PPN including root-knot nematode in green house experiments [23]. However, the existing literature has limited information on at what stage(s) of PPN are affected by EPN applications. Tests conducted in the field have demonstrated PPN population suppression for up to 8 weeks after application of EPN products [135].

Extensive research over the past decade has produced numerous effective isolates and strains as well as substantial advancements in mass production and formulation technology, all inspired by the need to reduce pesticide consumption [136]. Currently, *S. carpocapsae, S. feltiae, S. kraussei, S. glaseri, S. riobrave, H. bacteriophora* (CAB Reviews 2018 13, No. 058) http://www.cabi.org/cabreviews, and *H. megidis* are the most commonly used and successfully applied nematodes due to their easily production in liquid culture [112]. Today, EPN are mostly used in soil, galleries of boring insects, circumstances where insecticide resistance has occurred, or when dangerous pesticides are prohibited, conditions where chemical pesticides have failed [40].

In India, several scientists have used EPN in both lab and field conditions to attack cutworms, ragi pink borer, stem borer, white grubs, etc. [137]. Few EPNs, including *S. corpocapsae* (strain DD-136), *H. bacteriophora* (strain Burliar), and *Heterorhabditis* species, were found in field trails against *Amsacta albistrigata* larvae in their fourth instar on ground nuts [138]. Green Commandos (*S. corpocapsae*) and Soil Commandos (*H. bacteriophora*), two EPN-based formulations formed by Ecomax Compant in 1980 using exotic species, were both removed from the marketplace due to their lack of effectiveness against insect pests due to their poor adaptability to Indian environmental conditions or formulation issues.

4. Factors affecting survival and efficacy

Matching the optimal EPN species or strain to the target host and environment requires consideration of innate efficacy and suitability of environmental conditions. EPN population persistence is determined by the permanence of individual IJs and population recycling in host insect larvae and many factors that can influence both mechanisms [139, 140]. IJs in different EPN species vary in natural longevity from several months to over a year. Losses of 50% can be reached within hours after soil application until the IJ settles in the soil. Losses then range from 5 to 10% per day for 1 to 6 weeks and often only about 1% of the original inoculum survives. To compensate for these losses, a general rule of thumb for application rates is 25 IJ per cm2 of treated area, although higher or (rarely) lower application rates may be used depending on the pests targeted and the cropping system may be required. As a result, IJ populations in soil or similar substrates generally remain tall enough to provide effective control for 2–8 weeks. Post-application recycling is frequent, but not sufficient to achieve multi-seasonal control, as the distribution of IJs usually becomes too patchy over time. UV light can inactivate and kill his IJs within minutes of application, but the effects vary by his EPN strain and species [129]. If application is in the early morning or evening, IJ losses are minimized by adding a UV protectant to the IJ suspension and applying the soil with a large amount of carrier combined with immediate rinsing with sufficient water. Most EPN species perform optimally at 20–30°C [30], become sluggish below 10–15°C, and are inactivated above 30–40°C. Various *Steinernema* spp. have been isolated from other EPN species in cold, hot semi-arid, and even arid regions and may have promise for use in extreme environments. In soil, IJ travels in a film of water that covers crevices. Adequate substrate moisture is essential for good IJ activity. Dry conditions limited IJ activity, but gradual dehydration can cause IJs to enter and persist dormancy. Drought-intolerant *H. bacteriophora* IJ actively seek soil layers with high water content, whereas drought-tolerant *S. carpocapsae* IJ can survive better in drier conditions. In water logged soils, anoxic conditions and low surface tension can compromise IJ movement and even survival. Movement and survival of IJs are generally more restricted in microstructured soils than in sandy soils [141], although sandy soils may dry out faster and reduce IJ activity. EPNs are adversely affected by pH values <4 and >8. Various biological factors can also affect the survival of IJ or EPN populations in soil. Many species of arthropods and other invertebrates prey on IJs (e.g., mites, springtails, tardigrades, predatory nematodes, nematophagous fungi), or feed on EPN-infected hosts. Other entomopathogens (e.g., entomopathogenic fungi, bacteria, or viruses) or parasites compete with host EPNs [140].

5. Conclusion

The inordinate use of chemical pesticides in farming causes serious damage to soil, air, water, flora, fauna, and human beings. Thus, it is necessary to develop environmentally friendly druthers to control soil pests, similar as entomopathogenic nematodes (EPNs). Since EPN species are host specific, they can be used widely for the target organisms. Today, the use of entomopathogenic nematodes as biopesticides against agricultural insect pests has grown quickly in recent times. These bio-pesticides play a great part in producing organic crops and export goods. So, researchers and advanced institutions should have to give attention in producing, formulating, and storing environmentally safe biopesticides. Nowadays, it is insolvable to ensure sustainable growth of crop yields without the application of fertilizers. Still, when using soil emendations, it is necessary to select the once that will round the living terrain to help negatively affecting the crops but also the structure and agrochemical conditions of the soil and the soil biota. Overall, it is apparent that nitrogen negatively affects the bacterial symbionts, but it is unknown whether or not potassium counteracts the toxin, or if the bacterial growth was affected by commodity differently. Results suggest that organic diseases brace better with *Xenorhabdus nematophila* and *Photorhabdus luminescens*, but they may warrant too important nitrogen for the shops to be suitable to produce.

Conflict of interest

"The authors declare no conflict of interest."

Author details

Kajol Yadav*, Lovely Bharti and Ashok Kumar Chaubey
Nematology Laboratory, Department of Zoology, Chaudhary Charan Singh University Meerut, India

*Address all correspondence to: kajolyadav45@gmail.com

References

[1] Li D, Wang X. Nitric oxide emission from a typical vegetable field in the Pearl River delta, China. Atmospheric Environment. 2007;**41**:9498-9505

[2] Rashti MR, Wang W, Moody P, Chen C, Ghadiri H. Fertiliser-induced nitrous oxide emissions from vegetable production in the world and the regulating factors: A review. Atmospheric Environment. 2015;**112**:225-233

[3] Zhong WH, Bian BY, Gao N, Min J, Shi WM, Lin XG, et al. Nitrogen fertilization induced changes in ammonia oxidation are attributable mostly to bacteria rather than archaea in greenhouse-based high N input vegetable soil. Soil Biology and Biochemistry. 2016;**93**:150-159

[4] Meng L, Ding W, Cai Z. Long-term application of organic manure and nitrogen fertilizer on N2O emissions, soil quality and crop production in a sandy loam soil. Soil Biology and Biochemistry. 2005;**37**:2037-2045

[5] Wardle DA, Nicholson KS, Bonner KI, Yeates GW. Effects of agricultural intensification on soil-associated arthropod population dynamics, community structure, diversity and temporal variability over a seven-year period. Soil Biology and Biochemistry. 1999;**31**:1691-1706

[6] Bardgett RD, van der Putten WH. Belowground biodiversity and ecosystem functioning. Nature. 2014;**515**:505-511

[7] Mader P, Fliessbach A, Dubois D, Gunst L, Fried P, Niggli U. Soil fertility and biodiversity in organic farming. Science (New York, N.Y.). 2002;**296**:1694-1697

[8] Hartmann M, Frey B, Mayer J, Maeder P, Widmer F. Distinct soil microbial diversity under long-term organic and conventional farming. The ISME Journal. 2015;**9**:1177-1194

[9] Bonanomi G, De Filippis F, Cesarano G, La Storia A, Ercolini D, Scala F. Organic farming induces changes in soil microbiota that affect agro-ecosystem functions. Soil Biology and Biochemistry. 2016;**103**:327-336

[10] Lori M, Symnaczik S, Mäder P, De Deyn G, Gattinger A. Organic farming enhances soil microbial abundance and activity a meta-analysis and metaregression. PLoS One. 2017;**12**:0180442

[11] Yeates GW, Bardgett RD, Cook R, Hobbs PJ, Bowling PJ, Potter JF. Faunal and microbial diversity in three Welsh grassland soils under conventional and organic management regimes. Journal of Applied Ecology. 1997;**34**:453-470

[12] Atandi JG, Haukeland S, Kariuki GM, Coyne DL, Karanja EN, Musyoka MW, et al. Organic farming provides improved management of plant parasitic nematodes in maize and bean cropping systems. Agriculture, Ecosystems & Environment. 2017;**247**:265-272

[13] Coll P, Le CE, Blanchart E, Hinsinger P, Villenave C. Organic viticulture and soil quality: A long-term study in Southern France. Applied Soil Ecology. 2011;**50**:37-44

[14] Ferris H, Venette RC, Lau SS. Dynamics of nematode communities in tomatoes grown in conventional and organic farming systems, and their impact on soil fertility. Applied Soil Ecology. 1996;**3**:161-175

[15] Sanchez-Moreno S, Cano M, López-Pérez A, JMR. B. Microfaunal soil food webs in Mediterranean semi-arid agroecosystems. Does organic management improve soil health? Applied Soil Ecology. 2018;**125**:138-147

[16] Nielsen UN, Ayres E, Wall DH, Li G, Bardgett RD, Wu T, et al. Global-scale patterns of assemblage structure of soil nematodes in relation to climate and ecosystem properties. Global Ecology and Biogeography. 2014;**23**:96-8978

[17] Wu SM, Hu DX, Ingham ER. Comparison of soil biota between organic and conventional agroecosystems in Oregon, USA. Pedosphere. 2005;**15**:395-403

[18] Tsiafouli MA, Argyropoulou MD, Stamou GP, Sgardelis SP. Soil nematode biodiversity in organic and conventional agroecosystems of Northern Greece. Russian Journal of Nematology. 2006;**14**:159-169

[19] Koppenhöfer AM. Nematodes. In: Lacey LA, Kaya HK, editors. Field Manual of Techniques in Invertebrate Pathology: Application and Evaluation of Pathogens for Control of Insects and Other Invertebrate Pests. second ed. Dordrecht: Springer; 2007. pp. 249-264

[20] Grewal PS, Ehlers RU, Shapiro-Ilan DI, editors. Nematodes as Biological Control Agents. Wallingford: CABI Publishing; 2005

[21] Shapiro-Ilan DI, Gouge DH, Koppenhöfer AM. Factors affecting commercial success: Case studies in cotton, turf and citrus. In: Gaugler R, editor. Entomopathogenic Nematology. Wallingford: CABI Publishing; 2002. pp. 333-355

[22] Georgis R, Koppenhöfer AM, Lacey LA, Bélair G, Duncan LW, Grewal PS, et al. Successes and failures in the use of parasitic nematodes for pest control. Biological Control. 2006;**38**:103-123

[23] Ishibashi N, Choi DR. Biological control of soil pests by mixed application of entomopathogenic and fungivorous nematodes. Journal of Nematology. 1991;**23**:175-181

[24] Ehlers RU, Linau S, Krasomil O, Osterfield KH. Liquid culture of Entomopathogenic nematode bacterium complex *Heterorhabditis megidis/Photorhabdus luminescence*. Biological Control. 1998;**43**(1):77-88

[25] Campbell JF, Lewis EE, Stock SP, Nadler S, Kaya HK. Evolution of host search strategies in entomopathogenic nematodes. Journal of Nematology. 2003;**35**:142-145

[26] Lewis EE, Barbarosa B, Gaugler R. Mating and sexual communication by *Steinernema carpocapsae* (Nemata: Steinernematidae). Journal of Nematology. 2002;**34**:328-331

[27] Shapiro-Ilan DI, Bruck DJ, Lacey LA. Principles of epizootiology and microbial control. In: Vega FE, Kaya HK, editors. Insect Pathology. second ed. San Diego: Academic Press; 2012. pp. 29-72

[28] Hazir S, Kaya HK, Stock SP, Keskin N. Entomopathogenic nematodes (Steinernematidae and Heterorhabditidae) for biological control of soil pests. Turkish Journal of Biology. 2004;**27**:181-202

[29] Gaugler R, Campbell JF, Selvan S, Lewis EE. Large-scale inoculative releases of the entomopathogenic nematode *Steinernema glaseri*: Assessment 50 years later. Biological Control. 1992;**2**:181-187

[30] Grewal PS, Selvan S, Gaugler R. Thermal adaptation of entomopathogenic nematodes, niche breadth for infection, establishment, and reproduction. Journal of Thermal Biology. 1994;**19**:245-253

[31] Lello ER, Patel MN, Mathews GA, Wright DJ. Application technology for entomopathogenic nematodes against foliar pests. Crop Protection. 1996;**15**:567-574

[32] Majić I, Sarajlić A, Lakatos T, Tóth T, Raspudić E, Puškadija Z, et al. Virulence of new strain of *Heterorhabditis bacteriophora* from Croatia against *Lasioptera rubi*. Plant Protection Science. 2019;**55**:134-141

[33] Cruz-Martínez H, Ruiz-Vega J, Matadamas-OrtízPT, Cortés-MartínezCI, Rosas-Diaz J. Formulation of entomopathogenic nematodes for crop pest control – A review. Plant Protection Science. 2017;**53**:15-24

[34] Wright DJ, Peters A, Schroer S, Fife JP. Application technology. In: Grewal PS, Ehlers RU, Shapiro-Ilan DI, editors. Nematodes as Biocontrol Agents. Wallingford: CABI Publishing; 2005. pp. 91-106

[35] Laznik Ž, Trdan S. The influence of insecticides on the viability of entomopathogenic nematodes (Rhabditida: Steinernematidae and Heterorhabditidae) under laboratory conditions. Pest Management Science. 2014;**70**:784-789

[36] Laznik Ž, Tóth T, Lakatos T, Vidrih M, Trdan S. Control of the Colorado potato beetle (Leptinotarsadecemlineata [say]) on potato under field conditions: A comparison of the efficacy of foliar application of two strains of *Steinernema feltiae* (Filipjev) and spraying with thiametoxam. Journal of Plant Diseases and Protection. 2010;**117**:129-135

[37] Glazer I, Klein M, Navon A, Nakache Y. Comparison of efficacy of entomopathogenic nematodes combined with antidesiccants applied by canopy sprays against three cotton pests (Lepidoptera: Noctuidae). Journal of Economic Entomology. 1992;**85**:1636-1641

[38] Sher RB, Parella MP. Biological controlof the leafminer, *Liriomyza trifolii*, in chrysan-themums: Implications for intraguild predation between *Diglyphus begini* and *Steinernema carpocapsae*. Bulletin of the International Organization for Biological and Integrated Control of Noxious Animals and Plants: Integrated Control in Glasshouses. 1999;**22**:221-224

[39] Head J, Walters KFA, Langton S. The compatibility of the entomopathogenic nematode, *Steinernema feltiae*, and chemical insecticides for the control of the South American leaf miner, *Liriomyza huidobrensis*. Biocontrol. 2000;**45**:345-353

[40] Ehlers R-U. Mass production of entomopathogenic nematodes for plant protection. Applied Microbiology and Biotechnology. 2001;**56**:623-633

[41] Ehlers RU. Entomopathogenic nematodes – Save biocontrol agents for sustainable systems. Phytoprotection. 1998;**79**:94-102

[42] Glaser RW, Farrell CC. Field experiments with the Japanese beetle and its nematode parasite. Journal of the New York Entomological Society. 1935;**43**:345-371

[43] Akhurst R, Smith K. Regulation and safety. In: Gaugler R, editor. Entomopathogenic Nematology.

Wallingford: CABI Publishing; 2002. pp. 311-332

[44] Igbedioh SO. Effects of agricultural pesticides on humans, animals and higher plants in developing countries. Archives of Environmental Health. 1991;**46**:218

[45] WHO. Public Health Impact of Pesticides Used in Agriculture. Geneva: W.H.O; 1990. p. 88

[46] Environews Forum. Killer environment. Environmental Health Perspectives. Environ Health Perspect. 1999;**107**(2):A62–A63

[47] Hurley PM, Hill RN, Whiting RJ. Mode of carcinogenic action of pesticides inducing thyroid follicular cell tumours in rodents. Environmental Health Perspectives. 1998;**106**:437

[48] Liroff RA. Balancing risks of DDT and malaria in the global POPs treaty. Pestic Safety News. 2000;**4**:3

[49] Brouwer A, Longnecker MP, Birnbaum LS, Cogliano J, Kostyniak P, Moore J, et al. Characterization of potential endocrine related health effects at lowdose levels of exposure to PCBs. Environmental Health Perspectives. 1999;**107**:639

[50] European Commission. Monitoring of pesticide residues in products of plant origin in the European Union. Report. 1998;**15**:1

[51] Karunakaran CO. The Kerala food poisoning. Journal of the Indian Medical Association. 1958;**31**:204

[52] Kashyap R, Iyer LR, Singh MM. Evaluation of daily dietary intake of dichlorodiphenyltrichloroethene (DDT) and benzenehexachloride (BHC) in India. Archives of Environmental Health. 1994;**49**(1):63-66

[53] Kannan K, Tanabe S, Ramesh A, Subramanian AN, Tatsukawa R. Persistent organochlorine residues in foodstuff s from India and their implications on human dietary exposure. Journal of Agricultural and Food Chemistry. 1992;**40**:518-524

[54] Roberts TR. Metabolic pathway of agrochemicals. Part I. In: Herbicides and Plant Growth Regulators. Cambridge, UK: The Royal Society of Chemistry; 1998

[55] Andreu V, Pico' Y. Determination of pesticides and their degradation products in soil: Critical review and comparison of methods. Trends in Analytical Chemistry. 2004;**23**(10-11):772-789

[56] Savonen C. Soil microorganisms object of new OSU service. Good Fruit Grower. 1997. Available from: http://www.goodfruit.com/archive/1995/6other.html

[57] Kelley WD, South DB. In vitro effects of selected herbicides on growth and mycorrhizal fungi. In: Weed Sci. Soc. America Meeting. Auburn, Alabama: Auburn University; 1978. p. 38

[58] Chakravarty P, Sidhu SS. Effects of glyphosate, hexazinone and triclopyr on in vitro growth of five species of ectomycorrhizal fungi. European Journal of Forest Pathology. 1987;**17**:204-210

[59] MoormanTB A. Review of pesticide effects on microorganisms and microbial processes related to soil fertility. Journal of Production Agriculture. 1989;**2**(1):14-23

[60] Kole RK, Banerjee H, Bhattacharyya A. Monitoring of market fish samples for Endosulfan and Hexachlorocyclohexane residues in and around Calcutta. Bulletin of

Environmental Contamination and Toxicology. 2001;**67**:554-559

[61] Bortleson G, Davis D, U.S. Geological Survey, Washington State Department of Ecology. Pesticides in Selected Small Streams in the Puget Sound Basin 1987-1995; 1997. pp. 1-4

[62] US Department of the Interior. Pesticides in ground water: Current understanding of distribution and major influences. In: U.S. Geological Survey. National Water Quality Assessment. Factsheet Number FS. 1995. pp. 244-295

[63] U.S. Geological Survey, National Water-Quality Assessment, Pesticide National Synthesis Project. Pesticides in surface and ground water of the United States; Summary of results of the National Water Quality Assessment Program. Available from: http://water.wr.usgs.gov/pnsp/allsum/fig02.gif. 1998

[64] Wall GR, Riva-Murray K, Phillips PJ. Water Quality in the Hudson River Basin, New York and Adjacent States, 1992-95. Reston, VA: USGS. U.S. Geological Survey Circular; 1998. p. 1165

[65] U.S. Geological Survey. The quality of our nation's waters – nutrients and pesticides. Circular 1225. Reston VA: USGS. Available from: http://water.usgs.gov/pubs/circ/circ1225/.1999

[66] Waskom R. Best Management Practices for Private Well Protection. Colorado State Univ. Cooperative Extension (August). Available from: http://hermes.ecn.purdue. edu:8001/cgi/convertwq?. 1994;7488

[67] Kole RK, Bagchi MM. Pesticide residues in the aquatic environment and their possible ecological hazards. Journal of the Inland Fisheries Society of India. 1995;**27**(2):79-89

[68] Glotfelty S. In: Sawhney BL, Brown K, editors. Volatilization of Pesticides from Soil in Reactions and Movements of Organic Chemicals in Soil. Madison, WI: Soil Science Society of America Special Pub; 1989

[69] Majewski M, Capel P. Pesticides in the atmosphere: Distribution, trends, and governing factors. In: Volume One, Pesticides in the Hydrologic System. Ann Arbor Press Inc.; 1995. p. 118

[70] Straathoff H. Investigations on the phytotoxic relevance of volatilization of herbicides. Mededelingen van de Faculteit Land bouwweten-schappen, Rijksuniversiteit Gent. 1986;**51**(2A):433-438

[71] Dreistadt SH, Clark JK, Flint ML. Pests of Landscape Trees and Shrubs. An Integrated Pest Management Guide. University of California Division of Agriculture and Natural Resources. Publication; 1994. p. 3359

[72] Locke D, Landivar JA, Moseley D. The effects of rate and timing of glyphosate applications of defoliation efficiency, regrowth inhibition, lint yield, fiber quality and seed quality. In: Proceedings of the Beltwide Cotton Conferences. San Antonio, Texas, USA: National Cotton Council; 1995. pp. 1088-1090

[73] Kergunteuil A, Campos-Herrera R, S'anchez-Moreno S, Vittoz P, Rasmann S. The abundance, diversity, and metabolic footprint of soil nematodes is highest in high elevation alpine grasslands. Frontiers in Ecology and Evolution. 2016;**4**:1-12

[74] Helmberger MS, Shields EJ, Wickings KG. Ecology of belowground biological control: Entomopathogenic nematode interactions with soil biota. Applied Soil Ecology. 2017;**121**:201-213

[75] Campos-Herrera R, Pathak E, El-Borai FE, Stuart RJ, Gutiérrez C, Rodríguez-Martín JA, et al. Geospatial patterns of soil properties and the biological control potential of entomopathogenic nematodes in Florida citrus groves. Soil Biology and Biochemistry. 2013;**66**:163-174

[76] Spiridonov SE, Hominick WM, Brisco BR. Morphology of amoeboid cells in the uterus of *Steinernema* species (Rhabditida: Steinernematidae). Russian Journal of Nematology. 1999;**7**:39-42

[77] Nguyen KB, Smart GC Jr. *Steinernema scapterisci* n. sp. (Steinernematidae: Nematoda). Journal of Nematology. 1990;**22**:187-199

[78] Mráček Z, Bednarek A. The morphology of lateral fields of infective juveniles of entomogenous nematodes of the family Steinernematidae (Rhabditida). Nematologica. 1991;**37**:63-71

[79] Jackson GJ. Differentiation of three species of *Neoaplectana* (Nematoda: Rhabditida), grown axenically. Parasitology. 1965;**55**:571-578

[80] Akhurst RJ. Use of starch gel electrophoresis in the taxonomy of the genus *Heterorhabditis* (Nematoda: Heterorhabditidae). Nematologica. 1987;**33**:1-9

[81] Poinar GO, Kozodai EM. *Neoaplectana glaseri* and *N. anomali*: Sibling species or parallelism? Revue de Nematologie. 1988;**11**:13-19

[82] Reid AP, Hominick WH. Cloning of the rDNA repeat unit from a British entomopathogenic nematode (Steinernematidae) and its potential for species identification. Parasitology. 1993;**109**:529-536

[83] Mráček Z, Weiser J, Gerdin S. Head and cuticular structures of some species in the family Steinernematidae (Nematoda). Nematologica. 1981;**27**:443-448

[84] Selvan SP, Grewal PS, Gaugler R, Tomalak M. Evaluation of steinernematid nematodes against *Popillia japonica*: Species, strain and rinse after application. Journal of Economic Entomology. 1994;**89**:605-609

[85] Campbell J, Gaugler R. Nictation behavior and its ecological implications in the host search strategies of entomopathogenic nematodes (Heterorhabditidae and Steinernematidae). Behaviour. 1993;**126**:154-169

[86] Grewal PS, De Nardo EAB, Aguillera MM. Entomopathogenic nematodes: Potential for exploration and use in South America. Neotropical Entomology. 2001;**30**:191-205

[87] Hill DE. Entomopathogenic nematodes as control agents of developmental stages of the black-legged tick. The Journal of Parasitology. 1998;**84**:1124-1127

[88] Johnigk S, Ehlers R. Juvenile development and life cycle of *Heterorhabditis bacteriophora* and *H. indica* (Nematoda: Heterorhabditidae). Nematology. 1999;**1**:251-260

[89] Griffin CT, O'Callaghan KM, Dix I. A self-fertile species of *Steinernema* from Indonesia: Further evidence of convergent evolution amongst entomopathogenic nematodes? Parasitology. 2001;**122**:181-186

[90] Hominick WM, Reid AP, Bohan DA, Briscoe BR. Entomopathogenic nematodes: Biodiversity, geographical distribution and the convention on

biological diversity. Biocontrol Science and Technology. 1996;**6**:317-332

[91] Strauch O, Stoessel S, Ehlers R U. Culture conditions define automictic or amphimictic reproduction in entomopathogenic rhabditid nematodes of the genus *Heterorhabditis*. Fundamental and Applied Nematology. 1994;**17**:575-582

[92] Poinar GO Jr, Karunakar GK, David H. *Heterorhabditis indicus* n. sp. (Rhabditida:Nematoda) from India: Separation of *Heterorhabditis* spp. by infective juveniles. Fundamental and Applied Nematology. 1992;**15**:467-472

[93] Ebssa L, Dix I, Griffin CT. Female presence is required for male sexual maturity in the nematode *Steinernema longicaudum*. Current Biology. 2008;**18**:R997-R998. DOI: 10.1016/j.cub.2008.09.032

[94] Ebssa L, Borgemeister C, Poehling HM. Effectiveness of different species/strains of entomopathogenic nematodes for control of western flower thrips (*Frankliniella occidentalis*) at various concentrations, host densities and temperatures. Biological Control. 2004;**29**:145-154. DOI: 10.1016/S1049-9644(03)00132-4

[95] Bastidas B, Portillo E, San-Blas E. Size does matter: The life cycle of *Steinernema* spp. in micro-insect hosts. Journal of Invertebrate Pathology. 2014;**121**:46-55. DOI: 10.1016/j.jip.2014.06.010

[96] Sanda NB, Sunusi M. Fundamentalsof biological control of pests. International Journal of Chemical andBiochemical Sciences. 2014;**1**(6):1-11

[97] Brodeur J, Cory J, Harwood JD, Hoffmann JH, Jacobsen B, Lewis EE, et al. Biological Control Editorial Board. Elsevier B.V: Elsevier Journal website; 2013

[98] Eilenberg J, Hajek A, Lomer C. Suggestions for unifying the terminology in biological control. Biological Control. 2001;**46**:387-400

[99] Pickett CH, Simmons GS, Lozano E, Goolsby JA. Augmentative biological control of water flies using transplants. Biological Control. 2004;**49**:665-688

[100] Van Lenteren JC. Success in Biological Control of Arthropods by Augmentation of Natural Enemies. Binnenhaven 7, 6709 PD Wageningen, The Netherlands: Wageningen University; 2000 Chapter 3

[101] Chitwood BG, Chitwood MB. An Introduction to Nematology. Baltimore, Maryland: Monumental Printing Co.; 1937

[102] Poinar GO Jr. Description and biology of a new insect parasitic rhabitoid, *Heterorhabditis bacteriophora* n. gen. n. sp. (Rhabditida; Heterorhabditidae n. family). Nematologica. 1976;**21**:463-470

[103] Rahn O. New principles for the classification of bacteria. Zentralblatt für Bakteriologie, Parasitenkunde, Infektionskrankheiten und Hygiene, Abteilung. 1937;**II**(96):273-286

[104] Thomas GM, Poinar GO Jr. *Xenorhabdus* gen. Nov., a genus of entomopathogenic, nematophilic bacteria of the family Enterobacteriaceae. International Journal of Systematic and Evolutionary Microbiology. 1979;**29**:352-360

[105] Boemare NE, Akhurst RJ, Mourant RG. DNA relatedness between *Xenorhabdus* spp. (Enterobacteriaceae), symbiotic bacteria of entomopathogenic nematodes, and a proposal to transfer *Xenorhabdus luminescens* to a new genus, Photorhabdus gen. Nov. International

Journal of Systematic and Evolutionary Microbiology. 1993;**43**:249-255

[106] Bhat AH, Chaubey AK, Askary TH. Global distribution of entomopathogenic nematodes, *Steinernema* and *Heterorhabditis*. Egyptian Journal of Biological Pest Control. 2020;**30**:31

[107] Kaya HK, Gaugler R. Entomopathogenic nematodes. Annual Review of Entomology. 1993;**38**:181-206

[108] Smart GC. Entomopathogenic nematodes for the biological control of insects. Journal of Nematology. 1995;**27**:529

[109] Pye AE, Burman M. Pathogenicity of the nematode *Neoaplectana corpocapsae* (Rhabditida, Steinernematidae) and certain microorganisms towards the large pine weevil, *Hylobius abietis* (Coleoptera, Curculionidae). Annual Review of Entomology. 1977;**43**:115-119

[110] Dillon A. Biological Control of the Large Pine Weevil, *Hylobius Abietis* L., (Coleoptera: Curculionidae) Using Entomopathogenic Nematodes. Dissertation. Maynooth, Ireland: National University of Ireland–Maynooth; 2003

[111] Dillon AB, Downes MJ, Ward D, Griffin CT. Optimizing application of entomopathogenic nematodes to manage large pine weevil, *Hylobius abietis* L. (Coleoptera:Curculionidae) populations developing in pine stumps, *Pinus sylvestris*. 2007;**40**(2)

[112] Abate BA, Wingfield MJ, Slippers B, Hurley BP. Commercialisation of entomopathogenic nematodes: Should import regulations be revised? Biocontrol Science and Technology. 2017;**27**(2):149-168

[113] Lacey LA, Georgis R. Entomopathogenic nematodes for control of insect pests above and below ground with comments on commercial production. Journal of Nematology. 2012;**44**:218

[114] Grewal PS. In: Gaugler R, editor. Formulation and Application Technology, Entomopathogenic Nematology. Wallingford, UK: CABI Publishing; 2002. pp. 265-288

[115] Yukawa T, Pitt JM. Nematode storageand transport. In: Zhu H, Grewal PS, Reding ME, editors. (2011), WIPO Patent No. WO 85/03412Development of a Desiccated Cadaver Delivery System to Apply Entomopathogenic Nematodes for Control of Soil Pests. Vol. 27. USA: Applied Engineering in Agriculture; 1985. pp. 317-324

[116] Bedding RA. Storing third stage infective nematode juveniles by mixing with clay, placing between layers of clay or contacting with adsorbent. Int Patent WO 88:08668. 1988

[117] Raja RK, Hazir C, Gümüs A, Asan C, Karagöoz M, Hazir S. Efficacy of the entomopathogenic nematode *Heterorhabditis bacteriophora* using different application methods in the presence or absence of a natural enemy. Turkish Journal of Agriculture and Forestry. 2015;**39**:277-285

[118] Shapiro DI, McCoy CW, Fares A, Obreza T, Dou H. Effects of soil type on virulence and persistence of entomopathogenic nematodes in relation to control of *Diaprepesab breviatus*. Environmental Entomology. 2000;**29**:1083-1087

[119] Grewal PS. Enhanced ambient storage stability of an entomopathogenic nematode through anhydrobiosis. Pest Management Science. 2000;**56**:401-406

[120] Lewis EE, Campbell J, Griffin C, Kaya H, Peters A. Behavioral ecology of entomopathogenic nematodes. Biological Control. 2006;**38**:66-79

[121] Susurluk A, Ehlers R-U. Field persistence of the Entomopathogenic Nematode *Heterorhabditis bacteriophora* in different crops. Biological Control. 2008;**53**:627-641

[122] Koppenhöfer AM, Kostromytska OS, McGraw BA, Ebssa L. Entomopathogenic nematodes in turfgrass: Ecology and management of important insect pests in North America. In: Campos-Herrera R, editor. Nematode Pathogenesis of Insect and Other Pests. Cham: Springer International Publishing; 2015. pp. 309-327. DOI: 10.1007/978-3-319-18266-7_12

[123] Hussaini SS, Ansari MA, Ahmad W, Subbotin SA. Identification of some Indian populations of *Steinernema* species (Nematoda) by RFLP analysis of ITS region of rDNA. International Journal of Nematology. 2001;**11**:73-76

[124] Hussaini SS, Kavitha J, Satya and Hussain MA. Survival and pathogenicity of indigenous entomopathogenic nematodes in different UV protectants. Indian Journal of Plant Protection. 2003;**31**:12-18

[125] Gupta P. Entomopathogenic nematodes–work done at Allahabad agriculture institute, Allahabad. In: Hussaini SS, Rabindra RJ, Nagesh M, editors. Current Status of Research on Entomopathogenic Nematodes in India. Bangalore: Project Directorate of Biological Control; 2003. pp. 161-166

[126] Vyas RV, Pharindera Y, Ghelani YH, Chaudhary RK, Patel NB, Patel DJ. In vitro mass production of native *Steinernema* sp. Annals of Plant Protection Sciences. 2001;**9**:77-78

[127] Ganguly S, Anupama KA, Parmar BS. Nemagel - a formulation of the entomopathogenic nematode *Steinernema thermophilum* mitigating the shelf-life constraint of the tropics. Nematologia Mediterranea. 2008;**36**:125-130

[128] Divya K, Sankar M. Entomopathogenic nematodes in pest management. Indian Journal of Science & Technology. 2009;**2**(7):0974-6846

[129] Gaugler R, Bednarek A, Campbell JF. Ultraviolet inactivation of Heterorhabditids and steinernematids. Journal of Invertebrate Pathology. 1992;**59**:155-160

[130] Ishibashi N. Integrated control of insects pest by *Steinernema carpocapsae*. In: Bedding R, Akhurst R, Kaya HK, editors. Nematodes and Biological Control of Insects. East Melbourne, Australia: CSIRO; 1993. pp. 105-113

[131] Koppenhöfer AM, Grewal PS, Kaya HK. Synergism of imidacloprid and entomopathogenic nematodes against white grub: The mechanism. Entomologia Experimentalis et Applicata. 2000;**94**:283-294

[132] Nishimatsu T, Jackson JJ. Interaction of insecticides, entomopathogenic nematodes, and larvae of the western corn rootworm (Coleoptera: Chrysomelidae). Journal of Economic Entomology. 1998;**91**:410-418

[133] Koppenhofer AN, Cowles RS, Cowles EA, Fuzy EM, Baumgartner L. Comparison of neonicotinoid insecticides as synergists for entomopathogenic nematodes. Biological Control. 2002;**24**:90-97

[134] Koppenhöfer AM, Kaya HK. Additive and synergistic interaction between entomopathogenic nematodes

and *Bacillus thuringiensis* for scarab grub control. Biological Control. 1997;**8**:131-137

[135] Grewal PS, Martin WR, Miller RW, Lewis EE. Suppression of plant- parasitic nematode populations in turf grass by application of entomopathogenic nematodes. Biocontrol Science and Technology. 1997;7:393-399

[136] Askary T, Ahmad MJ. Entomopathogenic nematodes: Mass Production, Formulation and Application. In: Mahfouz MM, Elgawad A, Askary T H, Coupland J, editors. Biocontrol Agents: Entomopathogenic and Slug Parasitic Nematodes CAB International, Wallingford, UK; 2017. pp. 261-286

[137] Singh SP. Isolation of an entomophilic nematode from potato cutworms. Current Science. 1977;**46**:454-455

[138] Bhaskaran RKM, Sivakumar CV, Venugopal MS. Biocontrol potential of entomopathogenic nematodes in controlling red hairy caterpillar (*Amasacta albistriga*) (Lepidoptera:Archtiidae) a groundnut (*Arachis Hypogea*). The Indian Journal of Agricultural Sciences. 1994;**64**:655-657

[139] Shapiro-Ilan D, Arthurs SP, Lacey LA. Microbial control of arthropod pests of orchards in temperate climates. In: Lacey LA, editor. Microbial Control of Insect and Mite Pests. Amsterdam: Elsevier; 2017. pp. 253-267. DOI: 10.1016/B978-0-12-803527-6.00017-2

[140] Shapiro-Ilan DI, Hiltpold I, Lewis EE. Ecology of invertebrate pathogens: Nematodes. In: Hajek AE, Shapiro-Ilan DI, editors. Ecology of Invertebrate Diseases. Hoboken, NJ: John Wiley & Sons, Ltd.; 2018. pp. 415-440. DOI: 10.1002/9781119256106.ch11

[141] Portillo-Aguilar C, Villani MG, Tauber MJ, Tauber CA, Nyrop JP. Entomopathogenic nematode (Rhabditida: Heterorhabditidae and Steinernematidae) response to soil texture and bulk density. Environmental Entomology. 1999;**28**:1021-1035. DOI: 10.1093/ee/28.6.1021

Chapter 10

Technological Development in the Use of *Allium sativum* Aqueous Extracts in the Agricultural Field

Víctor Jesús Albores Flores, Julieta Grajales Conesa, Leopoldo Cruz López, José Alfonso López García and Eduardo Lozano Guzmán

Abstract

The advance in agricultural technology could increase their commercialization, being the agronomic management for each crop an alternative. The management of natural products is a relevant and responsible need, in order to improve the quality and production of food, and to protect the agro-ecosystem biodiversity. Therefore, the aim of this chapter is to present our five-year study advances in mango and rambutan agronomic management with aqueous extract of *Allium sativum* and the use of natural adherent such as *Melipona solani* honey that improves the function of the components in the biological processes of the crop. Our results showed that this aqueous extract promotes the emission of vegetative and floral shoots, increases flower development, works as an attractant for pollinators, promotes fruit set, stimulates fruit growth, acts as an insecticide to control thrips and mealybugs and stimulates the production of defense metabolites, such as polyphenol compounds. The use of stingless bee honey as an adherent and the aqueous extract of *A. sativum* could be a key to potentiate the function of its components in leaves, panicles, flowers and fruits.

Keywords: aqueous extract, flowering, growth, pest control, metabolites

1. Introduction

Agronomic management in agricultural maintenance and production varies depending on the type of crop, where improvements or the inclusion of new technologies are already an important need. Development is constant and from twenty years ago to our time, each of the areas of agricultural production has been automating. These constant improvements not only refer to machinery such as a seeder, a sprinkler system, a fruit sorting machine, greenhouse production, among others, it also includes new biological processes and products, such as bio-fertilizers, bioles, leachates, antimicrobial strains among others, that improve the properties of the crops and enrich physicochemically the tissues or increase the generation of fruits; which also given an extra plus to each agricultural product [1, 2].

All the new agricultural technologies with unprecedented innovations push the process of globalization and the integration of commercial blocs, generating new income in all trade areas: national and international. These natural and organic products have generated a continuous offer of renewed, innovative products with new expectations of use or worldwide application. These changes have modified the traditional ways of carrying out agricultural activity, in its production, transformation and commercialization phases [1, 3, 4].

Technological development, as a basis for improving and safeguarding the integrity of field products, is emerging as an alternative with potential use or application to increase production, jointly reducing costs and environmental impact [5–7]. Thus, the use of vegetable extracts from plants is regaining importance, in addition of already demonstrated its functionality for presenting a variety of proven properties in the medical, veterinary, food and cosmetic areas, for being of a biological, vegetable and non-vegetable nature to be generated synthetically, and not generate toxicity to plants, animals and humans [7–9].

1.1 Importance of aqueous extracts of *Allium sativum*

Medicinal plants are a generating source of plant extracts with different and important uses, due to their bioactive components. There is a variation in its concentration and in the type and variety of compounds contained in each elaborated extract. The *A. sativum* aqueous extract contains mainly sugars, nitrogenous mineral substances and essential oils, in which there are bioactive sulfur substances such as allicin and other allyl sulfides responsible for their chemical qualities [10].

The properties discovered in *A. sativum* are attributed to its components, such as amino acids, minerals, vitamins, pantothenic acid, folic acid, niacin, among other compounds that present specific activities and that have provided much interest in the scientific areas of the medical, nutritional, cosmetic, agricultural areas, etc. [11]. The properties that *A. sativum* are used depending on the extraction process, generally considering two types of solvents, polar and non-polar, considering water separately [12].

The components extracted by polar solvents differ from those obtained by non-polar solvents, including whether the process (heat or not). The variation obtained between these types of extract is presented in the quantities obtained, the color of the extract, odor and the activity it exerts on the applied biological models [12–14].

In alcoholic extractions or with non-polar solvents, it is essential to eliminate these solvents and obtain only the dissolved compounds in them, leaving a liquid with a higher viscosity, dark in color, called crude extract [15]. This type of procedure is common in the industrial area, for the generation of commercial products, where one of the disadvantages is that certain molecules are susceptible to high temperatures, drastic temperature variations, the presence of sunlight and a lack of water in the environment [16, 17]. Therefore, this is a disadvantage when used in healing processes in the presence of light, environments above 25°C and excess air currents, considering both animal skin, human beings and solutions and the pH value presented for each case.

Aqueous extracts had less importance due to the lack of use and information that exposes the properties that an aqueous extraction has and the properties, such as benefits, both in medicinal use and in other areas [12, 18, 19]. In addition, all the components that are reported with alcoholic extractions are normally polar and can be extracted with water by the infusion method, without losing their activity, because these molecules support within their chemical limits the heat being the maximum temperature, that of boiling, being called thermoresistant [11, 18, 20]. The importance at the agricultural level of the use of plant extracts is reconsidered and is increasing, due to the risk that agrochemical mean to agroecosystem biodiversity and in particular to pollinators, which are toxic also for human being [20–22].

The purpose of this chapter is to show the results of new technological development during five study years with *A. sativum* aqueous extracts and stingless bee honey as an adherent in two of the most economically important crops in the Soconusco region in Mexico: rambutan (*Nephelium lappaceum*) and mango Ataulfo (*Mangifera indica*).

2. Applications of the aqueous extracts of *A. sativum* in agricultural crops

Recent studies showed that aqueous extracts have a variety of uses in the agronomic management of an agricultural crop, due to their physicochemical properties. These extracts can be used in recent preparations or after having undergone a natural fermentation process. Its potential for use is also increased by being mixed with other synthetic, natural or microbial products.

In particular for fruit trees, specifically in mango cultivation and rambutan, this has become important for presenting biological action in these crops as: a) Floral inducer, b) Increase in fruit setting, c) Fruit growth and d) Pesticide.

The applications indicated in this chapter have been under research development at the field level in commercial farms in areas with two dominant annual seasons, hot and rainy. It was also accompanied by the evaluation of total phenolic compounds, number of panicles, number of flowers formed, number of fruits per panicle, growth dynamics of the fruits and the number of pest insects per panicle.

Location of commercial properties. The geographical locations of the properties provided are placed, with an area of 1 ha as a minimum and 2.5 ha as a maximum, according to the planting distribution of the trees.

In mango it was the "San Juan" orchard in the ejido de viva México on the coastal highway Km 227.5 (Lat. 14°54′25.0"N, Lon. 92°17′43.0"W) in Tapachula, Chiapas, Mexico.

In rambutan it was in the "Toluquita" Canton in the Municipality of Tapachula, Chiapas, Mexico (14° 58′13.001′′ N, 92° 14′2′′ W).

Preparation of the aqueous extract. The elaboration of *A. sativum* aqueous extract was carried out according to what was reported by Bustamante [23], considering a ratio of 1 kilograms of garlic in four liters of water (1: 5), and after extraction with hot water, the rest time of 24 hours at room temperature was adjusted, prior to its application in the manual spray pumps used.

Before applying any solution, the pH value was measured with a portable manual potentiometer for the field, of the undiluted extract and after making the application mixture.

Adherent used. In order to improve the effect of the aqueous extract, a test was carried out with two types of adherent: a synthetic (Inex A ®) and a natural, which was *Melipona solani* honey. The first adherent was used at a concentration of 1.5 mL per 20 liters of water and the second one was 15 mL per 15 liters of water, both separately. A treatment without adherent was used in each fruit crop.

Evaluation of variables in the field. The number of floral buds, number of flowers, number of fruits, fruit size (with vernier), total polyphenolic compounds and presence of pest insects (in the field with a rambután magnifying glass and by

Concentrations	Action	Crop	Control *
1.25% v/v	Floral inducer	Mango	Potassium nitrate
2.5% v/v	Floral inducer	Rambután	Calcinit
2.5% v/v	Fruit set	Mango	Comercial amino acids
2.5% v/v	Fruit set	Rambután	FusióN-H
1.25% v/v	Fruit growth	Mango	SpeleR-K + FusióN-H.
1.25% v/v	Fruit growth	Rambután	Fusión H
10% v/v	Pesticide	Mango	Malathión
10% v/v	Pesticide	Rambután	Cypermethrin

**information provided by each producer. The applications were made from 7:00 to 11:00 am, evaluating the pH of the water in each property.*

Table 1.
Concentrations of aqueous extract of A. sativum for each agricultural action in the agronomic management of the two crops studied.

the table technique dark for thrips), it was carried out by dividing the orchard of each tree into four zones, selecting three panicles per zone, according to Gonzales and Quiñones [24]. The evaluations were carried out every week, during the time that the study lasted and the permanence of production of the fruits until their harvest.

Efficient concentrations used. According to the type of action expected by the *A. sativum* aqueous extract, the minimum concentrations to achieve the desired objective were the following (**Table 1**):

Phenols content evaluation. To obtain phenols, the sample of the main branch of the panicles was rested in a methanol-water solution (1:1). The total phenol content was determined by using the Garrido [25] technique. The Folin-Ciocalteu reagent diluted 1:10 with water and 0.7 M sodium carbonate was used. It was kept in total darkness for 15 min and was read in a spectrophotometer at 765 nm. Methanol-water (1:1) was used as blank. A calibration curve was made with gallic acid at different concentrations in methanol-water solution: 0, 10, 50, 100, 150 and 250 mg L-1. Reported in mg EGA/gr.

Catching bees. The collections of the bees were carried out with aerial entomological nets by beating [26], the collection was carried out in the orchard zone of the mango tree and the collections were made from one day before the application and until six days after applying the treatments. The captured bees were mounted on pins, in an entomological box. Taxonomic identification was carried out in the ECOSUR Bee Team, San Cristóbal de las Casas unit, Chiapas.

Analysis of results. All the variables evaluated were analyzed with analysis of variance and a comparison of means was made by Tukey (0.05), accompanied by a Pearson correlation, in the Infostat program, 2019.

3. Results and discussions

3.1 Flowering stage

The application of the *A. sativum* aqueous extract with or without adherent promoted the induction of vegetative shoots and transformation to floral shoots in rambutan culture and emission of floral shoots in mango, one week before the application of Calcinit in rambutan and potassium nitrate in mango (**Figures 1–3**).

Using *A. sativum* aqueous extract accompanied by Inex A promoted a value of 29.8% of vegetative shoots transformed into floral shoots and this value was lower than that obtained when honey was used as an adherent (44.15% of vegetative shoots transformed into floral) or not used any type of adherent (41.6% transformation to floral bud). These last values were close to what was observed with Calcinit (44.2%). A maximum transformation value from vegetative shoots to floral shoots was observed in rambután when Inex A and honey were used, two weeks before the calcinit treatment.

In the mango crop, the emission of floral buds with *A. sativum* aqueous extract compared with the use of potassium nitrate, it was 58.77% lower when no type of adherent was used, 16.32% lower when using Inex A as adherent and 2.44% less when using honey. In this culture when the aqueous extract was used, the maximum emission of floral buds was observed one week earlier than that presented in the treatment with potassium nitrate.

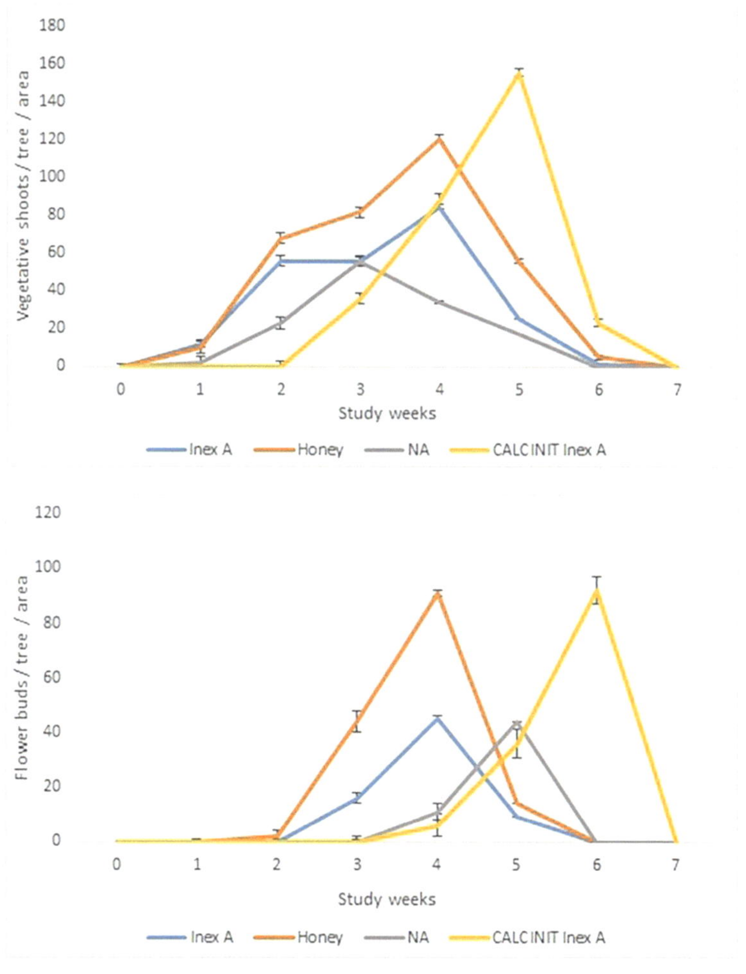

Figure 1.
Dynamics of vegetative shoots (A) and floral shoots (B) that were emitted in the rambután crop (NA: Without adherent).

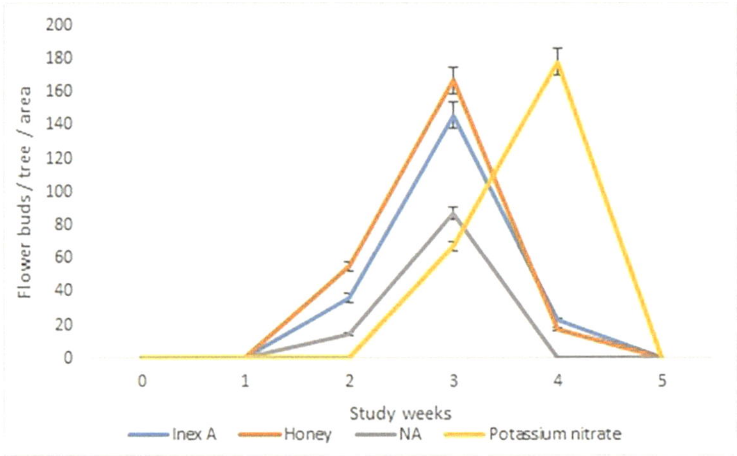

Figure 2.
Dynamics of flower buds emitted in the mango crop (NA: Without adherent).

Figure 3.
Images of vegetative to floral shoots in rambután and floral shoots of mango (A: Vegetative shoot in rambután, B: Vegetative shoot transformed into floral shoot in rambután, C: Shoot in mango, D: Growing mango panicle).

Statistically, there were significant differences between the treatments and between the study weeks, in both crops ($p < 0.05$).

These results suggest that when the *A. sativum* extract gets in contact with branches and leaves of both crops, acted as flower bud elicitors [8, 27]. Nonetheless, in mango crop this effect was rapid, unlike in the rambutan crop, where the generation of vegetative shoots was first stimulated before differentiation into floral shoots. The breaking of the abiotic stress in both crops was stimulated by the water factor which synchronized the vegetative and floral induction [28] and the biotic stress was influenced by the components of the plant extract, which were in sufficient quantities

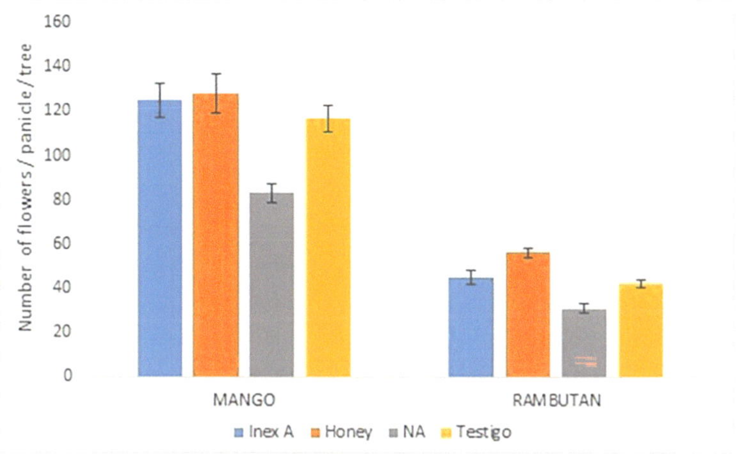

Figure 4.
Number of flowers formed in the mango and rambután culture (NA: Without adherent, control: in the mango culture it was potassium nitrate and in rambután it was Calcinit).

to reach and act as regulators of the metabolism and plant physiology of each tree after coming into contact with their tissue [29–31]. This differs from the nutritional imbalance induced by the agrochemicals used by producers at the foliar level, since they generated accumulation of nitrates in the leaves [28].

The results obtained in mango and rambutan when using *A. sativum* aqueous extract were a joint effect between the aqueous extract and the type of adherent incorporated into the applied solution. Both honey and Inex A acted as adjuvants and favored the penetration of the components of the *A. sativum* aqueous extract, by increasing cuticular permeability in the leaves [32]. When no type of adherent was used, the penetration of the extract components was less efficient, because the increase in permeability in the cuticle that protects the leaves was not facilitated [33], observing less shoot emission vegetative and floral.

After the stage of floral buds, the development of the flowers in the panicles of both mango and rambután is shown in **Figure 4**. A minimum of flowers was observed in the aqueous extract where no type of adherent was included, in both crops. Despite the fact that the values of the number of flowers formed with *Allium sativum* aqueous extracts in both crops were similar to those obtained with the chemical compound used to promote flowering, the percentage differences indicate that 6.8% and 9.4% more flowers can be generated of rambután and in mango of 7.14% and 33.33%, when including Inex A and honey to the solution, respectively. This is not related to the fact that there is more frequency of fruit set and it is necessary to study the fertility capacity and strength of the ovaries formed by the female flower in the future.

During flower formation, a variety of metabolic events occur that involve a varied production of secondary metabolites involved in the generation of floral hormones, osmotic pressure and ovary biogenesis [34–36]. The *A. sativum* extract stimulated the generation and formation of the rambután and mango flower, in the panicles produced by the components that were dissolved in the aqueous part, and according to Yakin [35] and Venegas-Gonzáles [37], these compounds provide energy, organic molecules that function as metabolic cofactors and elicitors, for the generation and maturation of the flower. Our results agree with the report by Ariza-Flores [36], who indicates that exogenous molecules act as biostimulants, where their rapid or delayed

Species	*Allium sativum* + Honey	*A. sativum*	A. sativum + Inex A	Potassium nitrate
Apis mellifera	8	8	1	4
Oxytrigona mediofura	0	0	0	1
Tetragonisca angustula	0	1	0	0
Trigona fulviventis	6	4	2	1
Trigona fuscipenis	8	2	3	1
Trigona nigerrimia	5	4	3	3

Table 2.
Average number of bee species found in the floral panicles of the mango crop, for each treatment of A. sativum extract (Honey: M. solani honey).

action depends on the concentration applied and absorbed in plant tissue. In this last aspect, the substance included in the final solution, before applying, plays an important role on the efficiency of the aqueous plant extract.

The application of the garlic extract in the mango crop, with Inex A, honey and without adherent, presented on average a higher number of pollinating bees compared to the number found in the treatment that only received potassium nitrate (**Table 2**). The commercial species *Apis mellifera* was the one that had the greatest presence in the mango flowers, observing a double value where honey was used and without adherent, compared to the chemical treatment. The aqueous extract added with Inex A, presented the lowest number of this bee species. The difference between treatments was highly significant ($p < 0.0001$).

The other bee species belong to the Meliponine group and are considered floral visitors, and are recognized as part of the ecosystem services that occur within agroecosystems with reduced environmental damage [38, 39]. The other species were also found in greater numbers in the extracts of *Allium sativum* with and without adherent, compared to the control treatment, standing out the extract of *A. sativum* with honey.

It has been reported that sulfide derivatives such as diethyl sulfide and propyl disulfide, including terpene derivatives, have an attractive action on bees [40, 41]. These compounds are similar to what is contained in the extract of *A. sativum*, according to what was reported by Yakin [35] and Duran [10]. The inclusion of honey in the plant extract strongly favored the attraction of bees or other pollinators, as reported by studies carried out by Kumari and Rana [38], Wankhede [42] and Pashte [43].

3.2 Fruit set

After flowering, the formation of fruits occurred and two stages were considered, one of them being before abscission and the second is after abscission.

In the first stage, in both crops a lower number of fruits was observed than in the treatments applied with a chemical product and in decreasing order according to the adherent included or without it (**Figures 5** and **6**). The percentages of differences calculated between the extract and the chemical product used, according to the binder included or without it, for each culture were: 6.25%, 12.5% and 35.4% in Inex A, honey and without binder, respectively in rambután and 3.4%., 18.9% and 37.9% of Inex A, honey and without adherent, respectively, in mango cultivation. The differences between the treatments were significant ($p < 0.05$).

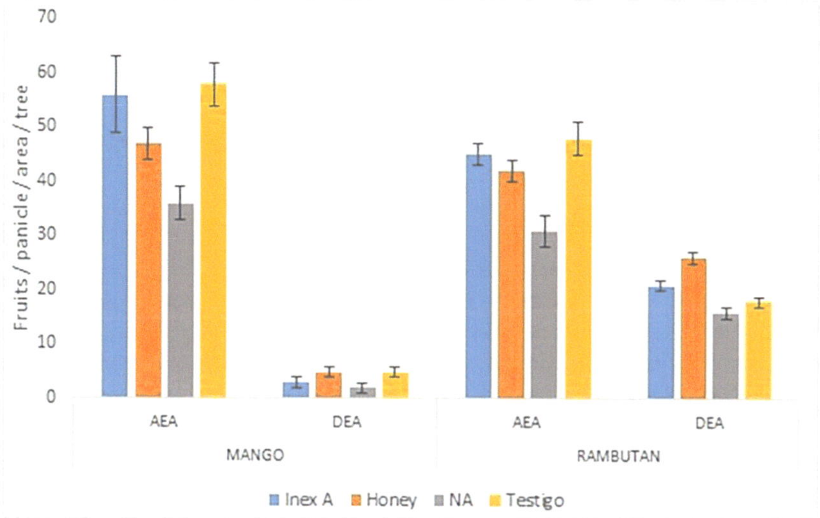

Figure 5.
Number of tied fruits in the mango and rambután crops (NA: Without adherent, control: in the mango crop it was with amino acids and in rambután it was Fusio N-H, AEA: Before the fruit abscission stage, DEA: After the stage of abscission of fruits).

Figure 6.
Fruit set before the abscission phase (A: Panicle with rambután fruits, B: Panicle with mango fruits).

In the second stage, after fruit abscission, it was observed that in the mango crop the number of fruits with definitive mooring was lower than that found in rambutan trees, regardless of the treatment. The percentage of definitive tied fruits per crop, calculated, were: 46.66%, 61.9% and 51.6% when Inex A, honey and no adherent were added, respectively, in the rambutan crop, and in the mango crop they were: 5.35%., 10.63% and 5.55% when Inex A, Honey and without adherent are added, respectively.

Only in the mango crop, when adding honey as an adherent, 2% more filled and harvested fruits were obtained, when compared with the control treatment (8.6%), where amino acids were applied. In the rambutan crop, a minimum of 9% (with Inex A) and a maximum of 24% (with honey) were obtained, more than tied fruits compared to the addition of the Fusio N-H product, in which the percentage of fruits produced was 37.5%.

In the fruit set, Ramírez-Luna [44] indicated that the components that stimulate the fertilization of the ovaries, the elongation of the pollen tube, the cell division and elongation that will generate the fruit are pantothenic acid, folic acid, macro- and micro-nutrients and amino acids. The *A. sativum* aqueous extract, with or without type of adherent, favored fruit set where possibly the components extracted in the elaboration process and those that penetrated the cuticle of the leaves of the trees were sufficient to stimulate metabolic activity and strengthen the process. The foregoing is sustained according to the report by Duran [10] and Ramírez-Concepción [9] who report the components that *A. sativum* has and those found in an aqueous extract, which are sugars, nitrogenous mineral substances, sulfur amino acids, vitamins, folic acid, pantothenic acid, niacin, polyphenols with antioxidant and antimicrobial activities, among others.

The abscission stage in fruit crops is a strategy that trees present to determine which fruits they can sustain, according to the nutritional capacity they present and the production of photoassimilates during the entire process of photosynthesis that supports the constant fruit growth [45, 46]. At this stage, the *A. sativum* extract received in the leaves and the type of adherent included in the mixture made it possible to amortize the effect of the abortion stage, where the components of the extract, received from the leaves, provided nutrients and inducing substances, among others, and including the nutritional capacity that the trees of each crop received before the anthesis stage, played a preponderant role in sustaining the growth of the fruits.

3.3 Fruit growth dynamics

The growth of the mango and rambutan fruits during a period of 16 weeks, without any growth stimulant, presented dynamics with lower growth and the size of the fruits was smaller than those obtained with the garlic extract, which received Inex A, honey and without adherent (**Figure** 7).

The application of SpeleR-K + Fusió N-H in mango and Fusió N-H in rambutan generated growth induction through cell division and fruit elongation, specifically in the third (21 days) and ninth week (63 days). These weeks are a key in the growth of both fruits, as they denote physicochemical changes and high activity in cell division that influence the elongation of the fruits. According to Caballero-Pérez [47], the first 68 days are a key for the longitudinal growth of rambutan fruits and Pérez and Barraza [48], indicated that the first 70 days of fruit growth after anthesis are key to obtaining good fruit size, and for this reason growth stimulation is recommended on these days, with inducing products that improve metabolism and promote cell division and elongation.

After each application of aqueous extract of *A sativum* in week 3 and 9, the effect was observed the following week, inducing elongation of the mango and rambután fruits. This response was similar to chemical treatment. From week 5, in both fruit trees, it was observed that the *Allium sativum* aqueous extract that did not include any type of adherent to the growth dynamics of the fruit presented lower values. Something similar was observed when Inex A was included, in the aqueous extract, in rambutan fruits. In mango fruits, a reduction in growth dynamics was observed two weeks after the second application. Only when honey was included in the aqueous extract solution, the growth dynamics was completely similar to the treatment that received the chemical product, reaching fruit sizes that differed by 0.5 cm between them. The difference between treatments was significant ($p < 0.05$).

The dynamics presented in the rambutan fruit, for each extract, allowed to see three growth stages (**Figure** 7) unlike what Caballero-Pérez [47] reports and in the mango fruits the dynamics are similar, for which follows that there are also three stages. It is proposed that the three stages are divided according to the stages reported by Martijn ten Hoopen [49], differentiated in times, where from day 1 to day 28 (week

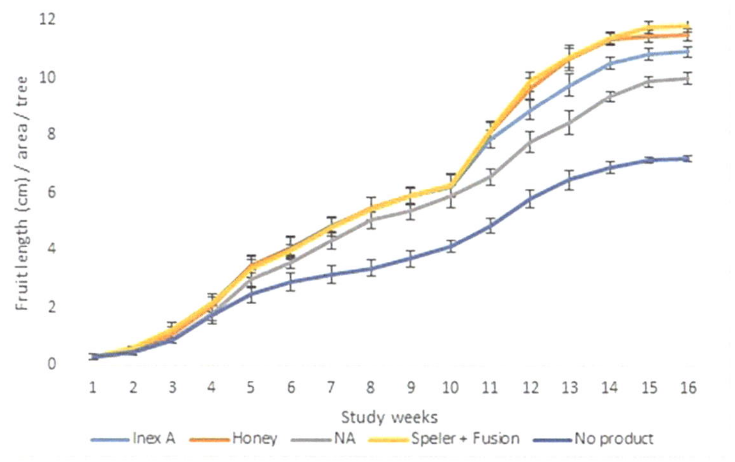

(a)

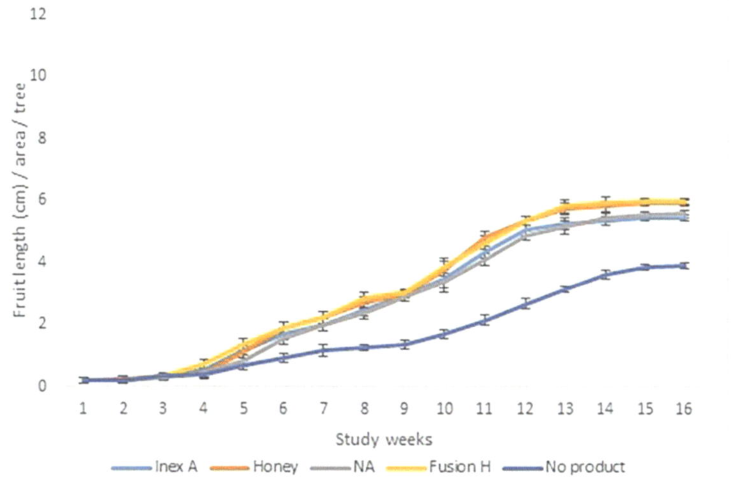

(b)

Figure 7.
Dynamics of fruit growth in the mango (A) and rambután (B) crop (NA: Without adherent, control: in the mango crop it was with SpeleR-K + Fusio N-H and in rambután it was Fusió N-H, none: non-product was applied).

4), is that of cell division and it is when the embryos are not yet growing, from day 29 to 70 (week 10) is the stage of cell elongation and the development of the embryo to generate the seed occurs, and the last stage from day 71 to day 112 or 120 (week 16 or 17), consists of the ripening of the fruit before harvest.

The components present in the aqueous extract stimulated the development of stage 2, which corresponds to cell elongation, and induce stage 3, which corresponds to cell differentiation and tree maturity. This effect of elongation at two different moments in the observed dynamics is related to the nutritive capacity of the tree, the production of photo-assimilated in the leaves and the transport of these to the fruits, to support their growth [45, 46].

The stimulation provided by the components contained in the *A. sativum* extract (sugars, nitrogenous mineral substances, sulfur amino acids, vitamins, folic acid, pantothenic acid, niacin, polyphenols, among others) [9], which was favored by the adjuvant that allowed these components to pass through the cuticle of the fruits, were inducers of cell elongation and division in stage two. According to Caballero-Pérez [47], in the growth stage that involves division and expansion, there is competition for carbohydrate sources, according to Pérez-Barraza [48], during the cell expansion of the fruit metabolic components are needed to synthesize hormones and compounds that allow the generation of tissue, and Elakbawy [50], indicates that organic compounds from plant sources, exogenous or endogenous to the tissue influence physiological processes that control cell division and the differentiation of the growth process, in which cell elongation is involved, finding the participation of sulfur amino acids, auxins and cytokinins.

3.4 Pest control

After the first application, the thrips population was reduced by 65.78%, 63.52%, 55.42% and 75% with *Allium sativum* extract mixed with Inex A, honey, without

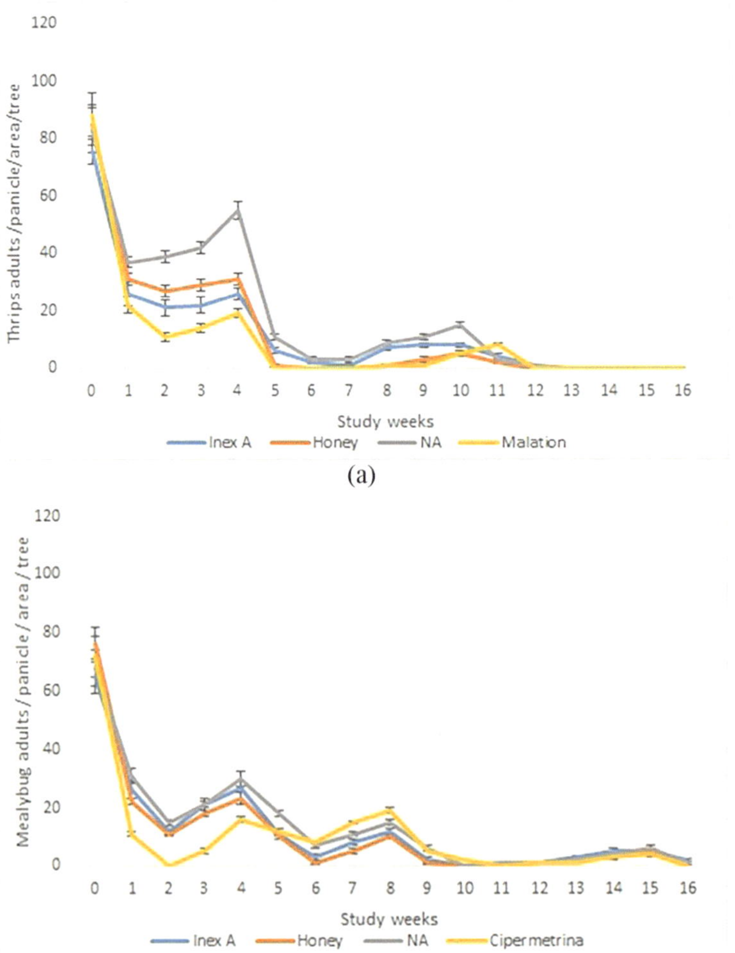

(a)

(b)

Figure 8.
Dynamics of adults of thrips in the mango crop (A) and of mealybugs in rambután (B) (NA: Without adherent, control: in the mango crop it was with malathion and in rambután it was cypermethrin).

adherent and chemical, respectively, in mango cultivation (**Figure 8a**). In rambutan, the reduction of the mealybug population was 60%, 71.05%, 54.41% and 84.7% with *A. sativum* extract including Inex A adherent, honey, without adherent and the chemical product, respectively (**Figure 8b**).

Subsequently, in both crops, the population of pest insects gradually increases in week 3 and 4, regardless of the adherent or without it. After the second application, in week 4, 31.57%, 36.47%, 40.96% and 25.0% more were reduced in the mango crop with the *A. sativum* extract mixed with Inex A, honey, without adherent and the chemical product, and in the cultivation of rambutan, a reduction of 35.38%, 27.63%, 35.29% and 4.16% more was with the *A. sativum* extract that was mixed with Inex A, honey, without adherent and the chemical product used by the producer.

The population dynamics in week 7 and 8 show an increase in the number of individuals of both pest insects. After the third application of the aqueous extract, in the mango crop, no thrips insect was found in the trees, unlike the rambután crop, where the population gradually increases until week 15 and after week 16, when it starts fruit harvest, no mealybugs were found.

The correlation found between the growth of the mango fruit and the number of thrips was $r = -0.65$ between the first 8 weeks, later the value increased, reaching $r = -0.92$, between weeks 9 and 16. The correlation found between the growth of the rambután fruit and the amount of mealybugs was $r = -0.58$ between the first 8 weeks, later the value increased, reaching $r = -0.90$, between weeks 9 to 16. The third application helps the rambután fruit no longer receive any damage from the mealybug larvae, and in mango the thrips larvae and adults are not affected significantly either, allowing a maximum induction of fruit growth. There were significant differences between the treatments ($p < 0.05$).

To control the two pests of both fruit trees, where the life cycle in time is similar and the arrival events for each crop begin a few days before the anthesis event occurs, attracted by volatile secondary metabolites emitted by mango trees and of rambután [51–53], is knowledge that should not be overlooked when designing control strategies. The application of the aqueous extract of *A. sativum* in time and form, for the control of mealybugs and thrips adults significantly reduced the population, with a minimum of 50% and a maximum of 70%, depending strongly on the type of adjuvant or not, similar to what was reported by Flores Villegas [54] and Jaramillo [55], both authors agreeing that the concentration used of plant extracts can reduce the population of pest insects by 50% by increasing the concentration of these or by reducing it, to which the insecticidal or repellent effect strongly depends on the extracted metabolites.

The control of pest insects, thrips and mealybugs, strongly depends on the flowering and fruiting periods of the crops, these stages being dependent on nutritional, physiological and environmental conditions [48, 51, 56]. Having started the applications at the beginning of the flowering stage and making three applications, every 4 weeks, kept the population of thrips low and prevented the number of individuals from increasing significantly, and this effect was maintained by including honey, Inex A in the mixture or without them, observing that in mango cultivation (**Figure 8a**), this effect is reduced when no adherent is applied. The reduction in the number of thrips at an extract concentration of 10% was similar to that obtained by Monje [57] who reported a mortality of adult thrips similar to chemical treatments, who used Spinosad, Imidacloprid and Thiamethoxan, without the use of adherent, being among the most efficient the extract of *A. sativum* and onion. Nava-Pérez [58] and Vázquez-Luna [59] indicated that the metabolites present in the extract of *A. sativum* present repellency, feeding and growth regulation of the larval growth faces of the insect plague. The extract of *A. sativum* has also had repellent effects against a variety of pest insects, where the mealybug is considered one of them [60]. Martinez and Rivera [61] and Marcano and Hasegawa [62] indicated that the lethal dose of the *A. sativum* extract depends on the population density that exists in the crop at the time of implementing its use as part of phytosanitary management, and in the same way the expected effect, where it can be biocidal or repellent.

3.5 Phenol production

The induction of phenols in mango panicles, regardless of adherent or non-adherent, started after the second application, unlike the treatment that received malathion (**Figure 9**). After week 8, when the third application was made, in the trees that received the aqueous extract, the production of phenols increased the concentration of phenolic compounds in the panicle. Only when Inex A was included in the application, after week 10, the increase in the concentration of these compounds decreases.

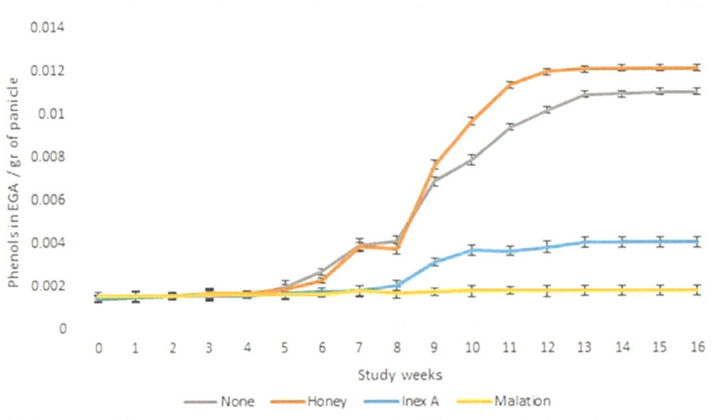

(a)

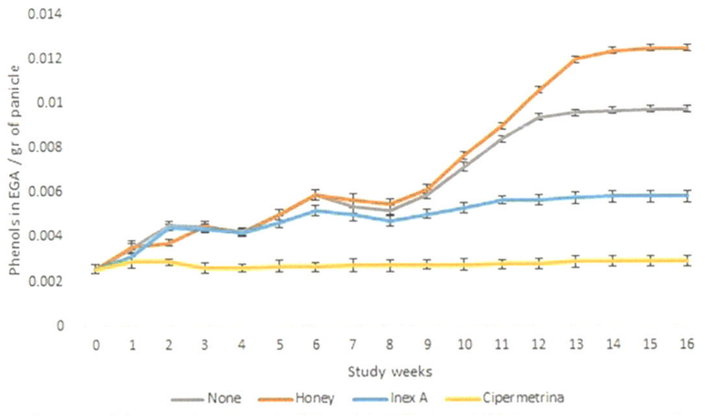

(b)

Figure 9.
Dynamics of phenol production in the main branch of the panicle in the mango crop (A) and in rambután (B) (NA: Without adherent, control: in the mango crop it was with malathion and in rambután it was cypermethrin).

After week 9, the extract that received honey and without adherent increases the concentration of phenols, observing that the dynamic that stands out with the highest production is the one that received honey as adherent. Only the aqueous extract that was mixed with honey reached concentrations higher than 0.011 mg EGA / g of panicle, equivalent to 8.94% more phenolic compounds than the extract without honey, 66.88% more phenolic compounds than the extract mixed with Inex A and 85.34% more phenols than the treatment that received malathion. A Pearson correlation value of r = −0.90 was found from week 4 to week 16, between the population density of thrips and the concentration of phenolic compounds found in the mango panicle.

In the rambutan crop, unlike the mango crop, the induction of phenolic compounds began one week after application with Inex A, honey and no adherent in the *Allium sativum* aqueous extract, reaching a maximum value at the second week and reducing its concentration in the third week. A similar dynamic was presented after the second application, observing that after the third application the dynamics of phenol production was different.

Only when Inex A was included in the application, after week 11, the increase in the concentration of these compounds decreases, similar to that observed in the mango crop after week 9, the extract that received honey and without adherent increases the concentration of phenols, observing that the dynamic that stands out with the highest production is the one that received honey as adherent. The aqueous extract that was mixed with honey reached concentrations higher than 0.0100 mg EGA / g of panicle, equivalent to 22.18% more phenolic compounds than the extract without honey, 53.53% more phenolic compounds than the extract mixed with Inex A and 76.84% more phenols than the treatment that received cypermethrin. A Pearson correlation value of r = −0.98 was found in all study weeks between the mealybug population and the concentration of phenolic compounds obtained from the rambután panicle.

The *A. sativum* aqueous extract contains a wide variety of secondary metabolites, among which are phenolic compounds, sulfur molecules and amino acids, which have been shown to have an eliciting action for secondary metabolism in cells or tissues where they are applied exogenously and penetrate to act endogenously [9, 10, 44]. According to Bailey [63] and Al-Oubaidi [64] the induction of phenolic compounds in biological models is a response to biotic or abiotic stress promoted by an organism or molecules that came into contact with the tissue or the cell of a host or receptive plant. The compounds of the extract of *A. sativum* generated induction in the production of phenolic compounds, which favored the control of the population density of both mealybugs and thrips, in both crops, being more efficient in rambután for observing a rapid response. Compared to that found in the mango crop, Delgado-Oramas [65] exposes something similar, who mentions that eliciting production of secondary metabolites is to avoid pest attack, it is a form of resistance induction, causing the host plant to present a different taste or smell, and on other occasions, they serve as repellents for phytophagous insects.

Based on the results obtained, fruit set, fruit growth and behavior of the pest, we propose that the production of phenolic compounds changes according to the physiological age or function that the organ is performing, which in this case are the panicles of both crops, according to what Albores-Flores [66] and Viveros-Legorreta [67] report, who agree that the concentration of phenols depends on the growth stage of the organ, varying in the types of compounds and their concentrations, depending on the function they perform, among these would be as a cell wall component, for biosynthesis of other molecules or as part of defense mechanisms, during cell division, cell elongation or cell differentiation.

This result will vary according to the crop, because in mango the induction of phenolic compounds was after the second application, unlike the rambutan crop where the phenol induction response was rapid. These results are related to the penetration capacity of the molecules in the cuticle of the tissues of each organ [33], and it is possibly lower in mango, because there was more hydrophobicity with its components, and faster in rambután, where the condition was less hydrophobic due to the types of compounds that constitute its cuticle.

4. Conclusions

The *Allium sativum* aqueous extract presents potential for use in agriculture as it shows a variety of functions in mango and rambutan fruit crops, to replace chemical products, be included in agronomic management and improve their production.

The extract of *A. sativum* with and without adherent can be used for flower induction, pollinator attraction, fruit set, fruit growth inducement, secondary metabolite inducement such as polyphenolic compounds and control pests in the fruiting stage.

The biological action of the aqueous extract of *A. sativum* is further potentiated when a chemical adherent or a natural product is included in the mixture, such as *Melipona solani* honey at a minimum concentration of 0.001%.

The extract of A. sativum can be used as an agricultural technology in the production of fruit trees without affecting the variety of pollinators, such as bees.

Acknowledgements

The authors thank the owners of the orchards, Ms. Maria Luisa Palacios y Palacios, Dr. Roberto Canell Aquino, Lic. Alejandra Lizet Ortiz Orozco and Engineer David Juárez Pichardo, for allowing access to the cultivated areas of their production lands commercial.

Conflict of interest

The authors declare no conflict of interest.

Acronyms and abbreviations

ECOSUR	The College of the Southern Border
NA	without adherent
EGA	gallic acid equivalents
AEA	before the fruit abscission stage
DEA	after the stage of abscission of fruits

Author details

Víctor Jesús Albores Flores[1*], Julieta Grajales Conesa[1], Leopoldo Cruz López[2], José Alfonso López García[1] and Eduardo Lozano Guzmán[3]

1 Institute of Biosciences, Autonomous University of Chiapas, Blvd. Príncipe Akishino S/N col, Tapachula Chiapas, Mexico

2 The College of the Southern Border, Tapachula de Córdova y Ordoñez, Tapachula Chiapas, Mexico

3 Laboratory of Pharmacognosy, Faculty of Chemical Sciences, Juárez University of the State of Durango, Durango, México

*Address all correspondence to: alboresflores@gmail.com

References

[1] Arús P. Agriculture of the future: Science and technology for sustainable agricultural development. Métode science studies journal. 2020;**10**:33-39. DOI: 10.7203/metode.10.12546

[2] Gautam HR, Kumar R. Agricultural development: The road ahead. Kurukshetra a Journal on Rural Development. 2014;**62**(8):3-6

[3] Kumar P. Technologies to boost agricultura production. Kurukshetra a Journal on Rural Development. 2014;**62**(8):16-19

[4] Reddy YS, Roy S, Tiwari SK. New technologies in vegetable production. Kurukshetra a Journal on Rural Development. 2014;**62**(8):20-23

[5] Ferraris G, Couretot L, Toribio M. Pérdidas de nitrógeno por volatilización a partir de dos fuentes nitrogenadas y dos métodos de aplicación. Siembra. 2021;**8**:19-22

[6] Pérez BMH, Osuna GJA, Sánchez LR, Vázquez V. El paclobutrazol como promotor de la floración en mango 'manila', aun sin condiciones ambientales inductivas. INIFAP, Nayarit. México, Nogales. Revista Chapingo Serie Horticultura. 2011;**17**(Especial 1):47-52

[7] Pérez G LL. Inducción de la floración en fresa (*Fragaria x ananassa*) variedad Albión, mediante la aplicación de extractos de Sauce (*Salix humboldtiana*) y agua de coco (*Cocos nucifera* L.). Universidad técnica de Ambato; 2018. p. 67. Available from: https://repositorio.uta.edu.ec/bitstream/123456789/28651/1/Tesis-212%20%20Ingenier%c3%ada%20Agron%c3%b3mica%20-CD%20605.pdf. Repository: https://repositorio.uta.edu.ec/jspui/handle/123456789/28651

[8] Sisa M. Evaluación de extractos vegetales como alternativa ecológica para accionar el enraizamiento de estacas de rosas (Rosa spp). Tesis de Ingeniería Agronómica. Universidad Técnica de Ambato; 2017. p. 87. Available from: https://repositorio.uta.edu.ec/bitstream/123456789/26376/1/Tesis-172%20%20Ingenier%c3%ada%20Agron%c3%b3mica%20-CD%20518.pdf. Retrieved From. http://repositorio.uta.edu.ec/jspui/handle/123456789/26376

[9] Ramírez-Concepción HR, Castro-Velasco LN, Martínez-Santiago E. Efectos Terapéuticos del Ajo (*Allium Sativum*). Salud y Administración. 2016;**3**(8):34-47

[10] Durán MA, González P, PA, Cardona PL. Obtención y caracterización de la oleorresina del ajo (*Allium sativum*). Scientia et Technica. 2007;**37**:551-555

[11] Loria Gutiérrez A, Blanco Barrantes J, Porras Navarro M, Ortega Monge MC, Cerda Vargas MJ, Madrigal Redondo GL. General aspects about *Allium sativum* – A review. Ars Pharmaceutica. 2021;**62**(4):471-481

[12] Juaréz- Segovia KG, Díaz-Darcía EJ, Méndez-López MD, Pina-Canseco MS, Pérez-Santiago AD, y Sánchez-Medina MA. Efecto de extractos crudos de ajo (*Allium sativum*) sobre el desarrollo in vitro de *Aspergillus parasiticus* y *Aspergillus niger*. Polibotanica. 2019;**47**:99-111

[13] Rodríguez Cepeda R, Álvarez Suarez NY. Actividad antimicrobiana del extracto hidroalcoholico de Calendula officinalis L. Revista Ion. 2021;**34**:97-110

[14] De Assis Cardoso AF, da Silva J, Queiroga PJ, Vicente de Paula FNA,

Cárdenas Olivier N, Rojas ABG. Eficiencia de extractos vegetales como insecticida sobre Sitophilus zeamais en granos de maíz almacenados. Revista Ciencias Técnicas Agropecuarias. 2014;**23**(2):57-62. http://scielo.sld.cu/scielo.php?script=sci_arttext&pid=S2071-00542014000200010&lng=es&tlng=es

[15] Sarria-Villa RA, Gallo-Corredor JA, Páez MI. Isolation of Catechin and Gallic acid from Colombian bark of Pinus patula. Chemical Sciences Journal. 2017;**8**(174):1-11

[16] Benítez Benitez R, Sarria Villa RA, Gallo Corredor JA, Pérez Pacheco NO, Álvarez Sandoval JH, Giraldo Aristizabal CI. Obtención y rendimiento del extracto etanólico de dos plantas medicinales. Revista Facultad de Ciencias Básicas. 2019;**15**:31-40

[17] Gonçalves FG, Chaves ILS, Fassarella MV, Brito AS, Silva ÉSG, da, López YM, & Oliveira REG de. Extracción de taninos de la corteza de Pinus spp tratada térmicamente – aplicación como dhesive. Madera y bosques. 2021;**27**(1):e2712041. DOI: 10.21829/myb.2021.2712041

[18] García-Pérez JA, Alarcón-Gutiérrez E, Torres Pelayo VR. Extractos acuosos de plantas como inhibidores de la germinación de urediosporas de *Hemiliea vastatrix*; la roya naranja del café. Alianzas y tendencias BUAP. 2021;**6**(21):45-60

[19] Gull I, Saeed M, Shaukat H, Aslam SM, Samra ZQ, Athar AM. Inhibitory effect of *Allium sativum* a *Zingiber officinale* extracts on clinically important drug resistant pathogenic bacteria. Annals of Clinical Microbiology and Antimicrobials. 2012;**11**:8. DOI: 10.1186/1476-0711-11-8

[20] Figueroa Gualteros AM, Castro Triviño EA, Castro Salazar HT. Efecto bioplaguicida de extractos vegetales para el control de *Spodoptera frugiperda* in corn crop (*Zea mays*). Acata Colombiana. 2019;**24**:58-66

[21] Monasterio Quintana AO. IV. Diagnóstico sobre políticas públicas, marco legal y atribuciones institucionales relacionados con la conservación y uso de polinizadores en México. In Gobierno de México Editores, Diagnostico. Situación actual de los polinizadores en México 2021. Agricultura, Medio Ambiente, Senasica, Conabio, Conanp; 154

[22] Lubell M, Hillis V, Hoffman M. Innovation, cooperation, and the perceived benefits and cost of sustainable agricultura practices. Ecology and Society. 2011;**16**(4):1-12. DOI: 10.5751/ES04389-160423

[23] Bustamante AG, García-López AM, Cervantes-Díaz L, Aíl-Catzim CE, Borboa-Flores J, Rueda-Puente EO. Estudio del potencial biocontrolador de las plantas autóctonas de la zona árida del noroeste de México-study of the autochthonous plants as a potential biocontroller in the northwest area of Mexico. Revista de la Facultad de Ciencias Agrarias. 2017;**49**(1):127-142

[24] Gonzales M, Quiñonez A. Efecto del extracto de ajo como compensador de frio 12 en la brotación del manzano red Delicious en la región de nuevo ideal durando México. Revista ingeniería y región IX CLEIA, Bogotá, Colombia. 2018;**20**:19-24. DOI: 10.25054/22161325.1935

[25] Garrido G, Ortiz M, Pozo P. Fenoles y flavonoides totales y actividad antioxidante de extractos de hojas de *Lampaya medicinalis* F. Philosophical Journal of Pharmacy & Pharmacognosy Research. 2013;**1**:30-38

[26] Ramírez FL, Alanís FG, Ayala BR, Velazco MC, Favela LS. El uso de platos

trampa y red entomológica en la captura de abejas nativas en el estado de Nuevo León, México. Acta zoológica mexicana. 2014;**30**(3):508-538

[27] Paredes SN. Ciclo biológico de *Oligonychus Coffeae* (Acari: Tetranychidae) en aliso (*Alnus Acuminata*) y café (*Coffea Arabica*) y el uso de extractos etanólicos complementarios para su control. Tesis de Ingeniería agronómica, Universidad Técnica de Ambato; 2017. p. 84. Available from: https://repositorio.uta.edu.ec/bitstream/123456789/26942/1/Tesis-184%20%20Ingenier%C3%ADa%20Agron%C3%B3mica%20-CD%20540.pdf

[28] Joo-Pérez R, Avendaño-Arrazate C, Sandoval-Esquivez A, Espinoza-Zaragoza S, Alonso-Báez M, Moreno-Martínez J, et al. Alternancy Study on Rambutan (*Nephelium lappaceum* L.) Tree in Mexico. American Journal of Plant Sciences. 2017;**8**:40-52. DOI: 10.4236/ajps.2017.81004

[29] Hayat S, Ahmad H, Ali M, Hayat K, Khan M, Cheng Z. Aqueous garlic extracts as a plant biostimulant enhances physiology, improves crop Quiality and metabolite abundance, and primes the Denfense responses of receiver plants. Review of Applied Sciences. 2018;**8**:1505

[30] Agustí M, Reig C, Martínez-Fuentes A, Mesejo C. Advances in Citrus flowering: A review. Frontiers in Plant Science. 2022;**13**:868831. DOI: 10.3389/fpls.2022.868831

[31] El- Rokiek K, Dawood M, Sadak M. The effect of the natural extracts of garlic or Eucalyptus on the growth, yield and some chemical constituents in quinoa plants. Bulletin of the National Research Centre. 2019;**43**:119. DOI: 10.1186/s42269-0190161-3

[32] Räsch A, Hunsche M, Mail M, Burkhardt J, Noga G, Pariyar S. Agricultural adjuvants may impair leaf transpiration and photosynthetic activity. Plant Physiology and Biochemistry. 2018, Nov;**132**:229-237. DOI: 10.1016/j.plaphy.2018.08.042

[33] Tafolla-Arellano JC, González-León A, Tiznado-Hernández ME, Zacarías García L, Báez-Sañudo R. Composición, fisiología y biosíntesis de la cutícula en plantas. Revista Fitotecnia Mexicana. 2013;**36**(1):3-12 http://www.scielo.org.mx/scielo.php?script=sci_arttext&pid=S0187-73802013000100001&lng=es&tlng=es

[34] Basavaraja PK, Yogendra ND, Zodape ST, Prakash R, Ghosh A. Effect of seaweed sap as foliar spray on growth and yield pf hybrid maize. Journal of Plant Nutrition. 2018;**41**(14):1851-1861

[35] Yakhin OI, Lubyanov AA, Yakhin IA, Brown PH. Biostimulants in plant science: A global perspective. Frentier en ciencias vegetales. 2017;7:1-35. DOI: 10.3389/fpls.2016.02049

[36] Ariza-Flores R, Barrios-Ayala A, Herrera-García M, Barbosa-Moreno F, Michel-Aceves MA, Sánchez O, et al. Fitohormonas y bioestimulantes para la floración, producción y calidad de lima exicana de invierno. Revista exicana de ciencias agrícolas. 2015;**6**(7):1653-1666

[37] Venegas-González A, Loewe-Muñoz V, Toral M. Influencia del uso de reguladores de crecimiento sobre brotes vegetativos y número de estróbilos masculinos en pinus pinea L. en chile. Ciencia Forestal. 2016;**26**(4):1087-1096

[38] Kumari S, Rana K. Efficacy of bee attractants in attracting insect pollinators in onion seed crop. Journal of pharmacognosy and phytochemistry. 2018;7(5):2239-2243

[39] De la Peña AE, Pérez MV, Alcaraz LJ, Larrañaga N, Hormaza I. Polinizadores y polinización en frutales subtropicales: implicaciones en manejo, conservación y seguridad alimentaria. Revista Ecosistemas. *2018*;**279**:1-101

[40] Alves TC, Rodrigues E, Lago JH, Prado CM, Girardi CEN, Hipólide DC. *Petiveria alliacea*, a plant used in Afro-Brazilian smoke rituals, triggers pulmonary inflammation in rats. Revista Brasileira de Farmacognosia. 2019;**29**(5):656-664

[41] Luz DA, Pinheiro AM, Silva ML, Monteiro MC, Prediger RD, Maia CSF, et al. Ethnobotany, phytochemistry and neuropharmacological effects of *Petiveria alliacea* L. (Phytolaccaceae): A review. Journal of Ethnopharmacology. 2016;**185**:182-201

[42] Wankhede HK, Kulkarni SR, Pawar SA. Influence of indigenous bee attractants on qualitative and quantitative parameters of cucumber (*Cucumis sativus* L.). Journal of Entomology and Zoology Studies. 2018;**64**(4):1241-1244

[43] Pashte VV, Shylesha AN, Bhat NS. Effectiveness of attractants and scents in enticement of *Apis cerana* on *Sesamum crop*. Environment and Ecology. 2015;**33**(4):1504-1507

[44] Ramírez-Luna E, Castillo-Aguilar C, Aceves-Navarro E, Carrillo-Avila E. Efecto de productos con reguladores de crecimiento sobre la floración y amarre de fruto en chile ‘habanero. Revista Chapingo Serie Horticultura. 2005;**11**(1):93-98

[45] Patiño-Torres AJ, Jaimez-Arellano RE. Relación exica – fuerza de la demanda en el aborto de estructuras reproductivas, tasa fotosintética y rendimiento en Capsicum annuun. Agrociencia. 2016;**50**:649-664

[46] Hernández-Maruri JA, Castillo-González AM, Pérez-Barraza MH, Avitia-García E, Trejo-Téllez LI, Osuna-García JA, et al. Fertilización con boro y su relación con la producción de frutos sin semilla en mango “Ataulfo”. Revista Mexicana de Ciencias Agrícolas. 2015;**6**(8):1757-1768

[47] Caballero-Pérez JF, Arévalo-Galarza L, Avendaño-Arrazate CH, Cadena-Iñiguez J, Valdovinos-Ponce G, Aguirre-Medina JF. Cambios físicos y bioquímicos durante el desarrollo y senescencia de frutos de rambután (Nephelium lappaceum L.). Revista Chapingo serie Horticola. 2011;**17**:31-38

[48] Pérez-Barraza MH, Álvarez Bravo A, Avitia García E, Pérez Luna AI, Santos Cárdenas MV. Temperatura y desarrollo floral en la formación de frutos partenocárpicos en mango “Ataulfo”. Revista mexicana de ciencias agrícolas. 2019;**23**:199-210

[49] Martijn ten Hoopen G, Deberdt P, Mbenoun M, y Cilas C. Modelling cacao pod growth: Implications for disease control. Annals of Applied Biology. 2012;**160**:260-272

[50] Elakbawy WM, Shanab SMM, Shalaby EA. Enhancement of plant growth regulators production from microalgae cultivated in treated sewage wastewater (TSW). BMC Plant Biology. 2022;**22**(1):377. DOI: 10.1186/s12870-022-03764-w

[51] Aliakbarpour H, Che Salmah MR, Dieng H. Species composition and population dynamics of thrips (*Thysanoptera*) in mango archards of Northern Peninsular Malaysia. Environmental Entomology. 2010;**39**(5):1409-1419. DOI: 10.1603/EN10066

[52] Moreno A, León DF, Giraldo GA, Rios E. Análisis del perfil de compuestos

volátiles del mango (Mangifera indica L. var Tommy Atkins) tratado por métodos combinados. Revista Colombiana de Química. 2010;**39**:61-72

[53] Marín-Loaiza JC, Céspedes CL. Compuestos volátiles de plantas: origen, emission, efectos, análisis y aplicaciones al agro. Revista Fitotecnia Mexicana. 2007;**30**(4):327-251

[54] Flores-Villegas MY, González-Laredo RF, Prieto-Ruíz JA, Pompa-García M, Ordaz-Díaz LA, Domínguez-Calleros PA. Eficiencia del extracto vegetal de *Datura stramonium* L. como insecticida para el control de la mosca sierra. Madera y bosques. 2019;**25**(1):e2511642. DOI: 10.21829/myb.2019.2511642

[55] Jaramillo Hernández D, González Reina A, Pedraza Castillo N, Sierra Acevedo JI, García Martínez GL, Jara AR. Evaluación del efecto acaricida de Momordica charantia, Megaskepasma erythrochlamys y Gliricidia sepium sobre Rhipicephalus microplus. Revista MVZ Córdoba. 2019;**2019**:25. DOI: 10.21897/rmvz.1951

[56] Mani M, Shivaraju C. Mode of spread of mealybugs. In: Mani M, Shivaraju C, editors. Mealybugs and Their Management in Agricultural and Horticultural Crops. Springer; 2016. pp. 1-655, 113-116. DOI: 10.1007/978-81-322-2677-2

[57] Monje AB, Delgadillo UD, Gómez CJC, Herney VE. Manejo de *Neohydatothrips signifer* Priesner (Thysanoptera: Thripidae) en maracuyá (*Passiflora edulis* f. flavicarpa Degener) en el departamento del Huila (Colombia) Corpoica. Ciencia y Tecnología Agropecuaria. 2012;**13**:21-30

[58] Nava-Pérez E, García-Gutiérrez C, Camacho-Baéz JR, Vázquez-Montoya EL. Bioplaguicidas: una opción para el control biológico de plagas. Ra Ximhai. 2012;**8**(3):17-29

[59] Vázquez Luna A, Pérez Flores L, Díaz SL. Biomoléculas con actividad insecticida: una alternativa para mejorar la seguridad alimentaria. Ciencia y Tecnologia Alimentaria. 2007;**5**(4):306-313

[60] Aguirre Yela V, Delgado V. Pesticidas naturales y sintéticos. Revista ciencia. 2010;**13**:43-53

[61] Martinez R, y Rivera M. Evaluación de acción repelente, insecticida y protectora de los extractos acuosos e hidroalcolico de *Allium sativum* (ajo) contra el *Zabrotes subfasciatus* (gorgojo común) de frijol almacenado. Tesis en química y farmacia. Universidad de El Salvador; 2008. p. 137. http//ri.ues.edu.sv/id/eprint/3031/1/1610024.pdf. Repository: https://ri.ues.edu.sv/id/eprint/3031/

[62] Marcano D, Hasegawa M. *Fitouimica organica.* Consejo de desarrollo científico y humanístico, Universidad central de venezuela (CDCH-UCV). 2002. http://saber.ucv.ve/omp/index.php/editorialucv/catalog/view/18/10/56-1.

[63] Bailey BAJ, Crozier RC, Sicher MD, Strem R, Melnicka MF, Carazzolle GGL, et al. Dynamic changes in pod and fungal physiology associated with the shift from biotrophy to necrotrophy during the infection of *Theobroma cacao* by *Moniliophthora roreri*. Physiological and Molecular Plant Pathology. 2013;**81**:84-96

[64] Al-Oubaidi HKM, y Kasid NM. Increasing phenolyic and flavonoids compounds of Cicer arietinum L. from embryo explant using titanium dioxide nanoparticle in vitro. World Journal of Pharmaceutical Research. 2015;**4**(11):1791-1799

[65] Delgado-Oramas BP. La resistencia inducida como alternativa para el manejo

de plagas en las plantas de cultivo. Revista de protección vegetal. 2022;**35**:e07 Link: http://scielo.sid.cu/scielo.php?script=sci_arttext&pid=S1010-27522020000100001&lng=es&tlng=es

[66] Albores-Flores VJ, García-Guzmán G, Espinosa-García FJ, Salvador-Figueroa M. Degree of domestication influences susceptibility of Theobroma cacao to frosty pod rot: A severe disease devastating Mexican cacao. Botanical Sciencies. 2018;**96**:84-94

[67] Viveros-Legorreta JL, Sarma SSS, Guerero-Zuñiga LA, Rodríguez-Dorantes A. Bioensayo del efecto de fenoles producidos por *Myriophylum aquaticum* en cultivo de Lactuca sativa. Hidrobiología. 2018;**28**:109-119

Chapter 11

Utilization of Moringa Leaves and Pods as Organic Fertilizers in Enhancing Soil Fertility and Crop Growth

Ebido Nancy Ekene and Ndubuaku Mabel Uchenna

Abstract

The use of Moringa extract as bio-fertilizers positively influences agriculture. The different parts of the plant have diverse functions. The extracts have proven to improve crop growth and yield when applied as foliar fertilizers or green manure. The growth and yield of two cultivars of cocoyam (Nce 001 and 012) were enhanced by the application of aqueous moringa leaf and pod extract (AMLE and AMPE). Also the use of moringa leaves as green manure increased the growth of maize. These effects could be traced to its potentials in improving the soil fertility status and also its phyto chemical properties. Therefore, the use of moringa as an organic fertilizer is highly recommended. This book chapter emphasizes the use of moringa leaf and pod extracts, as good alternative bio fertilizers for improved crop growth and yield.

Keywords: *Moringa oleifera*, aqueous moringa extract, organic fertilizer, crop growth and yield, soil fertility

1. Introduction

Moringa is also known as Horse-radish tree or drumstick tree. It is a shrub or tree belonging to the Onogenic family, known as *Moringaceae*. The tree is indigenous to Agra and Oudh in India's North West, south of the Himalayas. It is cultivated worldwide, particularly in Pakistan, Asia Minor, Africa, and Arabia, [1]. At the start of the 20th century, it was brought from India to Eastern Africa. The *Moringa oleifera* Lam. species is the most widely grown of the 13 species that make up the moringa family. Moringa grows fast into a short, slender perennial tree, about 7–9 m tall, and can grow up to 6–7 m within a year under low rainfall of at least 400 mm/annum [2]. The bark is grey and thick, and looks like cork, peeling in patches. Propagation is either by seed or vegetatively through cuttings. When the moringa plant is 8 months old, it starts to bloom, and the flowering season lasts from January to March. The triangular (30–50 cm long) pods of the fruit, which ripens between April and June, have oily, black winged seeds inside [3, 4]. It has long tuberous tap roots that grow deep into the soil in order to absorb nutrients from the subsoil during the dry season. The moringa tree is renowned for its ability to adapt to difficult growing conditions that most trees cannot handle, including resilience to diseases and drought [5]. In the sub-Saharan regions, moringa is a relatively new crop for farmers, and it is primarily produced in backyards for domestic consumption. However, there is little information available on farmers' perspectives of its production, processing, and use.

Moringa is quite interesting to scientists since it possesses beneficial qualities and traits. These include the high levels of protein found in the leaves and stem, the oil found in seeds, and the abundance of distinctive polypeptides found in seeds that can bind to various moieties. Since ancient times, its traditional, medical, and industrial uses have been promoted. The plant's varied sections are excellent providers of protein, vitamin B, amino acids, and numerous phenolic compounds. They also include profiles of significant minerals [3, 6]. Phytochemicals found in the moringa plant are abundant and unusually combined [7].

Owing to its numerous usages and advantages for agriculture and industries, *Moringa oleifera* Lam. has received a lot of attention recently [8]. All parts of the moringa plant are utilized for medicinal and other reasons, earning it the nickname "wonder tree." Moringa has antibacterial qualities in its bark, seeds, roots, stem, leaves, and flowers [9, 10]. Moringa has in recent times been studied for its potential to increase soil fertility, crop growth and production, making it a valued plant [11]. Because it includes several plant hormones that promote growth, moringa extract has demonstrated to boost crop development and productivity when used as foliar fertilizers or green manure. Due to their involvement in all phases of plant growth and development, plant hormones have the potential to increase yield. Auxins, gibberellins, abscisic acid, ethylene, and cytokinins are examples of growth-regulating hormones [12]. One type of cytokinin that naturally occurs in plants is zeatin. Fresh moringa leaves have a high zeatin concentration, according to studies [13]. Onions, pepper, soy beans, sorghum, coffee, tea, melon, and maize leaves were sprayed with moringa leaf extract, and it was discovered that doing so increased their yield [14]. Moringa has been reported to have high yielding and drought resistant qualities with its mean annual pod yield capacity of 37.69 tonnes/ha/yr. and seed yield capacity of 16.74 tonnes/ha/yr., and was also reported to have grown and yielded very well in the arid savannah zones [15]. The fertility of agricultural soils is increased by using moringa shoots as green manure. For this, moringa seeds are planted in well-prepared seed beds 10 cm apart and 1–2 cm deep. After 25 days of planting, the young plants are ploughed into the ground at a depth of 15 cm [3].

2. Moringa production

Moringa seedlings for transplanting are grown in seed bags of about 18 cm in height and 12 cm in diameter. The suitable soil medium is a mixture of topsoil, manure and fine sand in the ratio of 3:2:1. Three seeds are planted per bag at a depth of 2 cm. The seedlings are thereafter transplanted one month after sowing. Before transplanting, holes with dimension of 30 cm wide and 30 cm deep are dug at a spacing of 20 cm × 50 cm. The holes are then half-filled with the soil medium to sustain the seedlings at the early stage of their growth. The moringa seedlings are transplanted with their ball of earth and watered routinely. One month after transplanting, the moringa plants are trimmed using a pruning saw to promote branching, increase yield and facilitate harvesting. Six months after transplanting, when the pinnules would have been broad enough, the leave can be harvested for processing by snaping leafy stems from branches. Young shoot tips can be harvested to promote the development of side branches. The plant should be allowed to develop new shoots and branches before subsequent harvest.

3. Economic importance of moringa

Moringa is an all-purpose plant. There are numerous uses for every component of the plant. The leaves are highly nutritious. They are good suppliers of phenolic compounds, beta carotene, minerals, vitamins, amino acids, and protein. Zeatin, quercetin, beta-sitosterol, caffeoylquinic acid, and kaempferol are all present in them in rich and uncommon combinations. They act as cardiac and circulatory stimulants. They possess antibacterial, antifungal properties and some antioxidants. The leaves are ground and used for scrubbing utensils and for cleaning walls. The fresh leaves can be eaten raw or cooked like spinach or dried and made into powder that can be added to sauce etc., and young branches are eaten by livestock. It is planted as living fence tree.

The young green pods can be eaten whole and be comparable in taste to asparagus. The older pods can be used for their seeds, which can be prepared as peas or roasted and eaten as peanuts. The seeds yields about 40% of non-drying oil, known as Oleic or Ben oil, used for cooking, lubricating, cream and soap making etc. The oil is clear, sweet and odourless and also useful in the manufacture of perfumes and weavon oils in hair dressing [16]. The oil compares favourably with olive oil. The mature seeds can also be used to purify water. Seed cake is a good source of fertilizer, however it is not advisable to use the seed cake for livestock feed as it contains alkaloids and saponin [17].

The flowers bloom around 8 months after planting. They are present all through the year and serve as good source of nectar for honey producing bees. Thus, their presence enhances growth of other crops due to increase in pollination activities by bees. The flowers can be eaten fried and have the taste and texture of mushrooms. Moringa wood yields a blue dye and the bark can serve for tanning.

4. Agricultural benefits/application of moringa

There are also many other uses of moringa especially in the agricultural sector, among these are;

4.1 Animal feedstock

There are significant nutritional restrictions on ruminant feed in many tropical areas due to low-quality and insufficient natural foods, which can result in an energy and protein shortage [18, 19]. The problem of ruminant feed shortages is made worse by the dry seasons, when natural pastures are deficient in protein and energy. Moringa has a high nutritional content and might be a suitable source of feed supplement, according to several research [20, 21]. Therefore, using moringa as a source of protein will boost cattle performance and balance other available nutrients [16].

4.2 Alley cropping

With the rapid growth, long tap roots, few lateral roots, minimal shade and large production of high-protein biomass, moringa trees are well suited for use in alley cropping systems.

4.3 Green manure

Moringa plants can be cultivated intensively and then ploughed back into the soil, as green manure or natural fertilizers for other crops. The leaves could also be pruned and incorporated into the soil as green manure. The incorporation of fresh or dry moringa leaves has been reported to increase the soil organic matter, nitrogen and phosphorus contents and reduced exchangeable aluminium and hydrogen ions. Among fresh and dry moringa leaves incorporated into the soil, nutrient release was higher in the dry than the fresh leaves. The incorporation of 5–15 t/ha fresh and dry moringa leaves enhances Maize growth and production [22].

4.4 Foliar nutrients

The juice extracted from the different parts of the plant especially the fresh leaves and pods can be used to make foliar fertilizers capable of increasing crop yield by up to 30%.

In an experiment conducted by Ebido et al. [23], fresh leaves and pod husks of *Moringa oleifera* were collected and washed under tap water; rinsed with distilled water and dried in shade. The dried samples were powdered in a hammer mill grinder. Powders of the leaves and pod husks were mixed separately with distilled water in a ratio of 100 g: 1 lit and left overnight to allow the powder get dissolved in water. The mixture was then filtered through muslin cloth to get 100% moringa plant tissue extract. The extracts were shaken after vacuum filtration. The undiluted extracts were added to the Erlenmeyer flasks, blocked with cotton individually, and heated at 50°C for 15 min to prevent contamination. To further dilute the extracts for usage, two ratio concentrations 1:1 and 2:1 (i.e., 5 ml of extract to 5 ml of water and 10 ml of extract to 5 ml of water, respectively) of distilled water were used.

5. Use of aqueous moringa leaf and pod extracts (AMLE and AMPE) as foliar fertilizers and the effects on crop growth and yield

In a field evaluation trial, Ebido et al. [23] studied the effects of aqueous moringa leaf and pod extracts (AMLE and AMPE) on the growth and yield of cocoyam. The trial was done at the Cocoyam Experimental Farm of National Root Crop Research Institute, Umudike, Abia State, South-East Nigeria. The experimental site was located in the rainforest agro-ecological zone. The soil was a coarse-textured Ultisol. The processed moringa leaf and pod extracts were further diluted with distilled water at the ratios of 1:1 and 2:1 using 5 ml of extract to 5 ml of water and 10 ml of extract to 5 ml of water respectively. The crop used for the experiment was cocoyam (*Colocasia*-NCe 001 and NCe 012). A month after planting, the treatment application began, and it was repeated monthly. Plant stand at harvest (survival count) was the growth data measured. Number of corms, number of cormels, weight of corms (kg), weight of cormels (kg), total number and weights of corms and cormels after final harvest were the yield components measured.

The results, as displayed in **Tables 1–3**, demonstrated that the AMPE provided a higher yield than the AMLE. All of the treatments produced higher yields when compared to the control, with AMPE 2:1 having a higher yield of about 20% on NCe 012 and AMPE 1:1 having a higher yield of nearly 50% on NCe 001. There were notable yield variations between the two cultivars, with NCe 012 doing better.

Treatments	Survival counts	No. of corms	Weight of corms (kg)	No. of cormels	Weight of cormels (kg)	No. of corms and cormels	Total weight (kg)
T1 (Con)	14.0a	16.3b	0.9b	109.0a	2.2a	125.7a	3.1b
T2 (AMLE 1:1)	14.3a	17.0b	2.0b	143.7a	3.9a	187.3a	5.9a
T3 (AMLE 2:1)	16.7a	24.0ab	1.2ab	163.0a	3.4a	160.7a	4.6ab
T4 (AMPE 1:1)	17.7a	32.0a	2.1a	156.7a	4.1a	188.7a	6.2a
T5 (AMPE 2:1)	14.3a	16.7b	1.2ab	145.0a	3.3a	161.7a	4.5ab

T = treatments, AMPE = aqueous moringa pod extract, AMLE = aqueous moringa leaf extract, means with the same alphabets are statistically similar.
Source: Ebido et al. [23].

Table 1.
Effects of aqueous moringa extracts on cocoyam (NCe 001).

Treatments	Survival counts	No. of corms	Weight of corms (kg)	No. of cormels	Weight of cormels (kg)	No. of corms and cormels	Total weight (kg)
T1 (Con)	19.3ab	23.7b	2.9b	249.7a	6.0a	273.3b	8.9a
T2 (AMLE 1:1)	16.3b	32.7a	3.6ab	273.0a	6.5a	305.7ab	10.1a
T3 (AMLE 2:1)	19.0ab	31.7ab	3.3ab	270.7a	6.4a	302.3ab	9.6a
T4 (AMPE 1:1)	16.7b	30.3ab	2.9b	267.0a	6.2a	397.3a	9.1a
T5 (AMPE 2:1)	20.0a	38.3a	3.9a	316.0a	7.2a	354.3ab	11.1a

T = treatments, AMPE = aqueous moringa pod extract, AMLE = aqueous moringa leaf extract, means with the same alphabets are statistically similar.
Source: Ebido et al. [23].

Table 2.
Effects of aqueous moringa extracts on cocoyam (NCe 012).

This demonstrates that using moringa extract as an organic foliar fertilizer substitute can increase cocoyam yield. This can be likened to the presence of growth enhancing hormones, especially cytokinin (Zeatin) in AMLE and AMPE which increases growth and yield of crops [3, 7, 13]. It had been reported that foliar application of moringa leaf extract improved the growth and yield of tomatoes, peanut, corn and wheat during the vegetative growth stage of the crops. Fuglie [3] reported yield increase of 25–39% in onions, pepper, soya, maize, sorghum etc. following the application of moringa leaf extract. Similarly, Phiri [24] observed that *M. oleifera* leaf extract improved germination of sorghum and increased hypocotyl length of wheat.

Treatments	Survival counts	No. of corms	Weight of corms (kg)	No. of cormels	Weight of cormels (kg)	No. of corms and cormels	Total weight (kg)
NCE 001	15.7	22.2	1.5	150.8	3.5	173.1	4.9
NCE 012	20.6	33.2	3.4	276.8	6.5	**309.9**	**9.9**
LSD (0.05)	1.5	3.9	0.4	37.6	0.8	38.3	1.2

Source: Ebido et al. [23].

Table 3.
Yield performance of the two cocoyam cultivars (NCe 001 and 012).

Adekiya et al. [25] pointed out that the application of moringa leaf extract increased the yield of okra when compared with the control.

6. Conclusion

The use of plant extract as bio-fertilizers has proven to influence agriculture positively. Moringa, which is referred to as a Miracle tree, because every part of the tree has great potentials, is no exception of plants whose extracts are used as bio-fertilizers. It grows very fast and can survive unfavourable conditions. It is a global crop because of its adaptability to diverse environmental conditions. The valuable properties and characteristics of moringa have made it a crop of great scientific interest. The different parts of the plant, contain profiles of important minerals, nutrients, hormones and phyto chemicals. Moringa extracts has proven to improve crop growth and yield when applied as foliar fertilizers or even green manure. Reports have shown that the application of moringa aqueous leaf and pod extracts increased cocoyam yield by about 50%. The use of moringa leaves as green manure also increased the growth of maize. These effects could be traced to its potentials in improving the soil fertility status due to its high composition of chemical and phyto chemicals. This, therefore, concludes that moringa extracts, particularly AMPE and AMLE are good sources of alternative bio-fertilizers for enhanced crop yield. Therefore, the use of moringa as an organic fertilizer is highly recommended. Based on the dearth of research on the use of moringa pod extracts for improved crop yield, it is recommended that more research be conducted in this regard.

Author details

Ebido Nancy Ekene[1*] and Ndubuaku Mabel Uchenna[2]

1 Department of Soil Science, University of Nigeria, Nsukka, Enugu State, Nigeria

2 Department of Crop Science, University of Nigeria, Nsukka, Enugu State, Nigeria

*Address all correspondence to: nancy.ebido@unn.edu.ng

References

[1] Mughal MH, Ali G, Srivasta PS, Igbal M. Improvement of drumstick (M. pterygosperma Gaetn.)—A unique source of food and medicine through tissue culture. Hamdard Medicus. 1999;**42**:39-42

[2] Odee D. Forest biotechnology research in drylands of Kenya: The development of *Moringa* species. Dryland Biodiversity. 1998;**2**:7-8

[3] Fuglie LJ. The Moringa Tree: A Local Solution to Malnutrition. ECHO Development Notes. 5th ed. Dakar, Sengal: Yumpu Magazines; 2005. pp. 12-18

[4] Ndubuaku TC, Ndubuaku UM. *Moringa oleifera*- Medicinal and Nutritional Crop, its Botany, Production and Utilization. Nsukka: University of Nigeria Press Limited, Bookshop/Bank Complex, University of Nigeria; 2011. 47 pp

[5] Foidl N, Makkar HPS, Becker K. The Potential of *Moringa oleifera* for Agricultural and Industrial Uses. What Development Potential for Moringa Products? Dar Es Salaam, October 20th–November 2nd 2011. 2011

[6] Foildl N, Paul R. *Moringa oleifera*. In: The Encyclopedia of Fruit and Nuts. Oxfordshire, UK: CABI; 2008. pp. 509-512

[7] Ndubuaku UM, Ndubuaku TCN, Ike E, Ezeaku PI. Effects of *Moringa oleifera* leaf extract on morphological and physiological growth of cassava and its efficacy in controlling *Zonocerus variegates*. African Journal of Biotechnology. 2015;**14**:2494-2500

[8] Ashfaq M, Basra SMA, Ashfaq U. Moringa: A miracle plant of agro-forestry. Journal of Agriculture & Social Sciences. 2012;**8**:115-122

[9] Anjorin TS, Ikokoh P, Okolo S. Mineral composition of *Moringa oleifera* leaves, pods and seeds from two regions in Abuja, Nigeria. International Journal of Agriculture and Biology. 2010;**12**:431-434

[10] Dwivedi SK, Enespa A. Effectiveness of extract of some medical plants against soil borne *Fusaria* causing diseases on *Lycopersicon esculantum* and *Solanum melongena*. International Journal of Pharma and Bio Sciences. 2012;**3**(4):1171-1180

[11] Fahey JW. *Moringa oleifera*: A review of the medical evidence for its nutritional, therapeutic and prophylactic properties. Trees for Life Journal. 2005;**1**(5):1-15. http://www.TFLJournal.org

[12] Prosecus P. Biosynthesis-Plant Hormones and Growth Regulators: Chemistry and Biology. Switzerland: Biosynth AG. Co.; 2006

[13] El-Awady A. Moringa Tree: Nature's Pharmacy. Yumpu Magazines, in Dakar Sengal; 2003. Available from: http://www.islamonline.net/english/Science/2003/02article06.shtml. [Accessed: October 20, 2007]

[14] Fuglie LJ. New uses of Moringa studied in Nicaragua: ECHO's technical network site-networking global hunger solutions. In: Development Notes. 2nd ed. Nicaragua: ECHO; 2000. pp. 1-5

[15] Ndubuaku UM, Ndubuaku TCN, Ndubuaku NE. Yield characteristics of *Moringa oleifera* across different ecologies

in Nigeria as an index of its adaptation to climate change. Sustainable Agriculture Research, Canada. 2014;**3**(1):95-100

[16] Odeyinka SM, Torimiro DO, Oyedele JO, Asaolu VO. Farmer's awareness and knowledge of *Moringa oleifera* in Southwestern Nigeria: A perceptional analysis. Asian Journal of Plant Sciences. 2007;**6**:320-325

[17] CSIR. Council for Scientific and Industrial Research. Pretoria: CSIR; 1962. pp. 1-60

[18] Mannetjet L. Nutritive value of tropical and subtropical pastures, with special reference to protein and energy deficiency in relation to animal production. In: Gilchrist FMC, Mackie RI, editors. Herbivore Nutrition in the Subtropics and Tropics. Craighall: The Science Press; 1984. pp. 56-66

[19] Odeyinka SM, Ademosun AA. The effect of season on the yield and nutritive value of *Gliricidia sepium* and *Leucaena leucoephala*. Nigerian Journal of Animal Production. 1993;**20**:96-103

[20] Dash S, Gupta N. Effect of inorganic, organic and bio fertilizer on growth of hybrid *Moringa oleifera* (PKM 1). Academic Journal of Plant Sciences. 2009;**2**:220-221

[21] Moyo B, Masika PJ, Hugo A, Muchenje V. Nutritional characterization of Moringa (*Moringa oleifera* Lam.) leaves. African Journal of Biotechnology. 2011;**10**:12925-12933

[22] Ebido NE, Ezeaku P, Ndubuaku UM. Contributions of moringa (*Moringa oleifera*) tree foliage for enrichment of soil nutrient status. The International Journal of Science and Technoledge, India. 2014;**2**(4):350-355

[23] Ebido NE, Onuba MN, Okoye BC, Unagwu BO. Comparative effect of aqueous moringa extracts (leaf and pod) as foliar fertilizer on cocoyam yield. Nigeria Agricultural Journal. 2020;**51**(2):413-417

[24] Phiri C. Influence of *Moringa oleifera* leaf extracts on germination and early seedling development of major cereals. Agriculture and Biology Journal of North America. 2010;**1**(5):774-777. DOI: 10.5251/abjna.2010.1.5.774.777

[25] Adekiya AO, Agbede TM, Aboyeji CM, Dunsin O, Ugbe JO. Green manures and NPK fertilizer effects on soil properties, growth, yield, mineral and vitamin C composition of okra (*Abelmoschus esculentus* (L.) Moench). Journal of the Saudi Society of Agricultural Sciences. 2019;**18**(2):218-223. DOI: 10.1016/j.jssas.2017.05.005

Chapter 12

Synthesis of Thermophosphate Fertilizers by a Plasma Torch

Nelson Mauricio Espinel Pérez

Abstract

Phosphoric rock (PR) is the basic building block to produce animal feed, fertilizers, and industrial phosphates. Global demand for PR is estimated to grow from 207 Mtons in 2018 to 263 Mtons in 2035, of which Colombia contributes approximately 0.06 Mtons per year. A novel technology to carry out calcination is the plasma torch, where the electrical resistivity of the system is increased and ionized gas is produced that can reach temperatures above 10,000°C, which facilitates the transformation of PR into thermophosphates. Two samples of PR from the region central of the Boyacá department, Colombia, were subjected to calcination through a plasma torch and as a result, showed a maximum concentration of total phosphorus between 27 and 33% of P_2O_5 and assimilable phosphorus corresponding to the range between 3.0 and 4.8% of P_2O_5 respectively. Finally, the energy consumption for calcination is ≤1.14 kW-h/Kg, respectively.

Keywords: fertilizers, plasma torch, thermophosphates, phosphoric rock, calcination

1. Introduction

Phosphoric rock is one commodity type and is considered a strategic mineral of Colombia with 20 million hectares of reserves. Global demand for phosphoric rock is estimated to grow from 207 Mton in 2018 to 263 Mton in 2035, of which Colombia contributes approximately 0.06 Mtons/year [1]. The crucial phosphoric rock deposits in Colombia are in the departments of Boyacá (Sogamoso, Pesca, Iza, Cuitiva, Tota, Monguí, Úmbita, and Turmequé), North Santander (Sardinata, Lourdes, Cúcuta, Bochalema, Durania, Santiago and Zulio), Huila (Tesalia and Aipe), Cundinamarca (Zipaquirá and Pacho) and Tolima (Natagaima). According to previous reports [2], the production was 2016 of 66,324 tons/year, being the department of Boyacá the most prominent producer with 34,501 tons/year, Huila with 20,615 tons/, and North Santander with 11,208 tons/year, respectively [3]. **Figure 1** shows the primary phosphoric rock deposits in Colombia.

On the other hand, 75% of the phosphoric rock does using for the wet production of phosphoric acid as an intermediate product, required to obtain fertilizers and other products such as diammonium phosphate (DAP), monoammonium phosphate (MAP), triple superphosphate (TSP), dicalcium phosphate (DCP), sodium tripolyphosphate (STPP), thermal phosphoric acid (TPA), simple superphosphate, (SSP) and direct

 IntechOpen

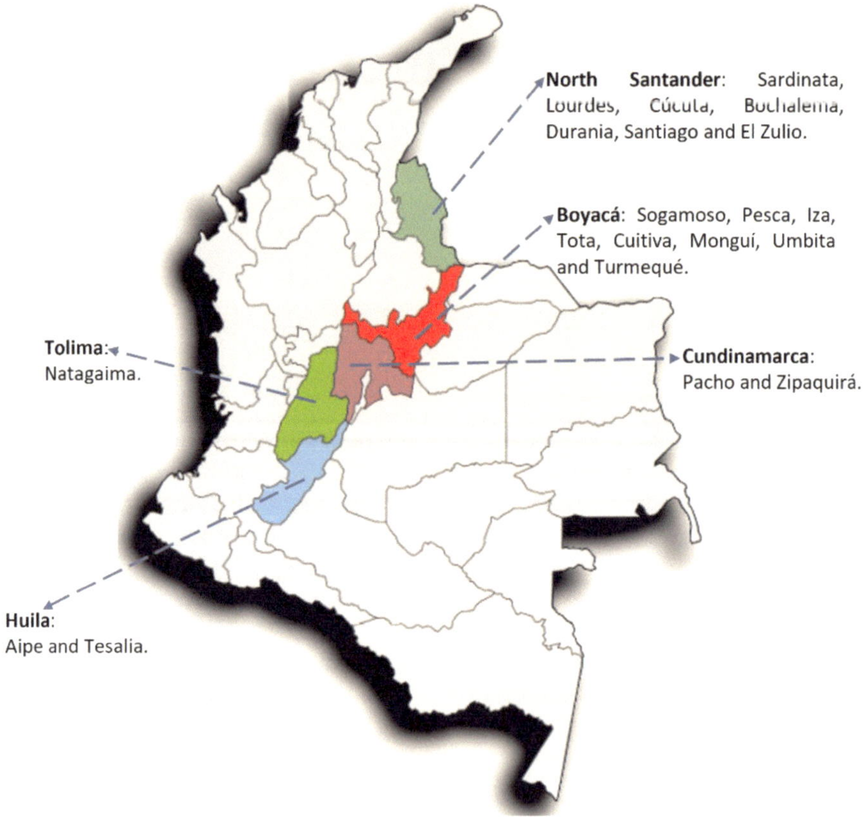

Figure 1.
Main phosphoric rock deposits in Colombia [4].

application phosphoric rock (DAPR) [1]. According to previous reports, more than 10% of the world fertilizer market produces by calcination [5], and a novel technology to carry out this operation is through the plasma torch [6]. It creates an electric arc through which gas passes, producing a stream of ionized gas or plasma. The system's electrical resistivity increases and ionized gas is created that can reach temperatures above 10,000°C, which facilitates the transformation of phosphoric rock into thermophosphates. The samples of phosphoric rock (PR) were taken from the localities of Iza and Sogamoso Boyacá – Colombia. Then the phosphoric rock beneficiation was carried out by drying, crushing, grinding, and sieving operations by obtaining particle sizes of 0.075 mm, because of a granulometric analysis that passed 200 meshes. To facilitate the agglomeration of the sieved mineral, a mixture of phosphoric rock, wheat flour, and water was prepared, forming a homogenized paste and pressed through a hydraulic system to obtain briquettes 2.5 cm in diameter and 1.2 cm thick. They were then calcined in a plasma reactor and the thermophosphates were obtained.

2. Importance of phosphoric rock in the world

Sustainable development goals (SDGs) seek to end poverty, defend the planet, and ensure prosperity for all people in 2030. Some goals like 1, 2, and 10, no poverty, zero hunger, and reduced inequalities [7], respectively, aim at improving living conditions worldwide. However, goal 2 is critical due to population growth, where every day needs to meet the demand for food. For food production, phosphoric rock is the primary source of phosphorus in fertilizers and can be obtained by a thermal or wet process; furthermore, your application, like soluble phosphorus, can increase the yield of crops and is a critical factor for the agriculture due to do not have a substitute for the growth of plants. Likewise, phosphorus in the shape of phosphate or its esters is involved in many biological processes, including relevant structural, metabolic, and transport functions, and it's a structural element in nucleic acids or phospholipids in biomembranes [8, 9]. In addition, nutrients like nitrogen are used less efficiently when soils contain less phosphate and potash [10]. Even if all other conditions and nutrients are plentiful, only phosphorus can make crops thrive [11]. On the other hand, the disponible of this raw material, nonrenewable in nature, depends on the extraction speed, and consumption determines the depletion rate [12]. Finally, the farmers must be conscientious about the efficient use of phosphorus fertilizers, and fertilizer producers should consider about developing materials that gradually release phosphorus and avoid dilution losses due to rainfall or other factors. Likewise, it must be carried on an equitable distribution of fertilizers that avoid social issues, favor economic growth, improve agricultural productivity, and improve food security worldwide.

3. Characteristics of phosphoric rock

In nature, phosphates are present in igneous, sedimentary, and metamorphic rocks, as well as marine and biogenic deposits. The most important group is the apatites, whose generic formula is $M_{10}(XO_4)_6Y_2$. Likewise, the most common natural deposits are fluorapatite ($Ca_{10}(PO_4)_6F_2$), hydroxyapatite ($Ca_{10}(PO_4)_6(OH)_2$), carbonate of hydroxyapatite ($Ca_{10}(PO_4,CO_3)_6(OH)_2$), francolite ($Ca_{10\text{-}x\text{-}y}Na_xMg_y(PO_4)_{6\text{-}z}(CO_3)_zF_{0.4z}F_2$), dahllite ($3Ca_3(PO_4)_2{\cdot}CaCO_3$) and collophane $3Ca_3(PO_4)_2{\cdot}nCa(CO_3,F_2,O){\cdot}xH_2O$ [5]. **Table 1** summary of the chemical composition of apatites.

4. Obtaining thermal phosphates

The thermal treatment of phosphate rock allows the production of thermophosphate fertilizers, sometimes with additives. At high temperatures, changes

Generalities	Cation/Anion/Properties	References
General Formula	$M_{10}(XO_4)_6Y_2$	[13]
M (Cations to replace)	Ca^{2+}, Mg^{2+}, Sr^{2+}, Ba^{2+}, Mn^{2+}, Fe^{2+}, Zn^{2+}, Cd^{2+}, Pb^{2+}, H^+, Na^+, K^+, Al^{3+}	
XO_4 (Anions to replace)	PO_4^{3-}, AsO_4^{3-}, VO_4^{3-}, SO_4^{3-}, CO_3^{2-} and SiO_4^{3-}	[13]
Y (Anions to replace)	OH^-, F^-, Cl^-, Br^-, O^{2-}, and CO_4^{2-}	[13]
Thermal stability	FAp > HAp > ClAp	[14]
Dissolution grade in tampon acid	ClAp > HAp > FAp	[15]

Table 1.
Generalities about apatites.

in the crystal structure of apatite occur, facilitating the availability of phosphorus to plants [16]. Three types of conventional thermal processes are applied to phosphoric rock [17]: calcination, sintering, and melting. Thermal plasma is a novel method, and its description is carried out later.

4.1 Calcination

It is a process that decomposes existing carbonates and eliminates CO_2.

4.2 Sintering

It is the process of agglomerating of small particles to form larger ones without reaching the melting point.

4.3 Melting

The raw ore is heated to above the melting point.

On the other hand, electric furnaces are widely used for calcinating phosphoric rock, and the thermal treatment carries on to specific temperatures, as described in **Table 2**.

4.4 Thermal plasma

Plasma is considered the fourth state of matter and consists of a mixture of electrons, ions, and neutral particles, which are generally electrically neutral. **Figure 2** shows the characteristics of process ionized plasma.

Characteristic	Temperature Range	Process	Reference
Water removal	120–150°C	Dry	[5]
Removal of organic matter	650–750°C	Calcination	
Carbonate dissociation	850–1000°C	Calcination	
Fluoride removal	>1350°C	Defluorinated	

Table 2.
Thermal treatments applied to apatite.

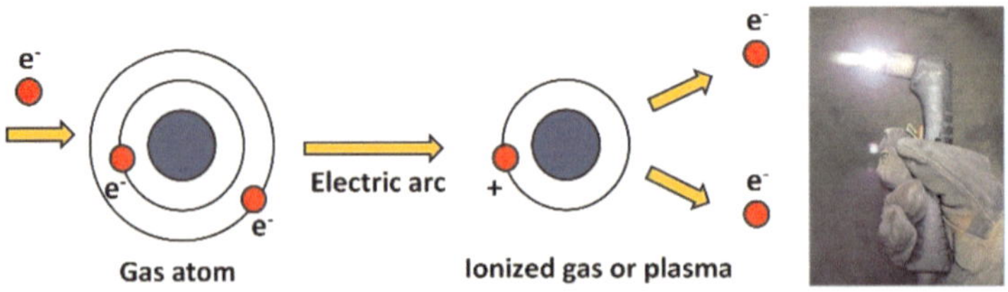

Figure 2.
Characteristic of process ionized plasma.

The creation of an electric arc sustained by the passage of electric current through a gas produces an increase in electrical resistivity throughout the system, generating heat, stripping electrons from the gas molecules, and resulting in a stream of ionized gas or plasma [18]. Plasmas are classified into thermal and cold plasmas: the former are known as hot, high-pressure, or equilibrium plasmas and are characterized by the fact that the temperature of the ions is very similar to that of the electrons. The second is low-pressure and non-equilibrium plasmas and is characterized by less frequent collisions between ions and electrons, generating a higher electron energy level (temperature) [19]. On the other hand, plasma torches can be classified into transferred, and non-transferred arcs. The difference is that the former has a wider physical separation between the cathode and the anode, which can vary between 1 cm and almost 1 m [18].

5. Material and methods

The phosphoric rock (PR) samples were taken in the central-eastern region of the Boyacá Department, Colombia, specifically in the municipalities of Iza (San Miguel) and Sogamoso (Pilar and Ceibita).

5.1 Benefit of minerals

Table 3 describes the benefits of minerals of raw phosphoric rock (PR).

5.2 Calcination process in plasma torch reactor

A plasma torch reactor was used to obtain thermophosphates by calcinating the phosphoric rock (RP) under the conditions described in **Table 4** and using the equipment Victor Cut Master A60™.

Unitary operation	Conditions	Equipment
Drying PR	50°C/8 h	Electric oven
Grinding	250 rpm/5 min.	Retsch planetary ball mill pm 400
Sieving	particle size ≤0.075 mm, passes 200 mesh/5 min.	Fritsch Analysette 3 Pro vibratory sieve and a sieve Tyler normalized series numbers 8, 16, 30, 50, 100, 200 and −200
Agglomeration	Proportions of 65 wt.% - 70 wt.% PR and 35 wt.% - 30 wt.% wheat flour in water paste	Laboratory material
Pressed	Pressure 119.84 Mpa, circular briquettes of 2.5 cm in diameter and 1.2 cm in height	Hydraulic press
Drying briquettes	105°C/2 days	Electric oven

Table 3.
Benefits of minerals and operation condition.

Feature	Description
Type of plasma	Thermal
Type of plasma torch	Transferred arc
Current output	20–80 A
Plasma gas	Air
Working pressure	4.1–6.5 Bar
Gas flow	142–235 L/min
Time calcination	30–40 s
Number of briquettes	4 briquettes/stainless-steel crucible

Table 4.
Conditions of calcination of phosphoric rock (PR) in plasma torch reactor.

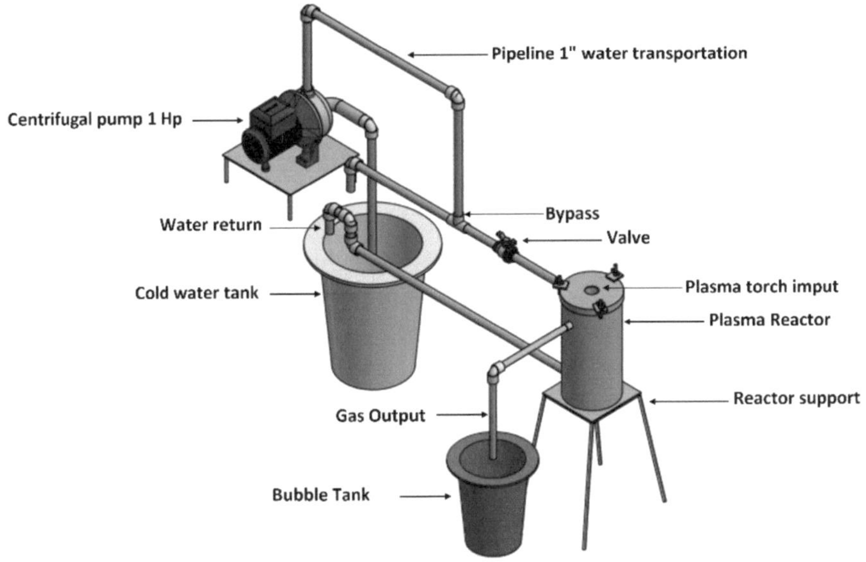

Figure 3.
Process for obtaining thermophosphates by a plasma torch.

The thermophosphates obtained were then cooled and ground in the Retsch planetary ball mill pm 400 for 5 min and sieved to get a particle size ≤0.075 mm. **Figure 3** is shown the process required to obtain thermophosphates fertilizers.

5.3 Design experiments

In a 2^4 – factorial design of experiments, two levels (high and low) and four factors were established: (**a**): current intensity (30–45 A); (**b**): mineral source, A: phosphoric rock Iza (PRIZA) and B: phosphoric rock Pilar and Ceibita (PRPC); (**c**): time (30–40 s) and (**d**): mineral mixture for briquette preparation (65–70 wt. % PR and 35–30 wt. % wheat flour in water paste). **Table 5** shows the design of experiments to obtain thermophosphates in a plasma reactor, likewise, through Minitab 17 software.

Levels	Factors			
	x_1^a	x_2^b	x_3^c	x_4^d
−1	30	A	30	65
+1	45	B	40	70

Table 5.
Design of experiments to obtain thermophosphates in a plasma reactor.

5.4 Analytic methods

Different analytical techniques were used to characterize the thermophosphates obtained by plasma torch, among which we find: X-ray diffraction (XRD) through equipment GNR XRD 600 of Cu Kα radiation (λ = 1.5418 Å), the diffraction angle range (2Θ) varying from *20° to 70°*, with a step of 0.02 and integration time 35 min. This technique facilitates the identification and quantification of the crystalline phases of samples (Rietveld refinement method). The X-Ray Fluorescence (XRF) carries out through the Epsilon 4™ equipment of Malvern PANalytical and the Omnian software. This equipment allows the quantification of the elemental chemical composition. The Shimadzu spectrophotometer UV-VIS 1601™ was used to quantify assimilable phosphorus, taking into account, as a reference, the AOAC standards: 963.03, 960.02960.03, and 993.31 [20].

6. Results and discussion

6.1 Chemical compositions analysis by XRF

The elemental analysis was performed by XRF on the Iza phosphoric rock feedstock (PRIZA), as well as its thermophosphates (TPPT-IZA), and the results are shown in **Table 6**. The concentration of the original raw material of Iza (PRIZA) is 27.05 wt. % P_2O_5, while the thermophosphates M1 to M8 report lower concentration ($\leq$27.00 wt. % P_2O_5) due to loss on ignition caused by thermal treatment. On the other hand, some samples of thermophosphates increase the concentration of CaO > 37.05 wt. % concerning PRIZA due to the thermal dissociation of carbonates, which is carried out at temperatures between 850 and 1000°C [5]. Likewise, this dissociation of carbonates required greater amounts of energy due to endothermic reaction, and as a result of calcination, there is low reactivity and lower relation between CaO/P_2O_5 [21]. According to previous reports, the original raw material of Iza (PRIZA) is composed of quartz-sandstone, and the highest concentration of SiO_2 was reported as 45.5 wt. % [22], respectively. This is the reason for the high SiO_2 concentration of M1 to M8 thermophosphate samples (>29 wt. %), exceeding the optimal concentration required for industrial use (5–15 wt. %) [23]. As a result, it could affect the solubility of the fertilizers. **Table 6** summarizes the composition of PRIZA oxides and their thermophosphates.

Table 7 shows the method for determining quality indexes of thermophosphates (TPPT-IZA), which include optimal values as $CaO/P_2O_5 \leq 1.6$; $MgO/P_2O_5 \leq 0.022$;

	Chemical composition of thermophosphates samples (wt.%)								
	PRIZA	M1	M2	M3	M4	M5	M6	M7	M8
Na_2O	0.32	0.00	0.00	0.00	0.24	0.35	0.00	0.00	0.19
MgO	0.14	0.19	0.18	0.19	0.17	0.18	0.19	0.19	0.21
Al_2O_3	3.29	3.25	3.27	3.19	3.36	3.33	3.22	3.31	3.14
SiO_2	29.93	30.26	30.42	31.06	31.21	30.62	29.67	29.87	29.46
P_2O_5	27.05	27.00	26.62	26.44	26.67	26.59	26.31	26.46	26.24
SO_3	0.16	0.24	0.20	0.19	0.18	0.21	0.25	0.21	0.20
K_2O	0.30	0.45	0.45	0.42	0.41	0.43	0.47	0.45	0.48
CaO	37.05	36.80	37.08	36.65	35.96	36.48	37.97	37.51	38.03
TiO_2	0.14	0.14	0.14	0.14	0.14	0.14	0.16	0.15	0.14
Fe_2O_3	1.28	1.41	1.40	1.46	1.39	1.41	1.50	1.42	1.49
Others	0.34	0.26	0.26	0.26	0.26	0.27	0.27	0.43	0.43
Total	100.00	100.00	100.00	100.00	100.00	100.00	100.00	100.00	100.00

Table 6.
Chemical composition of thermophosphates obtained by a plasma torch TPPT–IZA (30 A–45 A), copied from [6].

Sample	CaO/P_2O_5	MgO/P_2O_5	$R_2O_3(Al_2O_3 + Fe_2O_3)/P_2O_5$
M1	1.360	0.007	0.170
M2	1.390	0.007	0.180
M3	1.390	0.007	0.180
M4	1.350	0.006	0.180
M5	1.370	0.007	0.180
M6	1.440	0.007	0.180
M7	1.420	0.007	0.180
M8	1.450	0.008	0.180

Table 7.
Quality indexes of TPPT-IZA, copied from [6].

R_2O_3 $(Al_2O_3 + Fe_2O_3)/P_2O_5 \leq 0.1$ [23, 24]. Concerning relations CaO/P_2O_5 and MgO/P_2O_5, all samples of thermophosphates (TPPT-IZA) comply with quality indexes, facilitating the production of phosphoric acid. However, the ratio of R_2O_3/P_2O_5 exceeds these parameters due to high concentrations of Al_2O_3 and Fe_2O_3, affecting the solubility.

Table 8 shows the results of elemental analysis by XRF of the Pilar and Ceibita phosphoric rock feedstock (PRPC), and their thermophosphates (TPPT-PC). As a result, PRPC raw material reported a concentration of 32.07 wt. % P_2O_5, while their thermophosphates (M9, M10, M11, M13, M14, M15, and M16) showed concentration less than this due to loss on ignition, high temperatures of plasma torch reactor, an

	Chemical composition of thermophosphates samples (wt.%)								
	PRPC	**M9**	**M10**	**M11**	**M12**	**M13**	**M14**	**M15**	**M16**
Na_2O	0.52	0.74	0.00	0.39	0.91	0.53	0.70	0.45	0.50
MgO	0.10	0.16	0.15	0.15	0.16	0.17	0.18	0.14	0.15
Al_2O_3	1.22	1.10	1.23	1.32	1.18	1.29	1.37	1.18	1.18
SiO_2	9.76	9.57	9.91	10.51	10.27	10.18	10.19	9.74	9.50
P_2O_5	32.07	31.78	30.96	31.86	33.12	31.29	31.84	30.90	31.65
SO_3	0.38	0.49	0.44	0.42	0.41	0.43	0.42	0.43	0.45
K_2O	0.13	0.27	0.29	0.25	0.23	0.27	0.27	0.29	0.29
CaO	54.43	54.48	55.66	53.86	52.49	54.30	53.30	55.27	54.64
Fe_2O_3	0.85	1.04	0.98	0.91	0.89	0.91	0.87	0.97	1.02
Others	0.54	0.39	0.38	0.34	0.34	0.63	0.85	1.81	0.63
Total	100.00	100.00	100.00	100.00	100.00	100.00	100.00	100.00	100.00

Table 8.
Chemical composition of thermophosphates obtained by plasma torch TPPT–PC (30 A–45 A), copied from [6].

increase of speed of heat transfer and chemical reactions [25]. However, the sample of thermophosphates M12 showed a higher concentration of 33.12 wt. % P_2O_5, due to the decomposition of absorbed and combined water, burns organic matter and combination with carbon dioxide, increasing the concentration of calcined phosphate [26]. Regarding the concentration of CaO 54.43 wt. % of PRPC, there were fluctuations in their thermophosphates due to the dissociation of carbonates and the variation of the experimental conditions. On the other hand, SiO_2 concentration of M9 to M16 are within previously reported parameters (5–15 wt. %) and is the optimal concentration required for industrial use [23].

According to the optimal values ($CaO/P_2O_5 \leq 1.6$; $MgO/P_2O_5 \leq 0.022$; R_2O_3 $(Al_2O_3 + Fe_2O_3)/P_2O_5 \leq 0.1$) [23, 24], **Table 9** summary the quality indexes of thermophosphates (TPPT-PC) from M9 to M16. The relation CaO/P_2O_5 only met in

Sample	**CaO/P_2O_5**	**MgO/P_2O_5**	**$R_2O_3(Al_2O_3 + Fe_2O_3)/P_2O_5$**
M9	1.710	0.005	0.070
M10	1.790	0.005	0.070
M11	1.690	0.005	0.070
M12	1.580	0.005	0.060
M13	1.740	0.005	0.070
M14	1.700	0.005	0.070
M15	1.780	0.005	0.070
M16	1.730	0.005	0.070

Table 9.
Quality indexes of TPPT-PC, copied from [6].

sample M12, and the others samples are above the expected results due to higher concentrations of CaO coming from the original raw material. Therefore, more elevated amounts of acid could be required to produce fertilizers [24]. Nevertheless, all samples of thermophosphates (TPPT-PC) comply with quality indexes according to the relations MgO/P_2O_5 and R_2O_3/P_2O_5, facilitating the solubility of fertilizers as the final product.

6.2 Analysis of crystalline phases by XRD

According to a previous study, the Rietveld refinement of the thermophosphate samples of Iza (TPPT-IZA) and Pilar y Ceibita (TPPT-PC) is carried out to identify and quantify the phases with their respective adjustment criteria, such as R_{exp}, R_{wp} and x^2or GOF (goodness of fit) [6]. As a consequence of the analysis XRD of thermophosphates, the best results of assimilable phosphorus and solubility were taken as a reference to show results. Regarding database standards, the following samples and their phases are referenced, M3 (TPPT-IZA) quartz low ICSD (No 08–3849) and apatite–(CaOH) ICSD (No 08–1442); M9 (TPPT-PC) quartz low ICSD (No 08–3849), calcite ICSD (No 01–8166) and fluorapatite carbonate ICSD (No 07–1855); M10 (TPPT-PC) quartz low ICSD (No 06–2406), calcite ICSD (No 01–8165), hatrurite ICSD (No 08–1100), and fluorapatite carbonate ICSD(No 07–1854), and M12 (TPPT-PC) quartz low ICSD (No 20–0726), calcite ICSD (No 18–0349), and apatite–CaF ICSD (No 09–4082).

On the other hand, **Figures 4–7** show crosses representing the experimental data, and lines correspond to simulated XRD patterns. Likewise, Rietveld refinement

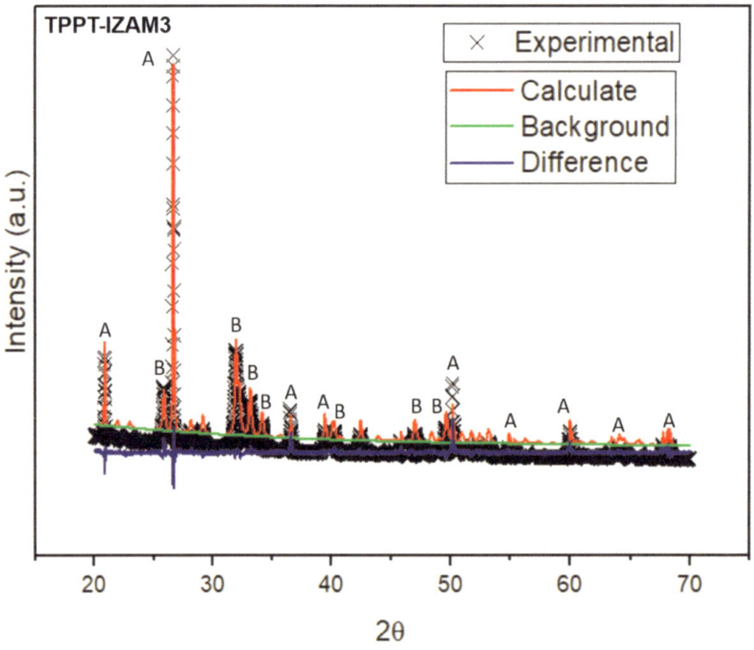

Figure 4.
XRD pattern of sample M3 TPPT-IZA (30 A, 70 wt. % PR, 30 s). Phases A: Quartz low; B: Apatite–(CaOH).

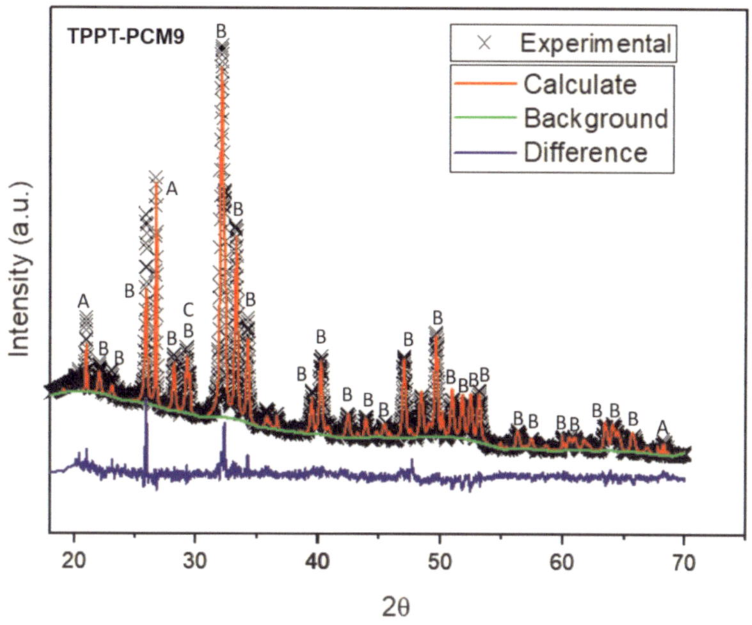

Figure 5.
XRD pattern of sample M9 TPPT-PC (30 A, 65 wt. % PR, 40 s). Phases, A: Quartz low; B: Fluorapatite carbonate; C: Calcite.

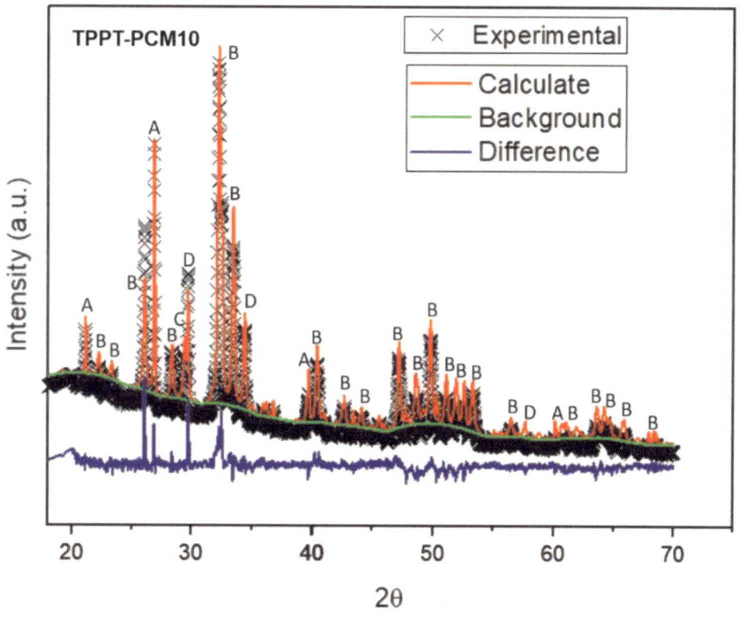

Figure 6.
XRD pattern of sample M10 TPPT-PC (30 A, 65 wt. % PR, 30 s). Phases, A: Quartz low; B: Fluorapatite carbonate; C: Calcite; D: Hatrurite.

permitted to establish that the samples of thermophosphates M3, M9, M10, and M12 presented a hexagonal crystalline system with space group P63/m (#176), and the lattice parameters are shown in **Table 10**.

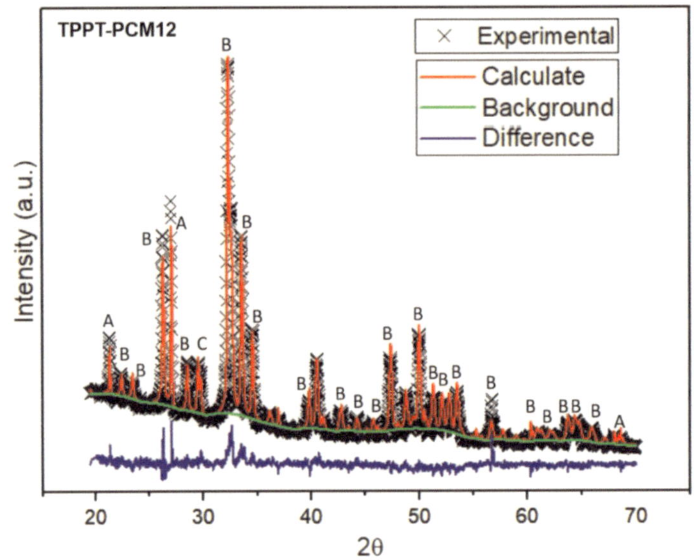

Figure 7.
XRD pattern of sample M12 TPPT-PC (30 A, 70 wt. % PR, 40 s). Phases, A: Quartz low; B: Apatite–CaF; C: Calcite.

Sample	Phase	Lattice parameters		
		a (Å)	b (Å)	c (Å)
TPPT-IZA-M3	apatite–(CaOH)	9.3500	9.3500	6.8917
TPPT-PC-M9	fluorapatite carbonate	9.3489	9.3489	6.8974
TPPT-PC-M10	fluorapatite carbonate	9.3536	9.3536	6.9000
TPPT-PC-M12	apatite–CaF	9.3612	9.3612	6.9027

Table 10.
Crystallographic parameters of thermophosphate, obtained from Rietveld analyses.

According to a previous study, overgrowth of substructures of the material could generate a variation of lattice parameters a and c [27] during plasma torch calcination. In addition, different experimental conditions of thermophosphates could affect the increase of lattice parameters and particle size.

Consequently, the Rietveld analyses allowed for establishing the strongest signals for angles 2θ, and their corresponding crystal lattice planes (hkl) or Miller indices, shown in **Table 11**.

Sample	TPPT-IZA-M3	TPPT-PC-M9	TPPT-PC-M10	TPPT-PC-M12
Phase	apatite–(CaOH)	fluorapatite carbonate	fluorapatite carbonate	apatite–CaF
hkl	2θ	2θ	2θ	2θ
200	21.93	22.05	22.15	22.33
111	22.96	23.02	23.23	23.36
201	25.49	25.55	25.77	25.94

Sample	TPPT-IZA-M3	TPPT-PC-M9	TPPT-PC-M10	TPPT-PC-M12
102	28.12	28.15	28.38	28.57
120	29.15	29.21	29.36	29.54
121	31.97	32.02	32.17	32.35
002	25.89	25.93	26.09	26.21
112	32.24	32.36	32.44	32.62
300	33.24	33.22	32.37	33.54
202	34.22	34.18	34.33	34.51
130	40.11	40.17	40.42	40.48
222	46.93	46.97	47.24	47.28
213	49.57	49.60	49.74	50.04
004	53.25	53.11	53.26	53.43
052	63.41	63.45	63.76	63.73
150	64.14	64.01	64.15	64.46
151	65.54	65.78	65.72	65.86

Table 11.
Crystal structure, lattice parameters, and miller indices of thermophosphates.

6.3 Analysis of assimilable phosphorus and solubility index by UV visible spectrophotometry

Table 12 summarizes the indexes of assimilable phosphorus and solubility obtained by the extraction method with neutral ammonium citrate solution (NAC) and visible UV spectrophotometry [6]. Regarding TPPT-IZA, we can observe that sample M3 subjected to thermal treatment of plasma torch 30 A, showed values of assimilable phosphorus and a solubility index of 3.07 wt. % P_2O_5 and 11.60 wt. % P_2O_5, respectively, overcoming the samples M5 to M8, subjected to 45 A current intensities. Likewise, the sample M10 (TPPT-PC), also subjected to thermal treatment of plasma torch 30 A, obtained the highest concentrations: assimilable phosphorus of 4.83% and solubility index of 15.59%, respectively. According to previous reports that used the method NAC, the solubility takes the following values: high >5.4 wt. % P_2O_5, medium 3.2–4.5 > wt. % P_2O_5 and lower <2.7 wt. % P_2O_5 [28]. Therefore, when comparing the solubility of samples M1 to M16, we found that all samples meet high solubility, while the assimilable phosphorus (an extractable fraction with weak acids) shown in **Table 12**, is low compared to sample superphosphate from the Cuban phosphoric rock (12.28 wt. % P_2O_5) [29]. However, the Colombian Agricultural Institute (ICA in Spanish) indicates that the minimum acceptable contents for solid fertilizers (NPK) for soil application must be at least 3.0 wt. % P_2O_5 as assimilable phosphorus [30]. In conclusion, the samples that comply with the parameters of assimilable phosphorus and solubility are M3 (TPPT-IZA) and M9, M10, M12, M13, M14, M15, and M16 (TPPT-PC), despite the non-use of additives to improve the quality indexes and the influence of novel plasma torch method in the results.

As a consequence of the above, it is shown that the use of the plasma torch reactor technology under conditions 30 A, 70% PRPC, and 40 s (TPPT-PC-M12) favors the

TPPT	Sample	P_2O_5 Total [wt.%]	P_2O_5 Assimilable [wt.%]	Solubility Index [wt.%]
IZA	M1	27.00	2.45	9.09
	M2	26.62	2.70	10.14
	M3	26.44	3.07	11.60
	M4	26.67	2.98	11.17
	M5	26.59	2.96	11.12
	M6	26.31	2.89	10.99
	M7	26.46	2.66	10.05
	M8	26.24	2.48	9.46
PC	M9	31.78	4.03	12.69
	M10	30.96	4.83	15.59
	M11	31.86	2.64	8.28
	M12	33.12	3.47	10.49
	M13	31.29	3.14	10.04
	M14	31.84	3.09	9.72
	M15	30.9	3.03	9.80
	M16	31.65	3.41	10.78
	PRIZA	27.05	3.62	13.38
	PRPC	32.07	4.02	12.54

Table 12.
Analysis of assimilable phosphorus and solubility index, copied from [6].

calcination and conversion of phosphoric rock and increases the concentrations of assimilable phosphorus, solubility index, and total phosphorus, as reported in the previous studies [25, 31].

6.4 Comparison of the composition of organic fertilizers with thermophosphates

Natural materials such as organic fertilizers from different animal or plant sources, including livestock manure, green manures, crop residues, household waste, and compost, improve soil fertility and increase water retention and NPK content [32]. According to certificate requirements of organic fertilizers, values of N-P-K must be 4%, respectively [33]. **Table 13** shows different sources of organic fertilizers, and all comply with this standard. However, total phosphorus (% P_2O_5 Total) in organic fertilizers is low (less than 10%) compared with thermophosphates (≥26%) in **Table 12** due to their origin from natural sources. At the same time, the thermophosphates were obtained from mineral deposits with high concentrations of phosphate and calcination process.

On the other hand, many factors affect the P dynamic in soils, such as pH, salinity, interaction with micronutrients, redox potential, soil structure and texture, and enzymatic activity [38]. However, the advantage of organic fertilizers is directly lead nutrients to plants, slow-release, increase organic matter and improve soil biological

Sources	Chemical characterization (%)			References
	N	P_2O_5	K	
Dried chicken manure	6.00	10.24	2.30	[34]
Meat and bone meal	7.88	10.69	0.34	[34]
Poultry compost	1.45	3.62	1.21	[35]
Cattle manure	1.70	6.92	1.82	[35]
Vermicompost	1.99	3.02	1.04	[36]
Mushroom compost	1.65	2.27	1.50	[36]
Farmyard manure	1.71	2.03	2.33	[36]
Beef cattle manure	2.22	5.70	1.08	[37]

Table 13.
Chemical characterization for different sources of organic fertilizers.

activity [32]. Finally, due to the growth of the world population and the need to obtain different sources of nutrients to produce food, organic fertilizers are a cheap, globally available, and environmentally friendly resource. However, inorganic fertilizers such as thermophosphates generated fast solutions for crops due to high concentration and production in scale.

6.5 Analysis of energy consumption

According to a study on energy consumption in an electric oven, the rank of the value is 12.0 kW-h/Kg-14.0 kW-h/Kg [39]. At the same time, thermophosphates shown in **Table 14** TPPT-IZA and TPPT-PC, submitted low energy consumption between 0.64 kW-h/Kg and 1.28 kW-h/Kg, respectively, for treating samples with a plasma torch to 30 A–45 A. Likewise, a previous study indicated that plasma energy consumption is 1.1 kW-h/Kg and 0.8–1 kW-h/Kg, values very close for this study [40, 41]. Plasma reactor technology also is used for tread waste of printed circuit board (PCB) and recovery metals, precious elements, and hazardous elements, with 2.0 kW-h/Kg electricity consumption [42]. Moreover, plasma technology has been used to simulate the production of biofuels starting from syngas with an electrical consumption of 0.48–2.2 kW-h/Kg for organic waste and 2.23 kW-h/Kg for wood sawdust [43, 44].

TPPT	Current intensity (A)	Plasma torch (KW)	Calcination time (h)	(KW-h)	(KW-h/Kg)
IZA	30	3.24	0.0083	0.027	0.64
	45	4.86	0.0111	0.054	1.28
PC	30	3.24	0.0083	0.027	0.64
	45	4.86	0.0111	0.054	1.28

Table 14.
Energetic consumption for thermophosphates produced by a plasma torch, copied from [6].

7. Conclusions

As a result of the thermal treatment of plasma torch for samples PRIZA and PRPC subjected to current intensities 30 A–45 A, thermophosphates fertilizers were obtained, likewise, shown the Rietveld analysis of experimental diffraction patterns.

The thermophosphates have a hexagonal crystalline system with space group P63/m (#176). Likewise, the Rietveld analysis shows that the most representative crystalline phases of thermophosphates are fluorapatite carbonate, apatite-(CaF), and apatite-(CaOH), respectively. Quality indices were determined as follows, total phosphorus, assimilable phosphorus, and solubility. The most representative results of TPPT-IZA sample M3 correspond to 26.44 wt. % P_2O_5, 3.07 wt.% P_2O_5, and 11.60 wt.% P_2O_5, respectively. On the other hand, the same analysis was carried out for TPPT–PC samples M9, M10, and M12, and the result was: 31,78 wt.% P_2O_5, 4.03 wt.% P_2O_5, and 12.69 wt.% P_2O_5; 30.96 wt.% P_2O_5, 4.83 wt.% P_2O_5, and 15.59 wt. % P_2O_5; 33.12 wt.% P_2O_5, 3.47 wt.% P_2O_5, and 10.49 wt.% P_2O_5, respectively. The best results shown above were obtained under current intensities of 30 A, while the percentage by weight and calcination times of PR can vary depending on the mineral source. TPPT-IZA and TPPT-PC thermophosphates showed low energy consumption between 0.64 kWh/Kg and 1.28 kWh/Kg, respectively. For samples treated with a plasma torch of 30 A–45 A, it was shown that a plasma torch technology is a viable alternative for obtaining thermophosphates at a low cost and using less calcination time for each sample. In addition, it makes the process more efficient and sustainable.

Acknowledgements

National Learning Service Sena, Mining Center and Sennova, Sogamoso, Colombia; Enterprise Phosphate Boyacá, Pesca, Colombia; Pedagogical and Technological University of Colombia IRME, Sogamoso, Colombia.

Conflict of interest

The author declares no conflict of interest.

Author details

Nelson Mauricio Espinel Pérez[1,2]

1 National Learning Service Sena, Mining Center, Sogamoso, Colombia

2 Faculty of Science, Student of Doctorate in Chemical Science, Pedagogical and Technological University of Colombia, Tunja, Colombia

*Address all correspondence to: nelespinel@gmail.com

References

[1] CRU Consulting. Roca Fósfórica Caracterización y análisis de mercado internacional de minerales en el corto, mediano, y largo plazo con vigencia al año 2035. Santiago de Chile: CRU; 2018. pp. 1-51

[2] CRU Strategies. Estudio para caracterizar el mercado nacional e internacional de los minerales estratégicos Reporte final preparado para UPME. Santiago de Chile: Upme; 2013. p. 957

[3] Ministerio de Minas y Energía República de Colombia. Plan Nacional de Desarrollo Minero con Horizonte a 2025. Bogotá D.C.: Minería responsable con el territorio; 2017. p. 174. Available from: www.upme.gov.co

[4] National Mining Agency. Roca Fosfórica. Bogotá D.C.: National Mining Agency; 2016. Available from: https://www.anm.gov.co/sites/default/files/ficha_roca_fosforica_es.pdf

[5] Abouzeid AZM. Physical and thermal treatment of phosphate ores - An overview. International Journal of Mineral Processing. 2008;**85**(4):59-84

[6] Espinel Pérez NM, Pazos MC, Parra E, Martínez D. Calcination of phosphoric rock by plasma torch, to obtain thermophosphates fertilizers. Revista Colombiana de Materiales. 2021; **18**:54-68. Available from: https://revistas.udea.edu.co/index.php/materiales/article/view/345653

[7] Departamento Nacional de Planeación R de C. Objetivos de Desarrollo Sostenible (ODS) Colombia 2030. Comisión ODS; 2015. Available from: https://www.ods.gov.co/es [Accessed: April 8, 2022]

[8] Richardson AE. Regulating the phosphorus nutrition of plants: Molecular biology meeting agronomic needs. Plant and Soil. 2009;**322**(1):17-24

[9] Yang XJ, Finnegan PM. Regulation of phosphate starvation responses in higher plants. Annals of Botany. 2010;**105**(4): 513-526

[10] Johnston AE. Potassium and Nitrogen Interactions in Crops. Harpenden: Potash Development Association; 2012. pp. 1-16. Available from: http://www.pda.org.uk

[11] Ridder M De, Jong S De, Polchar J, Lingemann S. Risks and opportunities in the global phosphate rock market: Robust strategies in times of uncertainty. 2012. 96. Available from: https://www.econbiz.de/Record/risks-and-opportunities-in-the-global-phosphate-rock-market-robust-strategies-in-times-of-uncertainty-ridder-marjolein/10009713558. [Accessed: October 18, 2013]

[12] Kooroshy J, Meindersma C, Podkolinski R, Rademaker M, Sweijs T, Diederen AM, et al. Scarcity of minerals. A strategic security issue. The Hague: The Hague Centre for Strategies Studies; 2009. pp. 23-28. Available from: https://hcss.nl/wp-content/uploads/2010/01/HCSS_Scarcity_of_Minerals.pdf

[13] Fahami A, Nasiri-Tabrizi B, Ebrahimi-Kahrizsangi R. Mechanosynthesis and characterization of chlorapatite nanopowders. Materials Letters. 2013;**110**:117-121

[14] Legeros RZ, Ito A, Ishikawa K, Sakae T, Legeros JP. Fundamentals of hydroxyapatite and related calcium phosphates. Advanced Biomaterials: Fundamentals, Processing, and Applications. 2010:19-52

[15] Kijkowska R, Lin S, LeGeros RZ. Physico-chemical and thermal properties of chlor-, fluor- and hydroxyapatites. Key Engineering Materials. 2002; **218–220**:31-34

[16] Pereira SCC, Cekinski E, Valarelli JV. Process for production of calcined phosphate in a grate furnace. Fertility Research. 1988;**16**(2):169-177

[17] Watti A, Alnjjar M, Hammal A. Improving the specifications of Syrian raw phosphate by thermal treatment. Arabian Journal of Chemistry. 2016;**9**:S637-S642. DOI: 10.1016/j.arabjc.2011.07.009

[18] Gomez E, Rani DA, Cheeseman CR, Deegan D, Wise M, Boccaccini AR. Thermal plasma technology for the treatment of wastes: A critical review. Journal of Hazardous Materials. 2009; **161**(2–3):614-626

[19] Bustinza NK. Aplicaciones Medioambientales del Plasma. DYNA. 1994;**6**:45-47

[20] AOAC. Official Methods of Analysis. 15th ed. Vol. 1. Arlington Virginia USA: Association Official of Analitycal Chemists; 1990. pp. 12-16

[21] Shariati S, Ramadi A, Salsani A. Beneficiation of low-grade phosphate deposits by a combination of calcination and shaking tables: Southwest Iran. Minerals. 2015;**5**(3):367-379. Available from: https://www.mdpi.com/2075-163X/5/3/367

[22] Cuy Buitrago HL, Martín G. Correlación estratigráfica y caracterización petrográfica de algunos niveles fosfáticos de la formación hermitaño (Kse) aflorantes en los municipios de Iza (vereda centro) y Sogamoso (vereda Pilar y Ceibita). Sogamoso, Boyacá-Colombia: Universidad Pedagógica y Tecnológica de Colombia; 2016

[23] Unión Temporal GI. Georecursos. Análisis de la Estructura Productiva y Mercados de la Roca Fosfórica. Bogotá D. C.: Unión Temporal G.I. Georecursos; 2005

[24] FAO. Utilización de las rocas fosfóricas para una agricultura sostenible. In: Boletín FAO Fertilizantes y Nutrición Vegetal. 13th ed. Vol. 13. Roma Italia: FAO; 2007. p. 177. Available from: http://www.fao.org/3/a-y5053s.pdf

[25] Venkatramani N. Industrial plasma torches and applications. Current Science. 2002;**83**(3):254-262

[26] Elgharbi S, Horchani-Naifer K, Férid M. Investigation of the structural and mineralogical changes of Tunisian phosphorite during calcinations. Journal of Thermal Analysis and Calorimetry. 2015;**119**(1):265-271

[27] Chaabouni A, Chtara C, Nzihou A, El FH. Textural And Mineralogical Studies Of Two Tunisian Sedimentary Phosphates Or Carbonated Fluorapatite Used In The Process Of Production Of Phosphoric Acid. 2016;**5**:05

[28] Ardalan M, Gholizadeh A, Tehrani MM, Hosseini HM, Karimian N. Solubility test in some phosphate rocks and their potential for direct application in soil. World Applied Sciences Journal. 2009;**6**(2):182-190

[29] Ordoñez Sánchez YC, Rodríguez Acosta C, Rodríguez SL. Determinación de las condiciones óptimas para la obtención de un fertilizante fosfatado a partir de la roca fosfórica cubana. Ingenieria y Competitividad. 1969;**19**(2):137-148

[30] Colombian Agricultural Institute. Reglamento Técnico de Fertilizantes y Acondicionadores de Suelos para Colombia [Internet]. Bogotá D.C.: "ICA", Colombian Agricultural Institute; 2003.

p. 71. Available from: https://www.ica.gov.co/areas/agricola/servicios/fertilizantes-y-bio-insumos-agricolas/resolucion-150-de-2003-1-1.aspx

[31] Mosse AL, Pechkovsky VV, Dzyuba ED, Krasovskaya LI. Production of condensed phosphate by plasma chemical treatment of phosphate raw material. Plasma Chemistry and Plasma Processing. 1985;**5**(3):275-282

[32] Roba TB. Review on: The effect of mixing organic and inorganic fertilizer on productivity and soil fertility. OALib. 2018;**05**(06):1-11

[33] Widyastuti S, Jupri A, Nikmatullah A, Kurniawan NSH, Kirana IAP, Abidin AS, et al. Analyses of organic matter and heavy metal composition in formulated macroalgae-based organic fertilizer. IOP Conference Series: Earth and Environmental Science. 2021;**913**(1):1-8. Article ID: 012024

[34] Bergstrand KJ. Organic fertilizers in greenhouse production systems – A review. Scientia Horticulturae. 2022;**295** (2021):1-8. Article ID: 110855

[35] De Morais RK, Zacharias MB, Della J, Forti VA. The potential of organic fertilizer application on the production and bromatological composition of millet (Cenchrus americanu). The Interne. 2022:1-13. Available from: https://www.researchsquare.com/article/rs-1486898/v1

[36] Cabilovski R, Manojlovic M, Bogdanovic D, Magazin N, Keserovic Z, Sitaula BK. Mulch type and application of manure and composts in strawberry (Fragaria × ananassa Duch.) production: Impact on soil fertility and yield. Zemdirbyste. 2014;**101**(1):67-74

[37] Eckhardt DP, Redin M, Santana NA, De CL, Dominguez J, Jacques RJS, et al. Cattle manure bioconversion effect on the availability of nitrogen, phosphorus, and potassium in soil. Revista Brasileira de Ciência do Solo. 2018;**42**(0):1-10

[38] Lizcano-Toledo R, Reyes-Martín MP, Celi L, Fernández-Ondoño E. Phosphorus dynamics in the soil–plant–environment relationship in cropping systems: A review. Applied Sciences. 2021;**11**(23)

[39] El Ouardi EM. Effect of temperature and residence time of calcination phosphate on the chemical reactivity: Application to the case of Bouchane phosphate (Morocco). International Journal of Innovation and Applied Studies. 2013;**4**(2):387-407

[40] Riegel ER, Kent JA. In: Kent PD JA, editor. Riegel's Handbook of Industrial Chemistry. Boston, MA: Springer Science & Business Media; 1991. pp. 1-871

[41] La T-g P. tecnología de plasma y residuos sólidos. La Tecnol Plasma y Residuos Sólidos. 2009;**13**(2):51-56

[42] Sanito RC, You SJ, Wang YF. Application of plasma technology for treating e-waste: A review. Journal of Environmental Management. 2021;**288**(200):112380. DOI: 10.1016/j.jenvman.2021.112380

[43] Tamayo-Pacheco JJ, Peña-Pupo L, Vázquez-Peña A, Brito-Sauvanell ÁL. Hydrogen-rich syngas production by plasma gasification of existing biomasses in Cuba. Rev Ciencias Técnicas Agropecu. 2020;**29**(4):53-63. Available from: http://scielo.sld.cu/scielo.php?script=sci_arttext&pid=S2071-00542020000400006&lang=pt

[44] Aliferov AI, Anshakov AS, Domarov PV. Plasma gasification of organic waste for production of motor fuel. Journal of Engineering Thermophysics. 2022;**31**(2):238-247

Chapter 13

The Earthworms: Charles Darwin's Ecosystem Engineer

Rahul Kumar, Renu Yadav, Rajender Kumar Gupta, Kiran Yodha, Sudhir Kumar Kataria, Pooja Kadyan, Pooja Sharma and Simran Kaur

Abstract

The term ecosystem engineering focuses on how organisms physically change the abiotic environment and how this feeds back to the biota. Charles Darwin was the first naturalist who studied the role of the earthworms and their ecosystem services. Darwin's last publication on earthworms gave the role of earthworms in global bioturbation. Darwin also used the word 'friend of farmer' and 'nature ploughman' for the earthworm because of its important role in the soil ecosystem. In modern ecological theory, bioturbation is recognised as 'ecosystem engineering'. They are called as ecosystem engineers due to their different ecosystem services which cause the physical, chemical and biological changes in the soil. This review highlights the different ecological services provided by the earthworms that make them ecosystem engineers as said earlier by Darwin.

Keywords: bioturbation, Charles Darwin, earthworms, ecosystem engineers, soil properties

1. Introduction

Charles Darwin (1809–1882) in his 45-year-long career studied the earthworm and gave many experimental results, observations, interpretations and theories. He published a book "*The Formation of Vegetable Mould, through the Actions of Worms, With Observations on their Habits*" almost six months before his death. His last publication on earthworms demonstrated their role in global soil bioturbation [1]. According to him [2] "*It may be doubted whether there are many other animals which have played so important part in the history of the world, as played by these less organized animals*." He carefully examined the activities of earthworms and reported that earthworms play important role in turning over large amounts of soil, maintaining of soil structure, breakdown of forest litter, plant and animal dead material in soil, aeration and fertility etc. All these activities ultimately promote the plant growth [3–5].

The complex interactions of earthworms with their environment make their study a challenging task [6] like modifying geochemical gradients, humus formation, buffering capacity of their cast, nutrient cycling [3], redistributing food resources, viruses, bacteria, resting stages of various microbes and eggs [7]. Ploughing of soil

IntechOpen

was a most valuable and ancient invention of man, but there is no doubt that earthworms were ploughing soil from millions of years ago and maintaining the physical condition of the soil. Earthworms are regularly ploughing the soil and continued to plough the soil in which they are present. That's why these worms are well-known as "farmer's friends" or "nature's ploughman" [2, 8, 9].

Bioturbation is the term used for the biological reworking of soil and sediments [10] and its importance for soil processes and geomorphology by all organisms including microbes, burrowing animals and rooting plants. In modern ecological theory, bioturbation is recognised as 'ecosystem engineering' [7]. All these pioneering observations and experimental results of Darwin on earthworm makes him founding father of soil science [11]. The concept 'ecosystem engineering' refers to modification in the physical environment that strongly affects the other organisms [7]. In the case of earthworms, it mainly affects the physical condition of soil which is an abiotic factor. Earthworms improve soil fertility by modifying the soil environment. Thus ultimately earthworms are capable of structuring, maintaining or restoration of degraded soils [4, 8, 12, 13]. This review paper highlights, how earthworms play an important role in different ecological services that make them ecosystem engineers.

2. General biology and classification of earthworms

Earthworms have dark brownish to red body, covered with cuticle and average body weight 1400–1500 mg after 8–10 weeks. They are soft-bodied, long, narrow, cylindrical, bilaterally symmetrical, metamerically segmented and soil-dwelling invertebrates [14]. Their body contains a large amount of protein approximately 65% (around 70–80% 'lysine-rich protein'). Their body also has 14% carbohydrates, 14% fats and 3% ash [15]. The gut of an earthworm is an almost straight tube starting from the mouth followed by a muscular pharynx, oesophagus, gizzard, intestine associated digestive glands and ending outside through the anus [3]. Their gut contains some important inorganic nutrients, protein, mucus, various polysaccharides forms and symbiotic microbes like bacteria, protozoa and microfungi. Earthworm's gut environment provides optimum conditions (increased total organic carbon, nitrogen and moisture) for activation of microbes from the dormant stage and also for germination of endospores [16, 17]. Various digestive enzymes such as cellulase, amylase, protease, chitinase, lipase and urease were identified from the alimentary canal of earthworms. Microbes present in their gut are responsible for cellulase and mannose activities [16, 17]. Their life span varies from species to species with a range of 3–7 years. These are hermaphrodite animals but cross-fertilisation occurs. The average rate of cocoon production was 7.23, 0.99 and 0.53/worm/week in the monsoon season, winter and the summer respectively [4, 18, 19].

Ronald and Donald [20] reported that the earthworms and microorganisms were associated symbiotically and thus enhance the decomposition of organic matter. Due to this relationship, they can break down a large amount of organic material within the time limit and return it to the nutrient cycle which is responsible for the introduction of vermiculture. Earthworms are highly sensitive to pollutants, so they can be used as bio-indicator tools [21].

Evolutionary date of earthworm is about 6 billion years back. They belong to phylum Annelida and class Oligochaeta. Worldwide around 4200–4400 species of oligochaetes and 20 families with 3200 species of earthworms are well known [4, 16, 22, 23]. Some groups are abundantly distributed in the whole world. Different taxonomic groups were found in different ecosystems except for Antarctica. Some species of earthworms can also be noticed in estuarine waters like *Pontodrilus bermudensis*. *Lampito mauritii* can be bred and cultured in sandy soils [3]. In the Indian subcontinent, earthworms represent the bulk of the oligochaete fauna. According to Julka [24], there are around 509 species and 67 genera of earthworms. The most abundant earthworms found in grasslands and agricultural ecosystems in the Palearctic region are belonging to the family Lumbricidae [25]. The greatest variety is found in tropical soils with progressively smaller numbers of species in the northern region. Therefore in France, there are about 180 earthworm species [26] and in the U.K. about 25 species [27]. Earthworms are found in almost every part of the world except the driest and the coldest regions because earthworms are sensitive to a range of environmental factors such as pH, temperature, water content, aeration and salinity levels [8, 22, 28]. Edwards [29] reported that earthworms belonging to temperate regions cannot tolerate high temperatures. *Eisenia fetida* is most productive at 20°C if both reproduction and growth rate are considered. We can say that earthworms are widely distributed creatures in nature and more than 80% of soil invertebrate's biomass is occupied by the earthworms [30].

2.1 Trophic classification of earthworms

Bouche [31] classified earthworms based on ecological habitat and their feeding habits, into three major groups (**Figure 1**):

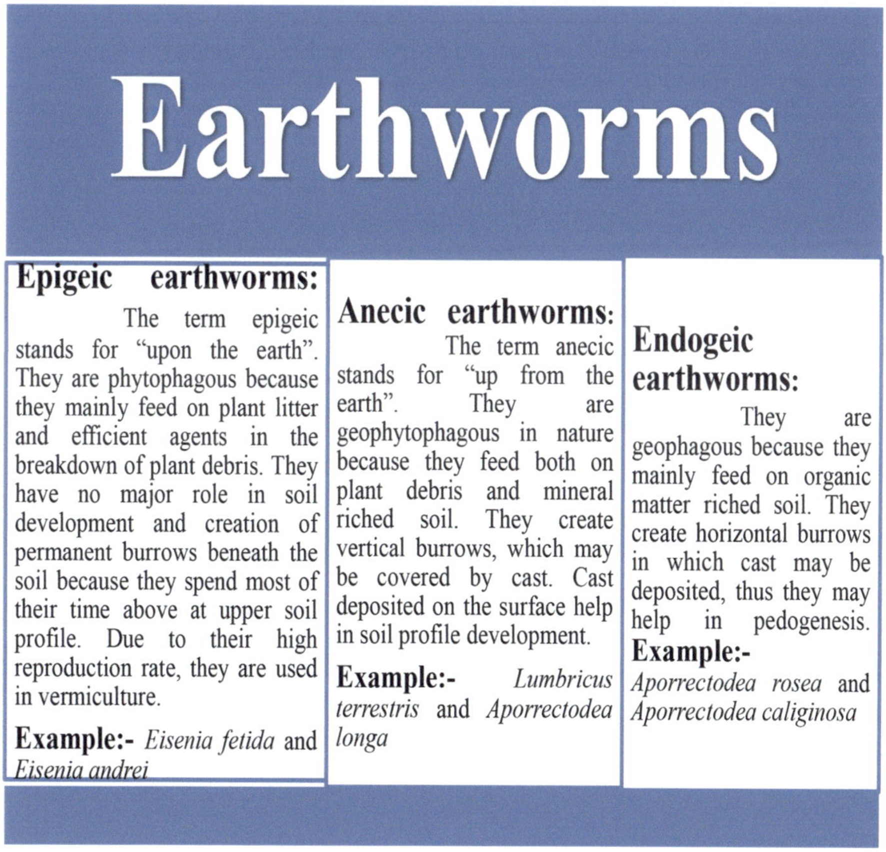

Figure 1.
Trophic classification of earthworms.

3. Ecosystem services of earthworms

The book "*The Formation of Vegetable Mould, through the Actions of Worms, With Observations on their Habits*" describes the role of earthworms in soil weathering, pedogenesis, nutrient cycling, organic degradation, soil aeration, burrow formation, development of vegetative mould (topsoil), soil profile differentiation, casting, interchange of the top soil profile, soil fertility, plant growth and protection of archaeological remains through their burial activity [12]. Earthworm's dry powder and its extract are also used as preventive agents and therapeutic agents for various diseases like bladder stone excretion promoting agents, bladder stone reducing agents, tonic agents, hair growth agents, jaundice, antipyretics, aphrodisiacs, therapeutic agents for convulsion, blood circulation promoters, therapeutic agents for hemiplegia, diuretics, indirect analgesics, antihypertensive and antiasthmatics agents [32]. Major ecosystem services of earthworms are: (**Figure 2**).

3.1 Effect of earthworms on soil properties

Darwin showed that earthworms are important ecosystem engineers in soil formation (pedogenesis), by mixing layers of soil, humus formation, affecting rock weathering rate and soil horizon differentiation. The influence of earthworms on soil properties depends upon the species present in that particular area and the life history of that species. For example, vertical burrows made by *L. terrestris* (anecic) facilitate the water flow through the soil profile. Thus they can increase

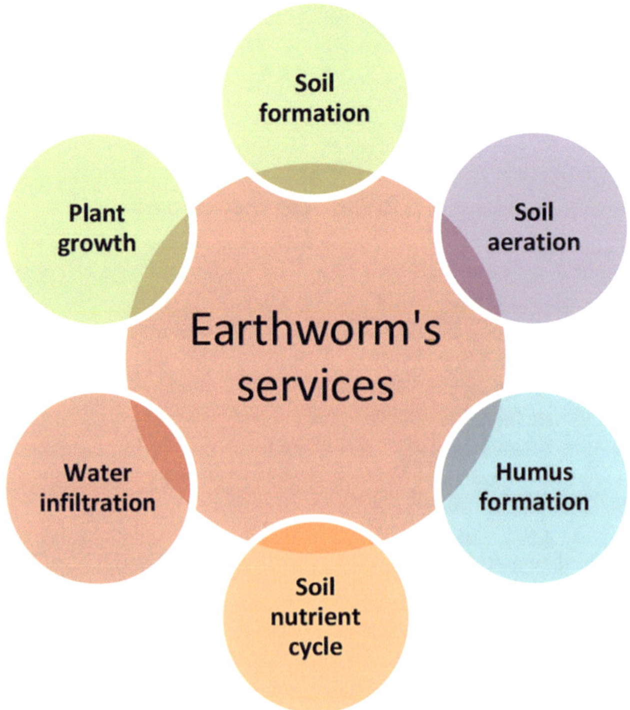

Figure 2.
Ecosystem services of earthworms.

the transport of water and nutrients in the deeper part of soil [4, 23]. It has been reported that surface cast play important role in soil profile development whereas cast deposited beneath the soil profile contributed to pedogenesis. Endogeic species *Octolasion tyrtaeum* lives in upper mineral soil layers and mainly consumes soil organic matter. Jones et al. [33] and Eisenhauer et al. [34] in their study reported that *O. tyrtaeum* and *L. terrestris* species act as keystone detritivores and ecosystem engineers.

3.1.1 Soil formation

The importance of earthworms in chemical weathering was first studied by Darwin [2] in an experiment where the red colour of red-oxide sand disappeared after passing through earthworm intestine, and dissolution of this red-oxide occur due to biochemical changes by specific enzymes released in gizzard and intestine [17]. According to Darwin [2], weathering of rock occurs due to physical and chemical changes in the nature of the rock. The physical weathering by earthworms is possible due to the breakdown of rocks in the gizzard of earthworms whereas chemical weathering occurs due to their intestinal enzyme and microflora. Darwin [2] studied that calciferous glands formed calcareous concretions which combine with small stone or sand present in the gizzard. At that time Darwin did not know exactly the role of calcite formed by the calciferous gland but know that it is a true excretory product. Recent studies showed that water balancing might be an important role by these glands [12, 35]. From the gizzard, grinded material (2–4 micron) pass through the intestine where chemical degradation occurs. Thus gizzard and intestine act as "bioreactors". However, the rest material is excreted out as cast [17].

Darwin [2] observed that earthworms produced almost 1140 kg/ha/year cast. Bertrand et al. [36] studied that 0–15 cm of soil layer drying was enhanced by endogeic species *A. calligenosa* and anecic species *L. terrestris* by increasing evaporation through their burrows. Erosion also helps in pedogenesis and earthworm's casts on soil surfaces have a significant role in this process. Aggregate size distribution is also affected by earthworm activities. For example, *Reginaldia omodeoi* increased the proportion of aggregate greater than 2 mm in diameter from 24.6 to 42.2% or 29.8 to 53.5% under maize and yam culture respectively [37–39].

Darwin [2] concluded that the process of ingestion, grinding and digestion in the earthworm's intestine, excretion of cast with mucus and microflora, mixing of organic nutrients and vegetative mould (topsoil) exposed the rock particles for chemical alteration as a result of which soil formation occur and ultmatelly soil amount increase.

3.1.2 Soil aeration

Air-filled pores are crucial to help the plant roots to thrive because the plant needs O_2 for photosynthesis and expel CO_2 from surrounding soil. Earthworms improve the exchange of these gases with the atmosphere (**Figure 3**).

Soil aggregation is improved by mixing of organic matter with soil in the earthworms' gut which is released as cast. These casts are highly stable aggregates which are deposited by some earthworms in their burrows and by others at the surface of the soil. Thus they may form permanent or temporary burrow [4]. Generally, vertical

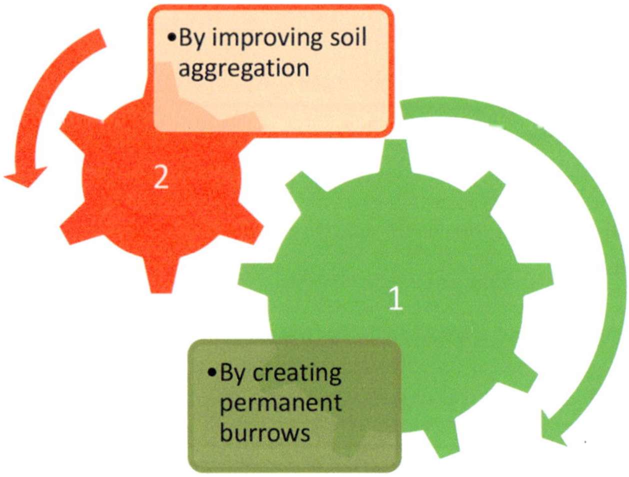

Figure 3.
Mechanism for increase soil aeration by earthworms.

burrows (more than 1 mm in diameter) are formed by anecic earthworm species that can extend upto 1 m depth in the soil. of endogeic species mainly orientated in the horizontal direction and their diameter are smaller than burrows made by anecic species whereas epigeic species are present in a few centimetres of upper soil and they mainly remain in plant litter [26, 36]. Wollny [40] reported that earthworms show an increase of soil volume from 8 to 30%.

3.1.3 Humus formation

Darwin was the first naturalist who recognised that earthworm's activities help in humus formation on the topsoil profile of the earth. He observed that earthworms were responsible for the breakdown as well as rapid incorporation of organic matter in different layers of soil. Coarse particles are continually used by the epigeic and anecic earthworms and triturated ingested organic particles in their gut [1, 12].

The short-term increase of mineral nutrients is well known however long term effects on soil organic matter (SOM) in the presence of earthworms are less clear [41]. Biodegradation of organic matter in the gizzard and the intestine occurs by proteases, lipases, amylases, cellulases and chitinases which bring about a rapid biochemical conversion of the cellulosic and the proteinaceous organic materials. Cast released by earthworms has many beneficial microbes which further help in the biodegradation of organic matter [17, 42]. Earthworm's gut alters the soil microflora and processes mediated by these microorganisms accelerate mineralisation and decomposition of soil organic matter. It was observed that soil inhabited by earthworms had more microflora than the soil devoid of earthworms [43].

3.1.4 Soil nutrient cycle

The cycling of nutrients is a critical function that is possible due to the decomposition of organic matter. The decomposition process is mediated by different types of microbes. Earthworms ingest plant debris and release it as casts. These casts have further beneficial microbes and important nutrients such as, nitrogen (N) and phosphorus (P) which can be used by microbes for their multiplication and vigorous action. Due to the symbiotic activities of earthworms and microbes, organic content which is unavailable to plants becomes available for plant growth in the form of inorganic nutrients. Earthworms also influence the supply of nutrients through their tissues but mainly through their burrowing activities [4, 17, 19, 44]. In earthworm's gut, casts and burrows; different groups of bacterial species were reported such as, *Aeromonas hydrophila* in *Eisenia fetida* [45], *Actinobacteria* in *L. rubellus* [46] and fluorescent *Pseudomonas* in *Lumbricus terrestris* [47].

During their function as ecosystem engineers; anecic and endogeic earthworms create semi-permanent burrows that permit the transport of O_2 into deeper soil profiles which may alter the principal nitrogen transformation reaction [48]. Haimi and Hutha [49] observed that *L. rubellus* stimulated microbial respiration by 15–28% whereas *Dendrobaena octaedra* stimulated it slightly. Both of these species increase nitrogen mineralisation within the soil environment. There are many cases in which carbon cycling occurs through the complementary mechanism of earthworms [41]. Decomposition of organic matter increases mineralisation of carbon upto 90% which is carried out by microorganisms such as bacteria and fungi [50].

3.1.5 Water infiltration

Water infiltration is regulated mainly by the soil porosity. Soil moisture and water infiltration rate decreased in a grass sward, when earthworms were experimentally removed [37, 51]. Soil mechanical and hydraulic properties are also affected by the earthworms, mainly through their burrowing activities which generate macropores. These macropores appreciably impact water infiltration and thus are important for supplying water to crops, in addition to controlling surface runoff and erosion [36]. Baker et al. [52] studied that sub soil properties like drainage and infiltration were improved in Australia with the introduction of *A. longa*.

The epigeic *L. rubellus* tends to favour water storage in the topsoil, because it leaves the litter at the soil surface rather than burying it, thus prevents evaporation. Compared to other species, *A. caliginosa* forms temporary burrows and continually rebuilt which cause a higher infiltration rate to the subsoil [36]. This infiltrated water can be a source of agricultural crop water or percolate through the soil horizon. The surface hydrological processes were also affected by water infiltration in the presence of earthworms [12].

In Ohio, anecic earthworm burrows greatly reduced soil erosion up to 50% due to an increase in infiltration rate [53]. Endogeic species *R. omodeoi* and *Dichogaster terraenigrae* improved weakly (+22 to 27%) water infiltration rate and in the case of *Hyperiodrilus africanus* it was strongly improved +77% [54].

3.2 Effect of earthworm on plant growth

According to Darwin [2] "earthworms help in preparation of soil in an excellent manner which enhances the growth of fibrous-rooted plants and seedlings of all kinds". Earthworms play a key role in nutrient rich manure production [55], mineral soil formation, soil porosity, water infiltration, amorphous colloidal humus formation and nutrient cycle which directly affect the plant's growth.

Earthworms are known for engineering seed bed conditions for plants. Some recent studies highlight direct and indirect interactions between earthworms and plant seeds which is responsible for plant community composition [4, 19, 34]. Krishnamoorthy and Vajranabhaiah [56] reported that earthworm's casts have some plant growth regulators such as cytokinins and auxins. Vesicular Arbuscular Mycorrhizae (VAM) population is enhanced due to the presence of earthworms and which is effective for the growth of wheat and other crops. VAM helps in the uptake of phosphate by plants [3]. Earthworms also help in controlling various types of pests in different crops. For example, *A. rosea* and *A. trapezoids* reduced the population of soil-borne fungal pathogens and the earthworm *R. omodeoi* reduced the damage caused by plant-parasitic nematodes *Heterodera sacchari* on rice plants [36, 57–59].

3.3 Effect of insecticides on growth, reproductive potential and avoidance behviour of earthworm

Insecticides have major effect on the reproductive, growth and avoidance behviour of earthworm. Due to use insecticides, benifical bacteria in earthworm's gut and its surrounding soil are decrease that ultimatelety effect the physiological activities of earthworm. Therefore the growth and reproductive potential of earthworm in contaminated decresed drastically [60] (**Figure 4**). Due to presences of protostomium, earthworms try to avoid the contaminated soil having the insecticides like chlorantraniliprole, fipronil, etc. (**Figure 5**). But in higly contaiminted, they are unble to avoid, because decrease in energy and die ultimately [4]. So we can say insecticides have decremental effect on trhe earthworm.

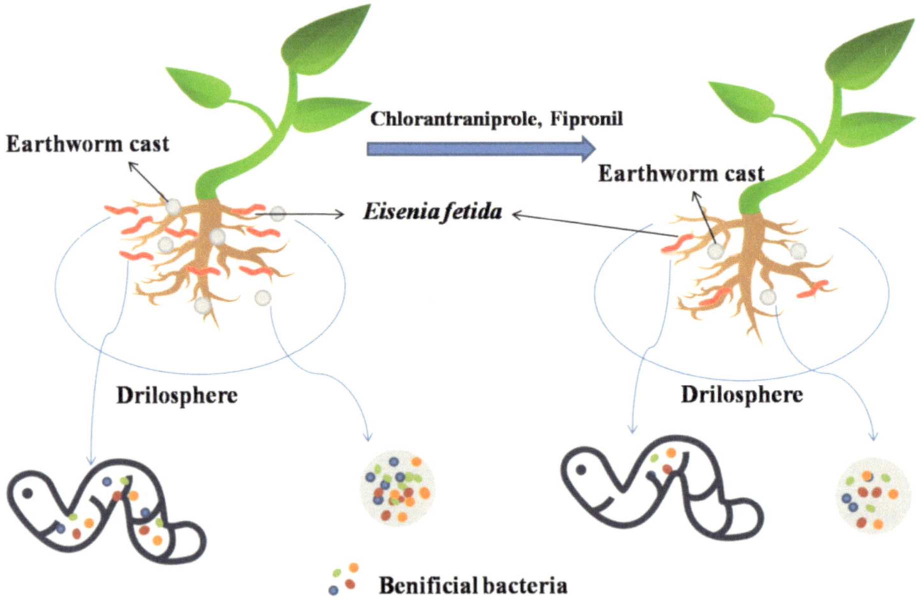

Figure 4.
Effect of insecticides on the beneficial bacteria on earthworms.

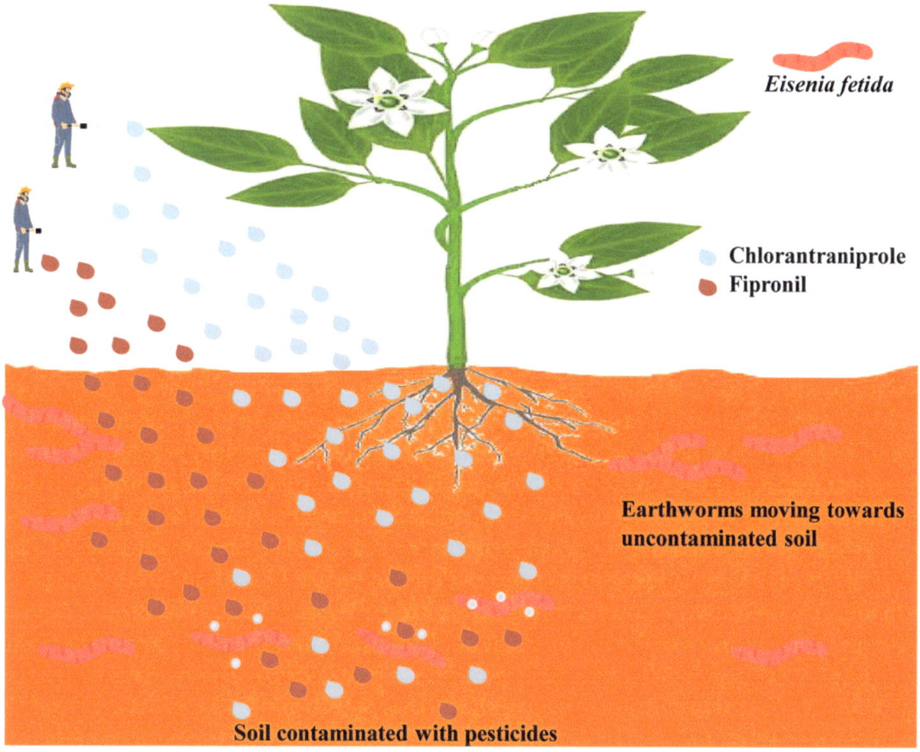

Figure 5.
Effect of insecticides on the avoidance behviour of earthworms.

4. Techniques used to identify the beneficial attributes of earthworms as ecosystem engineers

Earthworms can consume a wide range of unstable organic matter such as animal waste, industrial waste, sewage sludge, etc. [61, 62]. The burrowing activity of earthworms enhances decomposition, formation of humus, development of soil structure, and cycling of nutrients. Effect of these different kinds of processes done by earthworms are estimated by various techniques which are described below as:

4.1 HPLC-MS/MS

High-Performance Liquid Chromatography-Mass Spectrometry (HPLC-MS) is an excellent technique for component identification, quantification, and mass analysis, and it is frequently used to analyse chemical composition and purity. HPLC-MS has high sensitivity and is ideal for precise and repeatable quantitative analysis. For heat-labile chemicals in soil, such as insecticides and their metabolites, high-performance liquid chromatography techniques are suited. In HPLC-MS, the two techniques (HPLC and MS) are linked by an interface that transmits the separated components from the liquid chromatograph column into the mass spectrometer ion source, allowing the target compound to be identified and quantified. In a study, the enantioselective acute toxicity to earthworms of racemic fipronil and its enantiomers was done by Qu et al.,

[63]. For this reason, fipronil is released into the soil, where it is consumed by earthworms and degraded. HPLC was used to estimate the amount of pesticide ingested by the earthworm. For the determination of residues, HPLC analysis of extracts from earthworm samples collected at various times after fipronil application was performed.

4.2 GC-MS/MS

GC is a separation science technique that is used to separate the chemical components of a sample mixture and then detect them to determine their presence or absence and/or how much is present. GC detectors are limited in the information that they give; this is usually two-dimensional giving the retention time on the analytical column and the detector response. Identification is based on a comparison of the retention time of the peaks in a sample to those from standards of known compounds, analysed using the same method. However, GC alone cannot be used for the identification of unknowns, which is where hyphenation to an MS works very well. MS is an analytical technique that measures the mass-to-charge ratio (m/z) of charged particles and therefore can be used to determine the molecular weight and elemental composition, as well as elucidate the chemical structures of molecules. Data from a GC-MS is three-dimensional, providing mass spectra that can be used for identity confirmation or to identify unknown compounds plus the chromatogram that can be used for qualitative and quantitative analysis. Chang et al., [64] estimated the Bioaccumulation and enantioselectivity of type I and type II pyrethroid pesticides in earthworms by using GC-MS/MS. The earthworms treated with the two pyrethroids were processed and the residues were extracted in an n-hexane solvent. This extract was further cleaned up by using the adsorbents to remove any interfering substances coming at the retention time of the two pyrethroids. Then, this processed extract was subjected to residue analysis and the estimation of the residues of pyrethroids was done by using GC-MS/MS.

4.3 GC-ECD/FID/NPD

Most of the organic analyses that we conduct are for organic-priority pollutants such as pesticides, PCBs, PHCs, BTEX, and other petroleum products. Gas Chromatography (GC) is the standard methodology. From sample matrices such as soil, water, plants, fish, and earthworms, the analytes are extracted into organic solvents, separated from interfering substances (cleaned up), concentrated, and run on the GC. The detector was selected based on the type of analyte to be detected in the case of nitrogen and phosphorous-containing analytes, the NPD detector is used and in the case of analytes containing halogens, the ECD detector is used. For the estimation of hydrocarbons, an FID detector coupled with gas chromatography was used. Martinkosky et al., [65] demonstrated the bioremediation of soil by increasing the degradation rates of heavy crude oil hydrocarbons by the earthworms. This study demonstrated that earthworms accelerate the bioremediation of crude oil in soils, including the degradation of the heaviest polyaromatic fractions.

4.4 LC-MS/MS

It is a powerful analytical technique that combines the resolving power of liquid chromatography with the detection specificity of mass spectrometry. Liquid chromatography (LC) separates the sample components and then introduces them to the mass spectrometer (MS). The MS creates and detects charged ions. When a sample is injected, it is adsorbed on the stationary phase, and the solvent passes through the column to separate the compounds one by one, based on their relative affinity to the packing materials and the solvent. The component with the most affinity to the stationary phase is the last to separate. This is because high affinity corresponds to more time to travel to the end of the column. The difference between traditional LC and HPLC is that the solvent in LC travels by the force of gravity. In the application of HPLC, the solvent travels under high pressure obtained employing a pump to overcome the pressure drop in the packed column, which reduces the time of separation. As will be discussed, a continuous flow syringe pump is very useful in HPLC. Hu et al., [66] studied the behaviour of the imidacloprid herbicide in the earthworms by estimating the residues of this herbicide in soil via LC-MS/MS. Bioaccumulation and degradation of imidacloprid were estimating the concentration of this herbicide in the treated soil and earthworms after a few days of the treatment. Samples of the treated soil and earthworms present in it were taken at regular intervals of time and these samples were processed for further residue analysis using LC-MS/MS.

4.5 AAS

AAS is an analytical technique used to determine how many certain elements are in a sample. It uses the principle that atoms (and ions) can absorb light at a specific, unique wavelength. When this specific wavelength of light is provided, the energy (light) is absorbed by the atom. Electrons in the atom move from the ground state to an excited state. The amount of light absorbed is measured and the concentration of the element in the sample can be calculated. Ekperusi et al., [67] conducted a study to assess the levels of heavy metals present in crude oil-contaminated soil, and the application of the earthworm - *H. africanus* with an interest in the bioremediation of metals from the contaminated soil was investigated within 90 days under laboratory conditions. Selected heavy metals such as zinc, manganese, copper, nickel, cadmium, vanadium, chromium, lead, mercury, and arsenic were determined using AAS. Assessment of heavy metals indicated that heavy metals are present in crude oil at elevated levels beyond national regulatory guidelines. There was a significant ($P < 0.05$) decreasing trend in the percentage of heavy metals present in the soil after inoculation with an earthworm in zinc (57.66%), manganese (57.72%), copper (57.64%), nickel (57.69%), cadmium (57.57%), vanadium (57.68%), chromium (57.67%), lead (57.64%), arsenic (1.36%) and mercury (57.41%) after 90 days period. The bioaccumulation factor showed that zinc, manganese, copper, cadmium, vanadium, chromium, and lead had a factor of 1.36, while nickel, arsenic, and mercury had 1.37, 0.01, and 1.35 respectively. The results showed that the earthworms *H. africanus* can be effectively used to bioremediate heavy metals from crude oil-polluted soil.

5. Conclusion

Earthworms cause physical, chemical and biological changes in the soil and improve soil fertility mainly by increasing soil weathering, soil porosity, water infiltration and humus formation. All these activities of earthworms help in plant growth. Apart from these, they also protect the plants from various types of pathogens. Thus we can conclude that earthworms significantly modify their environment by improving the chemical, biological and physical properties of soil and because of these activities, Charles Darwin used the term "ecosystem engineer" for them.

Author details

Rahul Kumar[1*], Renu Yadav[1], Rajender Kumar Gupta[1], Kiran Yodha[1], Sudhir Kumar Kataria[2], Pooja Kadyan[2], Pooja Sharma[3] and Simran Kaur[4]

1 Department of Zoology and Aquaculture, Chaudhary Charan Singh Haryana Agricultural University, Hisar, Haryana, India

2 Department of Zoology, Maharshi Dayanand University, Rohtak, Haryana, India

3 Department of Chemistry, Chaudhary Charan Singh Haryana Agricultural University, Hisar, Haryana, India

4 Department of Animal Genetics and Breeding, Lala Lajpat Rai University of Veterinary and Animal Sciences, Hisar, Haryana, India

*Address all correspondence to: rahulrohila01@gmail.com; rkrohila22@gmail.com

References

[1] Kutschera U, Elliott JM. Charles Darwin's observations on the behaviour of earthworms and the evolutionary history of a giant endemic species from Germany, Lumbricus badensis (Oligochaeta: Lumbricidae). Applied and Environmental Soil Science. 2010;**2010**:1-11

[2] Darwin C. The Formation of Vegetable Mould through the Action of Worms, with Observations of their Habits. London: Murray; 1881

[3] Ismail A. Vermicology: The Biology of Earthworms. Hyderabad: Orient Longman; 1997

[4] Kumar R, Sharma P, Gupta RK, Kumar S, Sharma MMM, Singh S, et al. Earthworms for eco-friendly resource efficient agriculture. In: Resources Use Efficiency in Agriculture. Singapore: Springer; 2020. pp. 47-84

[5] Levinton J. Bioturbators as ecosystem engineers: Control of the sediment fabric, inter-individual interactions, and material fluxes. In: Linking Species & Ecosystems. Boston, MA: Springer; 1995. pp. 29-36

[6] Bartlett MD, Briones MJ, Neilson R, Schmidt O, Spurgeon D, Creamer RE. A critical review of current methods in earthworm ecology: From individuals to populations. European Journal of Soil Biology. 2010;**46**(2):67-73

[7] Meysman FJ, Middelburg JJ, Heip CH. Bioturbation: A fresh look at Darwin's last idea. Trends in Ecology & Evolution. 2006;**21**(12):688-695

[8] Chan KY. An overview of some tillage impacts on earthworm population abundance and diversity—Implications for functioning in soils. Soil and Tillage Research. 2001;**57**(4):179-191

[9] Russell RS. Plant Root Systems: Their Function and Interaction with the Soil. UK: Mc Graw-Hill Book Company Limited; 1977

[10] Kristensen E, Haese RR, Kostka JE. Interactions between Macro- and Microorganisms in Marine Sediments. USA: American Geophysical Union; 2005

[11] Johnson DL. Darwin would be proud: Bioturbation, dynamic denudation, and the power of theory in science. Geoarchaeology: An. International Journal. 2002;**17**(1):7-40

[12] Feller C, Brown GG, Blanchart E, Deleporte P, Chernyanskii SS. Charles Darwin, earthworms and the natural sciences: Various lessons from past to future. Agriculture, Ecosystems & Environment. 2003;**99**(1-3):29-49

[13] Lavelle P, Bignell D, Lepage M, Wolters V, Roger P, Ineson POWH, et al. Soil function in a changing world: The role of invertebrate ecosystem engineers. European Journal of Soil Biology (France). 1997;**33**:159-193

[14] Gajalakshmi S, Abbasi SA. Earthworms and vermicomposting. Indian Journal of Biotechnology. 2004;**3**:486-494

[15] Sinha RK, Bharambe G, Ryan D. Converting wasteland into wonderland by earthworms-a low-cost nature's technology for soil remediation: A case study of vermiremediation of PAHs contaminated soil. The Environmentalist. 2008;**28**(4):466-475

[16] Munnoli PM, Da Silva JAT, Saroj B. Dynamics of the soil-earthworm-plant relationship: A review. Dynamic Soil, Dynamic Plant. 2010;**4**(1):1-21

[17] Sinha RK, Agarwal S, Chauhan K, Chandran V, Soni BK. Vermiculture technology: Reviving the dreams of sir Charles Darwin for scientific use of earthworms in sustainable development programs. Technology and Investment. 2010;**1**(03):155

[18] Biradar VA, Amoji SD, Shagoti UM, Biradar PM. Seasonal variations in growth and reproduction of the earthworm Perionyxexcavatus (Oligochaeta: Megascolecidae). Biology and Fertility of Soils. 1999;**28**(4):389-392

[19] Le Bayon RC, Bullinger G, Schomburg A, Turberg P, Brunner P, Schlaepfer R, et al. Earthworms, plants, and soils. Hydrogeology, Chemical Weathering, and Soil Formation. 2021;**257**:81-103

[20] Ronald EG, Donald ED. Earthworms for Ecology and Profit. Vol. 2. Ontario, California: Earthworm and the Ecology; 1977

[21] Yasmin S, D'Souza D. Effect of pesticides on the reproductive output of Eisenia fetida. Bulletin of Environmental Contamination and Toxicology. 2007;**79**(5):529-532

[22] Edwards CA, Bohlen PJ. Biology and Ecology of Earthworms. Vol. 3. London, UK: Springer Science & Business Media; 1996

[23] Kooch Y, Jalilvand H. Earthworms as ecosystem engineers and the most important detritivors in forest soils. Pakistan Journal of Biological Sciences. 2008;**11**(6):819-825

[24] Julka JM. Earthworm Resources of India and their Utilization in Vermiculture. Calcutta: The Director, Zoological Survey of India (ed) Earthworm resources and vermiculture; 1993. pp. 51-56

[25] Perez-Losada M, Bloch R, Breinholt JW, Pfenninger M, Domínguez J. Taxonomic assessment of Lumbricidae (Oligochaeta) earthworm genera using DNA barcodes. European Journal of Soil Biology. 2012;**48**:41-47

[26] Bouche MB. Lombriciens de France. Eucologie at syste'matique. Annales de Zoologie—Eucologie Animale. 1972;**72**:1-671

[27] Sims RW, Gerard BM. Earthworms: Keys and Notes for the Identification and Study of the Species. Vol. 31. London: Brill Archive; 1985

[28] Lee KE. Earthworms: Their Ecology and Relationships with Soils and Land Use. London, UK: Academic Press Inc.; 1985. p. 411

[29] Edwards CA, Fletcher KE. Interactions between earthworms and microorganisms in organic-matter breakdown. Agriculture, Ecosystems & Environment. 1988;**24**(1-3):235-247

[30] Jouquet P, Plumere T, Thu TD, Rumpel C, Duc TT, Orange D. The rehabilitation of tropical soils using compost and vermicompost is affected by the presence of endogeic earthworms. Applied Soil Ecology. 2010;**46**(1):125-133

[31] Bouche MB. Strategies lombriciennes. In 'soil organisms as components of ecosystems'.(Eds U Lohm, T Persson). Ecological Bulletins. 1977;**25**:122-132

[32] Ishii Y, Okamoto T, Ishii S, Well Stone Co. Amyloid beta fibril decomposing agent, and therapeutic or preventive agent and therapeutic or preventive food composition for diseases caused by amyloid beta fibril formation. U.S. Patent Application 15/577,464. 2018. pp. 1-9

[33] Jones CG, Lawton JH and Shachak M. Organisms as ecosystem engineers. In:

Ecosystem management. Copenhage; 1994;**69**(3):373-386

[34] Eisenhauer N, Straube D, Johnson EA, Parkinson D, Scheu S. Exotic ecosystem engineers change the emergence of plants from the seed bank of a deciduous forest. Ecosystems. 2009;**12**(6):1008-1016

[35] Morgan AJ, Winters C, Williams AF, Turner M. The transformation of calciticspherites to excretory concretions in the earthworm calciferous glands: Did Darwin miss an ecophysiological function. In: Proceedings of the Seventh International Symposium on Earthworm Ecology. Wales: Cardiff; 2002

[36] Bertrand M, Barot S, Blouin M, Whalen J, De Oliveira T, Roger-Estrade J. Earthworm services for cropping systems. A review. Agronomy for Sustainable Development. 2015;**35**(2):553-567

[37] Blouin, M., Hodson, M.E., Delgado, E.A., Baker, G., Brussaard, L., Butt, K.R., Dai, J., Dendooven, L., Pérès, G., Tondoh, J.E. and Cluzeau, D., 2013. A Review of earthworm impact on soil function and ecosystem services. European Journal of Soil Science, 64(2), pp. 161-182.

[38] Gilot C. Effects of a tropical geophageous earthworm, M. anomala (Megascolecidae), on soil characteristics and production of a yam crop in Ivory Coast. Soil Biology and Biochemistry. 1997;**29**(3):353-359

[39] Gilot-Villenave C, Lavelle P, Ganry F. Effects of a tropical geophagous earthworm, Millsoniaanomala, on some soil characteristics, on maize-residue decomposition and on maize production in Ivory Coast. Applied Soil Ecology. 1996;**4**(3):201-211

[40] Wollny E. Studies on the influence of the fertility of the field crop by the activity of the earthworms. Research in the field of Agricultural Physics. 1890;**13**:381-394

[41] Lavelle P, Martin A. Small-scale and large-scale effects of endogeic earthworms on soil organic matter dynamics in soils of the humid tropics. Soil Biology and Biochemistry. 1992;**24**(12):1491-1498

[42] Dash MC. Role of Earthworms in the Decomposer System. Glimpses of Ecology. New Delhi: India International Scientific Publication; 1978. pp. 399-406

[43] Scheu S, Schlitt N, Tiunov AV, Newington JE, Jones HT. Effects of the presence and community composition of earthworms on microbial community functioning. Oecologia. 2002;**133**(2):254-260

[44] Singleton DR, Hendrix PF, Coleman DC, Whitman WB. Identification of uncultured bacteria tightly associated with the intestine of the earthworm Lumbricus rubellus (Lumbricidae; Oligochaeta). Soil Biology and Biochemistry. 2003;**35**(12):1547-1555

[45] Toyota K, Kimura M. Microbial community indigenous to the earthworm Eisenia foetida. Biology and Fertility of Soils. 2000;**31**(3-4):187-190

[46] Kristufek V. Actinomycete communities in earthworm gut and surrounding soil. Pedobiologia. 1993;**37**:379-384

[47] Devliegher W, Verstraete W. Microorganisms and soil physico-chemical conditions in the drilosphere of Lumbricus terrestris. Soil Biology and Biochemistry. 1997;**29**(11-12):1721-1729

[48] Costello DM, Lamberti GA. Non-native earthworms in riparian soils increase nitrogen flux into adjacent aquatic ecosystems. Oecologia. 2008;**158**(3):499-510

[49] Haimi J, Huhta V. Effect of earthworms on decomposition processes in raw humus forest soil: A microcosm study. Biology and Fertility of Soils. 1990;**10**(3):178-183

[50] Swift MJ, Anderson JM. Biodiversity and ecosystem function in agricultural systems. In: Biodiversity and Ecosystem Function. Berlin, Heidelberg: Springer; 1994. pp. 15-41

[51] Clements RO, Murray PJ, Sturdy RG. The impact of 20 years' absence of earthworms and three levels of N fertilizer on a grassland soil environment. Agriculture, Ecosystems & Environment. 1991;**36**(1-2):75-85

[52] Baker GH, Carter PJ, Barrett VJ, Kilpin GP, Buckerfield JC, Dalby PR. The introduction and management of earthworms to improve soil structure and fertility in South-Eastern Australia. Soil Biota: Management in Sustainable Farming Systems. 1994:42-49. Available from: http://hdl.handle.net/102.100.100/239744?index=1

[53] Shuster W, McDonald L, McCartney D, Parmelee R, Studer N, Stinner B. Nitrogen source and earthworm abundance affected runoff volume and nutrient loss in a tilled-corn agroecosystem. Biology and Fertility of Soils. 2002;**35**(5):320-327

[54] Guei AM Jr, Baidai Y, Tondoh JE, Huising J. Functional attributes: Compacting vs decompacting earthworms and influence on soil structure. Current Zoology. 2012;**58**(4):556-565

[55] Gaddie RE Sr, Gaddie Sr Ronald E. Habitat for earthworm cultivation. U.S. Patent 3,961,603. 1976. pp. 1-5

[56] Krishnamoorthy RV, Vajranabhaiah SN. Biological activity of earthworm casts: An assessment of plant growth promotor levels in the casts. Proceedings: Animal Sciences. 1986;**95**(3):341-351

[57] Blouin M, Zuily-Fodil Y, Pham-Thi AT, Laffray D, Reversat G, Pando A, et al. Belowground organism activities affect plant aboveground phenotype, inducing plant tolerance to parasites. Ecology Letters. 2005;**8**(2):202-208

[58] Stephens PM, Davoren CW. Influence of the earthworms Aporrectodea trapezoides and A. rosea on the disease severity of Rhizoctoniasolani on subterranean clover and ryegrass. Soil Biology and Biochemistry. 1997;**29**(3-4):511-516

[59] Stephens PM, Davoren CW, Doube BM, Ryder MH. Ability of the lumbricid earthworms Aporrectodea rosea and Aporrectodea trapezoides to reduce the severity of take-all under greenhouse and field conditions. Soil Biology and Biochemistry. 1994;**26**(10):1291-1297

[60] Kumar R, Singh D. Insecticide-tolerant bacterial population in Eisenia Fetida's gut and Vermicast exposed to Chlorantraniliprole and Fipronil. Centre For Advancement of Applied Sciences. 2022;**24**(3):273-279. DOI: 10.5958/0974-4517.2022.00033.7

[61] Lim SL, Lee LH, Wu TY. Sustainability of using composting and vermicomposting technologies for organic solid waste biotransformation: Recent overview, greenhouse gases emissions and economic analysis. Journal

of Cleaner Production. 2016;**111**:262-278. DOI: 10.1016/j.jclepro.2015.08.083

[62] Kumar R, Gupta RK, Yadav R, Saifi R, Yodha K, Pooja. *Eisenia fetida* as protein source for growth enhancement of *Heteropneustes fossilis*. Egyptian Journal of Aquatic Biology and Fisheries. 2022;**26**(2):577-588

[63] Qu H, Wang P, Ma RX, Qiu XX, Xu P, Zhou ZQ, et al. Enantioselective toxicity, bioaccumulation and degradation of the chiral insecticide fipronil in earthworms (Eisenia feotida). Science of the Total Environment. 2014;**485-486**:415-420. DOI: 10.1016/j.scitotenv.2014.03.054

[64] Chang J, Wang Y, Wang H, Li J, Xu P. Bioaccumulation and enantioselectivity of type I and type II pyrethroid pesticides in earthworm. Chemosphere. 2016;**144**:1351-1357. DOI: 10.1016/j.chemosphere.2015.10.011

[65] Martinkosky L, Barkley J, Sabadell G, Gough H, Davidson S. Earthworms (Eisenia fetida) demonstrate potential for use in soil bioremediation by increasing the degradation rates of heavy crude oil hydrocarbons. Science of the Total Environment. 2017;**580**:734-743. DOI: 10.1016/j.scitotenv.2016.12.020

[66] Hu M, Liu K, Qiu J, Zhang H, Li X, Zeng D, et al. Behavior of imidazolinone herbicide enantiomers in earthworm-soil microcosms: Degradation and bioaccumulation. Science of the Total Environment. 2020;**707**:135476. DOI: 10.1016/j.scitotenv.2019.135476

[67] Ekperusi OA, Aigbodion IF, Iloba BN, Okorefe S. Assessment and bioremediation of heavy metals from crude oil contaminated soil by earthworms. Ethiopian Journal of Environment Studies and Management. 2017;**9**:1036. DOI: 10.4314/ejesm.v9i2.9s

Section 3

Applications and Sustainable Use

Chapter 14

Reduction of Farmer's Attitudes and Value in the Implementation of Organic Agricultural Programs in Indonesia

Hamyana Yana, Kliwon Hidayat, Keppi Sukesi and Yayuk Yuliati

Abstract

Starting from the initiation of the "2010 Go organic" program which is full of various interests, placing the implementation of organic farming development at the crossroads between the project and the community movement. As a project, the values of organic agriculture are reduced in a sense that strays far from the true nature and philosophy of organic farming. Organic farming is nothing more than an approach to developmentalism that traps farmers in various dependence, powerlessness, and exploitation in other forms. However, the event of reducing the attitudes and values of organic farming on the one hand has also created anxiety for several actors who are concerned about the decline in the morality of farmers in treating land, water, and other resources. These concerns and concerns are then actualized in the resistance movement which seeks to reject the establishment, maintain ecosystem sustainability, and reorient local values and identities.

Keywords: value, organic farmer, awareness movement, organic agricultural programs, ecosystem sustainability

1. Introduction

As the world's response to agricultural modernization was carried out on a large scale in various parts of the world in the 1970s, organic farming in Indonesia was also a response to the green revolution policies of the same period. The first idea of organic farming (not organic, because organic is more technical) in Indonesia was initiated by Agatho Elsener, a Swiss organic practitioner who later became an Indonesian citizen in the 1980s, who dedicated almost his entire life (organic attitude) to running a farming system. Organic farming. This idea was followed by the emergence of the organic farming movement in Yogyakarta and its surroundings which was developed by G. Utomo, PR, and other civil society. At that time, organic farming had not yet become a concern of the government and developed as a response to an "alternative way of life" among farming communities. It is not easy to develop organic farming as an alternative because it was developed together with the green revolution policy which overhauled indigenous agriculture into an agricultural political-economic system that is integrated with the state and market, which dominates all instruments from the local to the national level and even links it to the global agricultural system [1].

This paper tries to describe the implementation of organic farming development based on the case of the Batu Go Organic program in Batu City, East Java, Indonesia. As a program, Batu Go Organic involves various actors or actors who have different interests which in the end result in a variety of actions taken. Differences in the interests of each actor lead to conflicts of interest. Apart from conflicts of interest, another thing that is also described in this paper is the reduction of farmers' values and attitudes in developing organic farming. The conflict between the spirit of morality, the spirit of spirituality, the spirit of globality, and the spirit of capital is the attraction presented in this paper. Hopefully, this article can provide benefits, especially in understanding other perspectives in the context of implementing organic farming when viewed from the perspective of marginal farmers.

2. Typology of actors and their importance in the implementation of organic farming development

Based on the results of interviews with key informants and informants as presented above, it is obtained an illustration that the parties involved as actors in the development of organic agriculture in the Research Location consist of the Government (regional and central), corporations, or agricultural entrepreneurs, environmental activists, Organic Certification Institutions (LSO), Academics and Practitioners of organic agriculture, farmers, and village community leaders. Each actor has different interests in the development of organic agriculture, socially, economically, and ecologically; so does the power it has. Differences in interests and power between actors are actualized in a variety of actors' behaviors and actions in running organic farming [2]. Therefore, to describe the typology of actors and their interests in the development of organic agriculture is framed using a genealogy analysis of actor actions.

The actions of actors in the context of this study do not place elements of agency and structure as an inseparable unit (duality). However, it does not mean that agency and structure cannot be identified for their dominance in the actions of actors. In certain conditions, the agency element may be very dominant and vice versa.

Gidens' view, is slightly different from Kinseng's which states that "I do not agree with Giddens that agency and structure elements cannot be separated (dualistic). In my opinion, even though these two elements are always present in every actor's action, and influence each other, the two can and must be separated analytically (dualistic in nature)" [3]. Apart from Kinseng, Layder et al. states that action (what is meant by agency) and structure are two aspects that can be separated and each has a degree of autonomy [4]. They said, "Thus we conclude that empirically structure and action are independent (and thus, deeply implicated in each other), but partly autonomous and separable domains."

The diversity of actors in interpreting the various experiences and knowledge they have is ultimately implemented in various patterns of behavior as the response they give. In the context of implementing organic farming, the results of observations and in-depth interviews with key informants and informants, identified at least four actor typologies in Bumiaji District, Batu City. To elaborate in depth on the involvement of various typologies of actors and their various interests identified in this study are presented in **Table 1**.

Typology actor	Orientation of the value	The importance of actor	The action of actor
Type A	Oriented to spiritual value	Worship in carrying out human nature as a caliph (leader)	Carrying out organic farming as a form of worship to carry out human nature as a leader, guardian of the continuity of life
Type B	Oriented to moral and cultural values	Maintain the sustainability and sustainability of social resources and natural resources	Carrying out organic farming as a strategy to improve soil fertility, biological balance and sustainability of local wisdom
Type C	Oriented to rational and market value	Complied market standardization, profit maximization	Carry out organic farming to meet market needs and increase income
Type D	Oriented to the value of power	Maintain power or consensus	Running organic farming as a media campaign and imaging

Source: Primer data, 2021.

Table 1.
Typology of actors and their importance in the implementation of organic farming.

Based on **Table 1**, it shows that the four typologies of actors and their interests have different value orientations and variations of interests. First, farmers who are classified as type A actors. The orientation of the values and interests of type A actors in carrying out organic farming is to worship Allah SWT. Type A actors carry out organic farming on the basis of their religious values and norms. Islam is the majority religion adhered to by the informants in this study. Adherents of Islam are obliged to maintain and safeguard the universe from actions that cause damage and destruction. The proportion of farmers who are classified as type A actors is very small in number, namely only four people.

The results of this study complement the results of research [5–7] that there is a relationship between spiritual values and biodiversity implemented in organic agriculture. Grim gives the example of the Ifugao Igorots as one of the indigenous tribes in the Philippines who perform rituals led by an indigenous priest to control rice pests, thereby preserving the plant species that the Igorots depend on for food. In addition, the Ifugao believe that "nature spirits" inhabit the trees and rocks of forests and watersheds, which are "centers of biodiversity," including more than 200 plant varieties. Other research results that support the findings in this study are Wilson, [8] who states that activities such as hunting and harvesting not only provide nutritional benefits, which support physical health, but also enable individuals to connect spiritually with Mother Earth, the Creator, and spirits, while on land. This is important because it allows individuals to simultaneously pursue a physical and spiritual connection to the ground that is essential for emotional and mental health.

Second, type B actors are oriented to moral and cultural values in implementing organic farming systems. Type B actors are more dominated by farmers who relatively act as traditional leaders or village or hamlet elders in the research location. The moral values passed down from generation to generation from the actors' ancestors are very strong in coloring their actions in implementing the development of organic agriculture. So for them, organic farming is more aimed at efforts to maintain local customs, habits, and wisdom passed down from their ancestors.

The results of this study confirm the research of Alhamidi et al. [9] which stated that the ability of farmers to integrate ethical values into their agricultural decisions and actions, has implications for managing natural resources in an analytical manner and not only for economic purposes. Moreover, making agriculture meaningful and sustainable. Loving farming is seen as a good job and a way of life, not just food production. This love is deeply rooted in the minds and hearts of the small scale farmers operating as custodians of the system. This makes farming a productive activity rather than an extractive one [9]. That is why the love of farming as a way of life is a constant theme in the alternative agriculture literature [9]. The cultural and spiritual dimension of farmer experience and knowledge underpins the relationship between farmer and farm. Likewise, research by Yazdanpanah et al. [10] states that the factors of morality and fear of disease are the dominant factors that affect the willingness of farmers to cultivate organic products in Iran.

Third, type C actors are oriented to the economic aspects of running organic farming cultivation systems. The economic aspect in question is more on the consideration of financial profit and loss. Referring to rational choice theory [11], there is the basic idea that people act deliberately toward a goal and that goal is shaped by values or choices. The actors will take action in order to maximize the benefits and satisfaction of their needs. The rational theory assumes that every human being is basically rational by always taking into account the principles of efficiency and effectiveness in carrying out every action. While still acknowledging the existence of determinant factors in the form of strong farming community solidarity, (material) economic subsistence, and pre-capitalist society's production relations, however, the influence of rationality always occurs in the context of the operation of the mechanism of the rational interests of individual members of the community. Humans do tend to maximize their rationality and always tend to calculate the value of something (utility) that they want to exchange, namely economic and moral utility. The theoretical facts are in accordance with the facts on the ground where most farmers prioritize the aspect of financial gain in choosing or not choosing an organic farming system.

Type C actors are the most vulnerable to return to conventional (non-organic) agricultural systems. The results of observations and interviews show that all actors belonging to type C have confirmed that they are no longer practicing organic farming. The reasons given varied. One of the informants said that the constraints faced by organic farming were product standards that were too high, so that farmers had difficulty fulfilling them. Apart from this, another reason is that although the price of organic products is high, the market segment is small, the demand is unstable, and the cost to produce organic vegetables is more expensive than conventional products.

The results of this study are in accordance with the results of research by Pham and Shively [12] which states that the adoption of organic production is significantly influenced by farm size, age of the head of household, yield differences, price differences, and input cost differences. Thus, in the context of type C actors, running organic farming is a rational choice if it can provide economic benefits financially, then they will implement this program. On the other hand, if in its development it turns out that this program is not profitable for them, then they will naturally stop.

The implementation of the Batu Go Organic policy as an intervention or stimulus for farmers to switch from conventional farming to organic farming, in fact, has not been fully successful. This was confirmed based on the statement of informant number 6 (identity withheld) that "The Batu Go Organic program is almost the same as government programs in general. It is very thick with the project approach where sometimes the involvement of local communities is very small. As a PPL, I'm also in the wrong position to stand on two sides. On the one hand, I am a government official, but I also have to be able to embrace farmers to be able to accept this program well. Finally, yes, I do what I can do."

The "Batu Go Organik" program is a project that is centralized, the implication is the lack of awareness of the farming community to be involved in a sustainable manner in the proposed program. A recurring event in almost all programs is the level of participation from program targets which is decreasing day by day. In the end, field officers are increasingly not considered by farmers. This is a bad sign for the government's existence as a partner of farmers.

Fourth, type D actors are farmers or community leaders who tend to be positioned as holders of power or farmers who are affiliated with political networks. So that the implementation of organic farming development is oriented toward efforts to perpetuate the power it has. The principles and standards referred to in organic farming are not considered as important, in fact, the main concern is how big the image is obtained from every action it takes, including the implementation of organic farming. On one hand, the rhetoric created by type D actors has given color to the development of organic farming. But on the other hand, environmental and sustainability issues that are juxtaposed in organic farming patterns are only mere rhetoric and jargon.

Agents and structures are seen as a dual entity and the determination of the agent or structure will determine the actions they take [3]. For example, the verses of the Koran and customary norms, in the terminology of structural theory are seen as structures that force individuals to act in accordance with these norms. Actors of type A and type B carry out organic farming systems because they are subject to religious norms and customary norms that they adhere to. Religious norms and customary norms are very strong, stable, and given structures in a society which will continue to be maintained and maintained along with individuals who maintain or practice them [3]. Likewise, the act of organic farming as an act of worship or a moral movement to maintain the preservation of the universe, will continue to survive and be implemented as long as there are actors who reproduce it. For this reason, internalization and crystallization of values that exist in religious norms and customary norms are needed in every organic farming activity. Although initially seen as something that is forced, over time the act of organic farming is interpreted as an activity of worship or a moral movement embedded in the actor's actions as a social practice.

Agree with Gidens that agent and structure are duality. However, that does not mean that agency and structure elements cannot be distinguished. Individual actions can still be identified which are more likely to be dominated by agency elements, and which tend to be dominated by structural elements. This slightly corrects what Kinseng said that agency and structure are dualistic, meaning that agency and structure are separated.

The agent's actions at one time are dominated by agency elements, but at other times they can be dominated by structural elements. This really depends on the actor himself and his environment. Therefore, the results of this study also yielded findings that agency and structure are not only dual in nature, but also dynamic Therefore, the results of this study also yielded findings that agency and structure are not only dual in nature, but also dynamic. Facts show that farmer behavior is not monotonous, static, but dynamic. They always modify their behavior in accordance with their own knowledge, experience, and environment. In other words, the dynamics of farmers' actions in carrying out organic farming are very high and tend to change quickly [13]. For example, type B actors, who tend to be dominated by structural elements, do

Typology	Actor	Determination agency	Determination structure
A	Religious leaders, senior farmers	Individual ways to strengthen confidence and gratitude for the grace of Allah SWT	Provisions in the Al-Quran oblige humans not to destroy nature
B	Village elders, traditional figures	Alternatif strategi dalam mengatasi kelangkaan dan mahalnya pupuk dan pestisida kimia	Moral values passed down from our ancestors to preserve the universe
C	Farmer, entrepreneur	Initiatives in taking advantage of market opportunities and healthy lifestyle trends	Process and product quality standardization
D	Local political elite, state apparatus	The drive for self-actualization, imagery and consensus tools	Legislative regulations and political ethics

Source: Data primer, 2021.

Table 2.
Determination of agency and structure in actions of actors.

not rule out switching to type C or D types, this really depends on the situation and determinant conditions that affect the type B actor.

Table 2 presents the determination of structure and agency in the actions of the following actors.

The structural determination that occurs in the typologies of actors A and B is inversely proportional to what occurs in the typologies of actors C and D. Actions carried out by actors of type C and type D are based on initiatives that arise from within themselves. The initiative to run an organic farming system as a strategy to increase income is purely an actor's decision free from the pressures of the structure that forces it. In this context, the actor or agency is autonomous, meaning that the actor's actions are not "dictated" by the structure, but are determined by the actor himself, who has the ability to think, judge, weigh, and choose what action is considered most appropriate at the time and place. Agencies are not only the ability to make changes, but also the ability to maintain existing conditions. Indeed, intrinsically every individual human is unique, no one is exactly the same as one another [14]. Therefore, it is not surprising that agencies also vary from one person to another; moreover, agency is also influenced by various other external factors.

3. Batu Go Organic Program: the intersection between the project and the community awareness movement for civilized and sustainable agriculture

Referring to the findings regarding the typology of actors and power relations as explained in the previous section, this research seeks to dig deeper into the question of whether organic farming is a project that is a transformation of modernity that hides behind the mask of wisdom, or an alternative route that seeks to get out of the shadow of modernity which has proven to have caused exploitation and erosion of environmental resources, dehumanization and injustice, as well as the marginalization of local wisdom and identity.

To understand the phenomenon of reducing farmers' attitudes and values in the implementation of organic farming development, the Batu Go Organic program was chosen as one of the observed contexts. The Batu Go Organic program is an intervention carried out by the Batu city government, East Java–Indonesia in response to the problem of decreasing soil carrying capacity or fertility on agricultural land and the level of continuous use of chemical pesticides by farmers. The results of observations at the research location showed that the behavior of farmers who always use chemical pesticides in every action to control pests and plant diseases is very dangerous for the safety of the food they produce. Even though in the last 3 years the results of pesticide residue tests for vegetable commodities in Batu City have shown values below the minimum residue limit set, but to move toward organic farming, the behavior of using pesticides and chemical fertilizers will be contrary to the established organic farming standards. The strategy used in the organic farming development plan by the Agriculture and Forestry Service of Batu City is essentially promoting the application of farming methods that lead to the application of organic farming and initiating the establishment of organic areas as pilot projects. The selection of this strategy refers to the goal to be achieved, namely changing the way of thinking of the farming community in Batu City from conventional (inorganic) agriculture to organic farming. However, in reality, the four identified typologies of actors at least provide an illustration that the development of organic farming through the Batu Go Organic program is at a crossroads between the project and the movement of community awareness. This is in stark contrast as seen from the interests of the actors who in general can be categorized into two dimensions, namely the moral-spiritual dimension (actor types A and B) and the rational dimension (actor types C and D). It is these two dimensions that have conflicting interest dimensions that place the implementation of organic farming development at the crossroads between projects and community awareness movements.

3.1 The "Batu Organic" program in project perspective

In order to describe the Batu Go Organic program as a project, this study explored primary data and secondary data related to the planning and implementation process of the program from 2012 to 2021. Based on the results of in-depth interviews with key informants, the Batu Go Organik program has not been fully implemented as a program. Pro farmer. This is supported by facts in the field which show that since program planning, farmers as the program's main target have not been involved intensely. The activities were arranged by elite bureaucrats consisting of government, practitioners, and academics. There are several important things that were not carried out in planning the Batu Go organic program such as the elaboration of macro policies into a more detailed program of activities, because a detailed and sustainable strategic planning document has not yet been prepared for the development of organic farming in Batu City within a certain period of time. Conceptually, the development of organic farming in Batu City, it can be identified that the planning process is carried out through a political, technocratic, quasi-participatory approach, top-down and bottom-up. Meanwhile, if viewed from the aspect of program implementation, the results of in-depth interviews and observations as well as document studies show that there are often various differences in the organic cultivation process. These differences are caused by the following factors:

a. The standardization has not been fully implemented by the farmers in Sumberbrantas village, so that each organic farming group or actor can set their own standard.

b. Market orientation, with standards that have been set by the group and if you can convince the market that the product is of high quality and deserves more respect, then it will suffice to use those standards.

c. The farmers in Sumber brantas village, with the green revolution, are accustomed to seeing their plants always in green condition. To carry out organic farming as it should, it often does not have a 100% determination so that in practice it still uses chemical fertilizers as basic fertilizers and has not completely abandoned the use of chemical pesticides

3.2 The "Batu Go Organic" program in the perspective of the awareness movement for organic farming communities

The Batu Go Organic program, which was launched in 2011 by the Mayor of Batu through the Agriculture and Forestry Service, is an effort to support a program of resilience and independence in agriculture. Batu City has very good potential for the implementation of organic farming such as its geographical conditions which are in the highlands, productive human resources, and policy support from the local government which is very supportive for the agricultural sector. As an area located in the highlands, it is possible to develop highland vegetable plants that have economic value; the many sources of springs support the availability and purification of water as one of the main factors in organic cultivation of plants. One of the local wisdoms of the people in Batu City is familiar with the world of agriculture. This provides convenience in terms of conveying information on organic farming cultivation technology. Technical matters conveyed through extension are easier for farmers to understand, and farmers can even innovate with the information provided.

Mr. Ma, Chairperson of the Anjasmoro II Poktan of Sumberbrantas Village, Bumiaji explained that the declaration of organic farming in Sumberbrantas Village, stems from the condition of the production pattern, especially for horticultural commodities, which have greatly exceeded the threshold for the use of chemical pesticides and chemical fertilizers. This causes degradation of soil fertility and biota on farmers' fields. Efforts to remind farmers of the dangers of what they have been doing have been carried out through outreach activities. However, it still does not show a significant change. For this reason, in 2011 the city of Batu launched a movement called the Batu Go Organic Program.

The next informant who was asked for information about the beginnings of organic farming in Sumberbrantas Village was Pak Jo. A farmer as well as a community leader who has been involved in agriculture for quite a long time in Sumberbrantas. Based on information from Mr. Joni, organic farming has actually been around since this idea was voiced, the problem has not been answered until now. Starting from planning, implementation, and even evaluation, they have not found an ideal and inconsistent form. Organic farming as a program or a project has many indicators which are very relative to determine its success. But organic farming as an awareness movement to liberate farmers from shackled modern narratives, only started after fertilizer scarcity occurred, pesticide prices were high, and agricultural product prices fell.

Furthermore, Mr. Jo said that organic farming must start from the movement to improve the soil as a place for plant life. Without real action to improve the physical, chemical, and biological properties of the soil where plants live, then the concept of organic farming will only be discourse and rhetoric. Pak Jo's statement regarding where organic farming should start, is also in accordance with some literature which states that organic farming is a pattern of agricultural production that aims for long-term ecological health, such as biodiversity and soil quality [15, 16]. The results of research by [17] also show that organic farming systems in Europe and Russia have been proven to be able to significantly increase soil organic matter in the last 10 years. Improvements in soil organic matter content will be followed by improved ecosystems in the soil.

The next informant who was asked for his opinion on the origins of organic farming in Sumberbrantas Village was Mr. Pu. Based on Mr. Pu's statement, the decline in fertility and the explosion of pest attacks in the last 15 years have made him move to find solutions to solve the problem. Discussions with fellow farmers, agricultural extension workers, students, lecturers, and other practitioners have made Mr. Pu believe that a cultivation pattern that only prioritizes production and planting acceleration is a mistake. He finally realized that the satisfaction of farming is not only a matter of abundant production, but besides that there is the issue of responsibility to pass down positive things and goodness for the future of children and grandchildren.

Mr. Pu is of the view that modern agriculture which is actualized in a high production pattern in the use of hybrid seeds, chemical fertilizers, and chemical pesticides has slowly but surely eroded the local wisdom left by their ancestors. One clear example is that traditionalism, which is synonymous with life in the countryside, has disappeared, and now it has merged with globalism, which has implications for the difficulty of distinguishing between localism and globalism, local knowledge and foreign knowledge, traditional technology, and modern technology. Misinterpretation of local potato seeds, for example, that they are considered hybrid potato seeds that are planted repeatedly. Even ironically, some farmers do not know the names of the local commodities in Sumberbrantas village. This confirms that the modernization of agriculture that has taken place in recent years has not only reduced locality-specific commodities, but has also taken away all elements of the soul and institutional farmers in Sumberbrantas.

In line with what was conveyed by Pak Pu, Pak Jo also said that recently the values adopted by farmers, especially young farmers in Sumberbrantas, have been very fast. The feeling of love for water, affection for the land and plants, slowly but surely begins to erode from the mentality, initiative, and soul of the farmers. The view that land and plants as living things are no longer instilled in the souls of most farmers today, so that exploiting land is something that is natural and common. The use of chemical fertilizers and chemical pesticides continues to be doubled in order to pursue productivity targets. In the name of efficiency, we accelerate production by growing plants without a time lag. The production process that is safe, correct, and wise is no longer the main thing because the most important thing is the satisfaction and interests of the market. Generational sustainability is not the main thing that becomes an orientation, but maximizing profits is the main goal. Reducing the value of gotong royong, community participation and kinship becomes an instant project that increasingly separates farmers from reality.

Pak Jo's statement stated that the struggle of the farmers was not only dealing with external parties such as bureaucrats, capital owners, and the market, but they were also dealing with their own instincts and conscience. On the one hand, they are faced with economic pressures and power pressures that demand the accumulation of rupiah and the actualization of power, but behind their deepest hearts are hidden rebellious consciences because they contradict the values inherited from their ancestors. Although the impact of modern technology has encouraged increased production, business efficiency, and increased farmers' income, lately it has become increasingly clear that the symptoms of the loss of humanity are implemented in farming activities.

Various efforts that have been made to resolve the uncontrollable impact of implementing the green revolution through modern agriculture 2.0 and then being refined into modern agriculture 3.0 are unable to heal the wounds caused by modernization and capitalization of agriculture. Ironically, instead of treating modern agriculture 3.0, it is actually contaminated by new creations and modernization formulations such as reform and transformation [18]. One form of this effort is the Batu Go Organic program which gives birth to a more environmentally friendly and sustainable farming system which is then packaged in organic farming terms. On the one hand, this program aims to atone for the sins of exploitation of land and other environmental resources, but on the other hand, its meaning has been reduced and trapped in the grip of the market.

4. Organic Agriculture: is a metamorphosis of modern agricultural imperialism or a post-modern agricultural model?

Referring to the previous sub-chapter, that the development of organic agriculture is at the junction between the project and the community awareness movement, leads us to the next question that I want to analyze in this paper. It is still being debated that organic farming is a step against modernization, or just a camouflage of modernism hiding behind the issues of sustainability and health. In order to dig up this information, the data search began by asking for information from the key informant, namely Mr. Jo.

Organic farming, whose concept is adopted from sustainable development, substantively prioritizes social, economic, and ecological sustainability. A creation of capitalism that at first glance offers friendliness and sustainability, while true ism is still development. The characteristics of developmentalism are very clearly reflected in its implementation which is dominated by the use of external inputs produced by outside industries, processes, practices, and results are more expensive, so that they are not affordable by the weak (marginal farmers). According to the informant of this study, Mr. Bb, said that internal inputs such as seeds, fertilizers, and pesticides which should be fulfilled independently from the implementation of integrated agriculture, in practice are still predominantly imported from outside. The high cost of internal inputs is not supported by local culture, for example, the culture of animal husbandry, growing plants for raw materials for vegetable pesticides, and local technology. So that the tendency is partial implementation, biased toward ecological sustainability and neglect of economic, social, and political sustainability. The implication is that land conversion is not controlled, regeneration does not occur, urbanization remains high and economic inequality is getting worse.

Mistakes in the implementation of sustainable agriculture lead to unsustainability in ecological, economic, social, and political aspects. Propaganda of sustainable agriculture that is seen to be more humanist and ecological is inversely proportional to the reality. Several critical thinkers actually labeled this sustainable agriculture as the green revolution volume 2. It is said that because although it is environmentally friendly it still depends on various external inputs. The biological fertilizers and pesticides used are factory-produced (such as liquid fertilizers and pesticides, granules) as well as home-made products from outside. This causes production costs to become more expensive, another implication of this high production cost is the reluctance of farmers to continue participating in sustainable agriculture like this. Its implementation is only limited when there is a project, after the project is completed the agricultural activities are also completed and switch to the conventional system.

At the global level, modern imperialism, which began with the industrial revolution, has not gone away. Since then, imperialism's camouflages have always transformed from one form to another featuring varied creations, from the visible to the virtual, from the materialist to the ideological. The transformation of physical imperialism into economic imperialism, innovation, information, technology, culture, and ideology carried out by developed countries for developing countries is evidence that imperialism entities already have long-term plans. According to Setiawan [18] there are 10 camouflage modes that are operated sequentially by developed countries against developing countries including Indonesia, namely, spatial imperialism, commodity imperialism, ideological imperialism, industrial imperialism, innovation imperialism, technological and information imperialism, market imperialism, standard imperialism, investment imperialism, and education imperialism.

The threat of imperialism from developed countries, such as China, the United States of America, and European countries, is real in the context of state life in general. As a concrete example that has gone viral in the last 10 years in Indonesia is the threat of modern Chinese imperialism. China's movement in building neocolonization with the mode of placing Chinese residents throughout the world, including Indonesia. The concentration of the placement of the Chinese population or their descendants in strategic business cities in Indonesian territory cannot be seen as a mere coincidence or business motive, but there could be geopolitical motives that Indonesia as a sovereign country must be aware of. Even if paying attention to the development of the trade war between China and the United States in the last 5 years, it has implications for changes in China's geopolitical strategy which seeks to relocate citizens of Chinese descent from the United States to Asian and African countries including Indonesia with the aim of sticking ideological and its impact on the country.

In the context of the agricultural sector, the control of strategic sectors such as rice, meat, strategic vegetables such as chili, is starting to be controlled by Chinese, Indian, and US investors. The various potential resources began to be mapped, in terms of their number, type, location and capacity, and then they invested to further control and exploit them. The Chinese investment tendency is not only to cram technology and capital, but also to include manpower in their every investment. This is what distinguishes the investment patterns made by Europe and the United States. Observing the aggressiveness of Chinese investment in Indonesia cannot only be seen from the aspect of spurious economic growth alone, but one must also look at the motives behind the propaganda and the agenda of the hidden geospatial political setting behind it.

Investments in the agricultural, fishery, plantation, forestry, and other vital sectors carried out massively by Chinese investors in collaboration with local residents of Chinese descent can be called a form of modern imperialism in this century. Indications of China's control movement are not only targeting the market sector, but also the production sector. This is very dangerous for the sovereignty of Indonesia as a sovereign country. Of course, it still remains in our memories regarding the case of plastic rice which shocked this country in 2017 and artificial eggs or synthetic eggs which also came from China which made people in this country worry about buying eggs.

The case of Chinese farmers who were caught in the Cianjur area several years ago is proof that the invasion by China was real and is happening in this country. Awareness to counter various acts of global imperialism, including Chinese imperialism, must be built and grown within the social strata, including the farming community. The hope is that in the future, farmers as part of the community in this country will also have good awareness and vigilance to ward off various intimidations and colonialism from outside themselves.

Specifically, the implementation of global imperialism in a real context is rooted in the implementation of modernization and industrialization of agriculture. Forms of imperialism in each era, for example, forced cultivation imperialism in the colonial era, agricultural innovation imperialism in the green revolution era, market imperialism in the globalization era. All forms of imperialism always start from the agricultural sector as the main foundation and then continue to imperialism in other fields [18].

One of the efforts to ward off these forms of imperialism can be done through the growth of self-reliance. Thus, if we want to replace imperialism in the field of agriculture, we must replace modernization and industrialization of agriculture with a new, civilized, independent, sovereign, beneficial, locally specific agricultural model, all of which are based on maximizing local civilization. To get to the agricultural model, various attempts have been made with various approaches and empirical studies. Organic farming can be seen as a hope that will create local self-reliance that is able to maximize local potential and reduce dependence on agricultural inputs from external parties.

Referring to Rigby and Cáceres [19], the goal of organic farming is to prioritize long-term ecological health, such as biodiversity and soil quality, rather than short-term productivity gains. Thus, the implementation of organic farming should be measured by how much it achieves in realizing the above goals. Organic farming is a technology or premodern technology for today's world [20]. Proponents of organic farming argue that the organic farming model is an innovation that defies some forms of modernity, with the vision of returning agriculture to a certain premodern structure, as well as an innovation that provides solutions to current agricultural problems [21].

In fact, we should question whether it is true that the implementation of organic farming is a way to fight modernity or that organic farming is a new incarnation of modernity hiding behind a mask of sustainability. When who want trace some common facts that occur, there are indeed indications in that direction that could have occurred. When organic farming is approached economically, it can almost be said that its implementation will tend to exploitation with other packaging. Market standard-oriented organic farming is one proof that this is actually an act of colonization through homogenization [22]. For the implementation of organic farming like this, it is certainly not the option referred to in postmodern agriculture in this study. The response of farmers to organic farming programs varies greatly. This is very dependent on the attitude and value orientation of farmers in running organic farming. As has been discussed in Chapter V of this dissertation about the typology of actors in the implementation of organic farming, each type of actor has different orientations of interests and values between one type and another. In addition to the typology of farmers who only place organic farming as a means of gaining projects, there are also farmers who base their orientation on cultural values or moral values.

The activities of organic farmers in Sumberbrantas Village reflect an awareness movement that was built on the motivation to fight against imperialism and the structured marginalization of the rulers and capitalists against the reality of farmers. The results of investigations at the research locations found that they were well aware that there was no power to fight imperialism from the grip of outsiders. However, the spirit to fight and consolidate the movements of the grassroots community never stops. The implementation of organic farming, is not only a matter of price and market interest which is quite high for organic products, but also for organic farmers in the village of Sumberbrantas, planting crops with reference to the principles of environmental and generational sustainability and sustainability is a calling. Planting crops, especially horticultural commodities such as potatoes, cabbage, and carrots, is almost impossible if you apply organic farming standards as standardized. However, the spirit to be free from the grip of fertilizer, pesticide and seed monopolies is the main thing.

Interpreting the statements of some of the informants above, there is an essential meaning that can be obtained that the choice to implement organic farming is motivated by the spirit of improving the quality of sustainable generations and environmental health rather than just economic issues. This does not mean that economic value is not important, but the correct term is probably that economic value is not everything. This can be interpreted that the capitalist approach which always emphasizes economic value which ultimately encourages exploitative actions must be stopped with movements to internalize the spirit of togetherness, the spirit of sustainability, the spirit of independence, and the spirit of spirituality. This condition is expressed by Pak Jo and Pak Pur as key informants in this study. In the point of view Pak Jo and Pak Pur, that organic farming is not merely a matter of a better price than non-organic, or a problem of residues and degradation of land fertility, but that organic farming is a form of human responsibility to God, responsibility to the environment, and responsibility to fellow humans. The implementation of the spirit of independence that is applied in the farming community in Sumberbrantas is reflected in the activities of farmers to independently prepare potato seeds. Pak Jo and other farmers at Sumberbrantas have been breeding potato seeds independently since the 1990s. Potato seeds produced by the farming community in Sumberbrantas are not only able to meet the needs of local farmers, but in recent developments, they have been able to sell potatoes to farmers outside Sumberbrantas Village, such as Ngantang sub-district, Ngadiwono sub-district, and have even reached Lembang, West Java. This condition shows that the desire to be independent in terms of seeds is a value that must be transmitted to other farmers. Farmers do not always have to be carried away by the mainstream, which sometimes the farmers themselves do not understand the agenda setting behind it. Pak Joni and several other farmer leaders always struggle to convince farmers that organic farming is not just a matter of organic certification for the products they produce. But what is most important is the will and awareness of the farmers not to submit to modernist narratives that seem to be a single truth.

5. Organic farming: between market traps, reduction of meaning, and moral identity

Since it was initiated in early 2000 which then strengthened into a policy "Go Organic 2010", the direction and orientation of organic farming policies seem to increasingly indicate a development design toward industrialization of agriculture and world trade [1]. This condition has provided great opportunities and opportunities for organic business actors with legal entities to take a role in a larger organic farming system. The Indonesian government's policies in implementing organic farming are outlined in various regulations issued. Until 2020, regulations issued by the government regarding organic farming include: Regulation of the Minister of Agriculture (Permentan) 20/2010 concerning the Quality Assurance System for Agricultural Products; Minister of Agriculture no. 70 of 2011 concerning Organic Fertilizers, Artificial Fertilizers, Soil Improvers; and Minister of Agriculture no. 64 of 2013 concerning Organic Farming Systems. These three regulations have a very important role in strengthening the implementation of organic agriculture in Indonesia.

The transition of the ruling regime from the government of Susilo Bambang Yudoyono (SBY) to the government of Joko Widodo (Jokowi) in 2014 has also had an impact on the orientation of policies in the development of organic agriculture. The "Go Organic 2010" program was changed to the "Thousand Organic Farming Villages" program. This program at least marked a change in policy direction from the previous one increasing production, quality competitiveness, and competition at the global level to achieve industrialization and world trade toward the development of organic agriculture based on food sovereignty at the village level. In addition to being colored by a development strategy that has the nuances of "building from the periphery", the new direction of "A Thousand Organic Farming Villages" also seems to emphasize the importance of village development as mandated by the Village Law [1].

To implement the "Thousand Organic Farming Villages" program, based on the Decree of the Minister of Agriculture No. 58 of 2015, the Minister of Agriculture formed a Working Group for the Development of a Thousand Organic Agriculture Villages. The decision stated, among other things, that this working group was tasked with coordinating, monitoring, and evaluating the implementation of the program at the village level. Even though this program is based on the spirit of food sovereignty, the various regulatory instruments used to implement this program are products of policies during the previous government which of course have different directions and orientations. In other words, the program which is based on the spirit of food sovereignty and the strategy of "building from the periphery" is trapped into various policy instruments that have been made during the previous administration which have directions and orientations toward increasing production, quality competitiveness and market competition at the global level to achieve industrialization and world trade development. Policy instruments that are biased toward one side and tend to be trapped in the clutches of the market have had implications for the reduction of the meaning internalized within farmers. This raises the question whether organic farming is a metamorphosis of modern agriculture that hides behind the issue of ecological sustainability? Or is organic farming really anti-modernism? The fact is that organic farming is more of a creation of capitalism which at first glance seems to offer friendliness and sustainability, whereas the realism is still development. Agricultural driving actors who have stronger discourse power, dominate other actors in imposing ideas, images, even beliefs that are not necessarily correct in terms of cultivation methods, even marketing.

Organic agriculture is used as a rhetorical agenda that is full of political interests, capitalist interests and lacks meaning and moral movements. The results of this study confirm research [1] that there has been a reduction in the meaning of organic to a partial and pragmatic tendency in the implementation of policies for developing organic agriculture in Indonesia. Organic farming development programs only give meaning to the meaning of organic limited to the acquisition of a label, logo, or stamp, not to give the meaning of a completely organic farming cultivation system. In fact, according to [23], organic agriculture has a role that is considered safe for the environment and the formation of high-quality food ingredients. Therefore, organic farming policies must consider environmental practices, the willingness of consumers to pay for products, and the social aspects of organic farming.

As a result of research [24], if organic practice conforms to the rhetoric associated with it from its inception as a social movement, then it will have much to offer in the present and the future in terms of its contribution to the possible adaptation pathways and flexibility that he offered. Thus, as a form of adaptation, the development of organic farming must be carried out within the framework of a social movement or moral movement. However, what was stated by [23, 24] has not been fully implemented in the development of organic farming in Indonesia. The color of the construction is obvious in practice. Apart from still being dominated by the use of external inputs (external inputs) produced by the industry, processes, practices, and results are also becoming more expensive, so they are not affordable for the weak (peasant). Internal inputs (seeds, fertilizers, pesticides) which should be met independently from the implementation of integrated farming, in practice are still predominantly imported from outside. The high cost of internal inputs occurs because they are not supported by local culture, both the culture of raising livestock, planting crops, and even biological pesticides and local technologies. The tendency is that organic farming is applied partially, biased ecologically sustainable and ignores economic, social, and political sustainability. The implication is that land conversion is not controlled, regeneration does not occur, urbanization is still high and economic inequality is getting higher. Organic farming is operational, but the needs of the present generation are still not being met, imports are increasing, the needs of future generations are being forgotten and employment opportunities or rural entrepreneurs are still not created. Even though it is seen as environmentally friendly, because it is still strong with the development perspective, organic farming deserves to be labeled as modern agriculture which hides under the mask of ecological sustainability.

Winnett, stated that from an environmental awareness approach the organic farming community in Kaliandra–East Java, has a bigger mission than the organic farming community in Milas–Central Java. However, this has an impact on the loss of public awareness of the importance of community independence as farmers because what is implemented is only limited to diverting the issue from looting forests by the community to organic farming. It is in stark contrast to what Milas has done by focusing on self-reliance and awareness of the community as organic farmers [25].

It is limited to diverting the issue from forest plunder by the community to organic farming. It is very contrary to what Milas has done by focusing on the independence and awareness of the community as organic farmers.

The debate in the context of the implementation of organic farming as camouflage for modern agriculture is not entirely true. At least, there are some farmers who consistently implement organic farming as a resistance movement against the various structural pressures that have been holding their daily lives as farmers. This research at least found some farmers who still apply organic farming as a moral identity that differentiates conventional farmers from organic farmers in morality. Even though the number of farmers who apply the principles of organic farming is limited to a few individuals, in this dissertation it is very important to describe how the struggle is being carried out to maintain their existence in the midst of the onslaught of modern tools, especially on the cultural aspect.

As a moral struggle, organic farming is used as a cultural identity that distinguishes it from others. In analyzing the phenomenon of moral struggle of organic farmers, it is analyzed from the perspective of postmodern theory. Postmodernism is a term that refers to various meanings, various terms, various versions, various disciplines, and various objects. As a term, postmodern has been used in various fields, from literature, art, film, sociology, philosophy, economics, communication, and culture. The first figure who pioneered the birth of the term postmodern was Federico de Onis, then developed and popularized by several theorists such as Lyotard, Nietzsche, Foucoult, Vatimo, Giddens, Derrida, Baudrillard, Capra, and others. Although explicitly not found in the field of communication, postmodern concepts can be traced in the works or thoughts of Habermas, Heidegger, Gadamer, Ricoer, Marry Hese, and McCarty both as the antithesis of modernism, criticism, and mainstream (heurmeneutics) [26].

The term postmodern in this research uses a philosophical perspective that views postmodernism as a distrust of grand narratives that are nothing more than mainstream metaphors. Postmodern is defined as a period in which everything is delegitimized. Borrowing from Liyotard's term, postmodern is an intensification of dynamism, an effort to seek sustainable novelty from modernity, experimentation, and revolution of sustainable life. Postmodern rejects liberalism, Marxism, and even subverts foundationalist epistemology (dominantly positive). Postmodern is also understood as any criticism of universal knowledge, criticism of metaphysical traditions. The conclusion in postmodern philosophy is all forms of critical reflection on modern epistemology and paradigms on metaphysics in general.

Departing from the postmodern concept in a philosophical perspective, the term postmodern agriculture was adapted from this concept of thought as a response, criticism and even an antithesis to the stagnation or, more precisely, the failure of modernization of agriculture with a positivistic paradigm. Postmodern agriculture in this study is positioned as a movement of radical thought and behavior, not a theory of social change that is contaminated with modernization. Postmodern agriculture is offered not only to replace modern agricultural terminology, but to radically dismantle modernity, radically dismantle the contradictory meaning hidden behind the reality of modern agriculture to find new meanings, new attitudes, and values that prioritize the spirit of spirituality, morality, and community spirit.

The spirit of spirituality is built on spiritual values internalized in each individual farmer. The internalization of one's spiritual values is inseparable from the process of individual interaction as an agent with a structure that becomes an inseparable part of their life. In the context of postmodern society, the value of spirituality becomes a contradictory phenomenon on one hand and becomes very sacred on the other. The sacredness that begins to fade due to the eroding of the depth of meaning on the one hand and the crystallization of meaning on the other hand caused by the values of hedonism that become the narrative of modernism is clearly a big disaster for human life in the future. The loss of sacredness in rituals, as actualized in prayers, prayers, clean village rituals, rituals at harvest or rituals just before harvest is a sign that there is a shift that is very far from the standard that should be. Even though the postmodern tendency always rejects all forms of grand narratives with all their foundations (spirit, logos, being, system, state, authority) and prefers to explore emotional, irrational, mystical, and magical dimensions excavated from the spirits of the past, it cannot be denied that the depth of meaning of spirituality needs to be re-articulated, reoriented and even reconstructed.

As one of the oldest activities in the world, farming requires an orderly and balanced mindset so that the abundance of natural resources does not drown and narrowness or crisis does not imprison. Human exploitation and greed cannot be fulfilled from the layers of the earth and layers of the sky. Because the desire is unlimited, while the resources are available in limited quantities. Lust for exploitation, hedonic nature, consumptive behavior, kufr pleasures is a spiritual disease that leads humans to the brink of destruction.

Referring to Mr. Jo's statement above, that the low level of awareness of farmers in general to preserve the abundance of resources as a gift, as a blessing from God, as a mandate that must be accounted for in the future. That farming is not only a question of how to harvest or how to take, but also how to plant or how to give to generations. Pak Joni, as an elderly figure and the oldest farmer in Sumberbrantas, also said that the tendency of farmers to only take it and forget or not know how to return it was a behavior that led to the deteriorating condition of Sumberbrantas agriculture in the last 10 years. How can it not fall, in the mid of a sluggish market, coupled with declining land productivity, an explosion of pests, and increasingly expensive production costs, is the brink of collapse that farmers must accept at this time. These consequences must be borne as a payment for cultivation behavior that does not pay attention to the benefit of others. Exploitative cultivation patterns that completely ignore ecosystem sustainability. What is being accelerated is only production, production and production to fulfill that unlimited wish.

Implicitly from Mr. Joni's statement, it can be interpreted that farming is not just a culture and activity that fulfills the stomach (worldly—secularistic) but has spiritual meaning and values that are transcendental. There are meanings of worship, blessing, benefit, civility, and piety. That agriculture is a word or order of the creator, Allah SWT, which must be lived on the basis of order, balance, and justice, which is not exploitative, sustainable, regenerative, and rights (there are other rights in the results and there are rights of future generations). Another spiritual value that can be interpreted from Mr. Joni's statement is that agriculture in the future must be able to minimize and even avoid actions that violate soil, water, plants, and other physical and non-physical environments. A good relationship must be built between humans and the creator, human-human relations, and human-nature relations. This human relationship with the creator is the foundation or foundation of spiritual values. The creator, Allah SWT. Firmly instructs mankind without exception to build relationships in the divine dimension, the dimensions of fellow human beings and fellow creatures—His creation. Humans are ordered to maintain, protect and use it and learn from the applicable natural laws.

In line with what was conveyed by Mr. Jo, Mr. Pur also said that the tendency of farmers in general to carry out farming or cultivation tends to be exploitative. The actualization of spiritual values as outlined in the form of responsible cultivation behavior is still a problem in the agricultural pattern at Sumberbrantas. Very few farmers really actualize agriculture as a vehicle to worship Allah SWT.

The statement above can be interpreted that economic orientation is the first thing in the life of farmers in general. In fact, the essence of farming is not merely an economic matter. In Islamic teachings, it has been explained that agriculture is not singular, but is involved with all aspects of life and all systems, be it ecosystems, geosystems, or sociosystems. Pak Purnomo's statement confirms to us that so far farming activities have been implemented only partially, whether the orientation is toward the economy, or other aspects. Agriculture is supposed to be an integrated activity. Geographically or geosystems, agricultural implementation is related to climate conditions, soil, water, environment, biodiversity, etc. Demographically, agriculture is related to the diversity of population, markets, policies, institutions, preferences, etc. The estuary of all the synergies that are built in agriculture is happiness, prosperity, justice, equality, social responsibility, sustainability and benefit for all.

The term used by Pak Pur in describing the condition of contemporary society which is currently in a crisis of guidance has a very deep meaning. The term guidance when referring to the terminology and definition refers to the guidelines and instructions that are believed by an individual in behaving and acting. In the context of the life of the Islamic community, guidance should originate from the Bible (Al-Qur'an) as a revelation sent down by Allah SWT. God of the universe. This means that agriculture must also be guided by the Koran as the absolute source of truth. The Islamic teachings contained in the Qur'an are comprehensive, including integrated, sustainable agriculture, both what has happened, what is happening, and what will happen in the future.

Another term that must be underlined from the interview excerpt with Mr. Pur is related to fading faith, bland worship, and rituals that have lost their sacredness. This term refers to various factual conditions that occur in the life of rural communities, especially farming communities at this time. What is meant by the waning of faith in Pak Purnomo's terms refers to the phenomenon of society that seems to have begun to doubt the fortune and destiny line that has been determined by the creator. People are competing to collect as much wealth as possible, competing in planting area, competing in crop yields, but they forget that there is God's hand, there is God's decree, there is destiny and Allah SWT's Irodhat. behind it all. It is as if all achievements, all successes and even failures are derived from the process they do, derived from their strengths and knowledge. In other words, in this condition, Allah SWT. The lord of the universe has been exiled, killed, even in the absence of existence. As a replacement, the gods of modernity such as markets, technology, the internet, and others were presented. Worship has shifted, dislocation and disorientation of rituals and worship are clear evidence that what Mr. Purnomo said about fading faith, bland worship and loss of the sacredness of spiritual values is undermining the life of contemporary society these days. The farming community worships Regent, Lipor, Ciyodane, and so on pesticides rather than introspection on farming activities that trigger pest explosions. Awareness, which is another form of faith in the supreme controller of life, has been eroded by synthetic gods that manifest in the technology of chemical pesticides, hybrid seeds and chemical fertilizers. What must be contemplated is why now, it seems as if nature is no longer friendly to humans? Why in the past was the balance of nature able to control problems that occurred, for example, pests, soil fertility, and other climates, but now it is not? The answer to all of these questions is that it could be because we are not introspective. It is not nature that causes our lives to be less harmonious, but we are the ones who make nature angry, angry and even reluctant to be friends with humans anymore.

Pak Jo and Pak Pur implicitly conveyed the value and meaning of spirituality built within each farmer that needed to be reoriented so that farming activities would be more stable in the future. If we look closely and study it, it is very clear that the meaning and value of postmodern agriculture are reflected in the verses of the Koran, both explicitly and implicitly. Postmodern agriculture which bases its activities on spiritual values will free it from desire, exploitation, domination, greed, and lust or desires. Farming must be emphasized that producing is not for exploiting humans, plants, land, water, and other resources. Producing (agriculture) is not to maximize production, not to sell as much as possible or satisfy the market, but to meet consumption needs. Farming must be adaptive and anticipatory in order to maintain sustainability in the future.

Farming must be emphasized that it produces not for exploiting humans, plants, land, water, and other resources. Producing (agriculture) is not maximizing production, not selling as much as possible or satisfying the market, but meeting consumption needs. Farming must be adaptive and anticipatory to maintain sustainability in the future. Agriculture that is patterned on nature's way of producing, which is orderly, fair, balanced, ethical, location-specific, unique, based on local self-sufficiency, which prioritizes local knowledge and technology, which is civilized and beneficial. Postmodern agriculture is agriculture based on diversity of localities, based on diversity of commodities, and based on diversity of communities that do not create competition but side by side, thus building exchanges, mutually reinforcing and complementing each other.

In addition to the value of spirituality, postmodern agriculture is supported by the value of morality as a spirit that fills every agricultural activity. According to Pak Joni, since he has been in the farming profession for more or less 65 years, indeed various phases of change in farmer culture have shifted from generation to generation. The thing that grieves Pak Joni the most as a farmer figure in Sumberbrantas Village is related to the moral degradation of farmers. As an example, he pointed out how love for the land, love for the land and plants, is slowly but surely being reduced from the soul of the current generation. It is as if land and plants are no longer considered as creatures, or at least the sense of moral responsibility for the quality and quantity that must be maintained for the next generation has almost completely disappeared in today's farmers. Individual moral sensitivity or sensitivity is eroded and replaced by being (becoming) from the various choices available.

The aspect of morality has a very important role in the midst of globalization that is currently sweeping the world, including rural areas in Indonesia. Borrowing Bauman's term which says that "The world is full" when looking at the world that is increasingly global. The world is full not in the physical or geographical sense, but in the perception of a sense of closeness. Furthermore, Bauman [27] uses the metaphor of liquid modernity in illustrating the fragility of relations between individuals and their culture. The term "liquid modernity" is used by Bauman to refer to the phenomenon of change from a solid, controllable, predictable, rational form into a liquid state, that is, an inevitable condition of mortality, a condition in which its members act into change more quickly than is necessary and integrate it into the habits of every individual. Individuals, due to the weakness of the state system in the era of fluid modernization. Individuals become free with their choices because they follow the speed of changes that occur in this era (Elliot 2007). The conditions as described by Bauman, in almost the same textuality also occur in Sumberbrantas Village. The exposure to globalization in a slightly different context is also happening in the lives of farmers. Even though the levels are not as high as in urban areas, the term liquid modernity is also found by the authors in the results of in-depth interviews with informants.

The exposure to globalization has made it increasingly difficult to have space for a sense of closeness, empathy, ethics, and courtesy, and ironically, a feeling of *getun* (an action that makes a person feel regretful, disappointed) appears when doing good for others. As stated by Pak Joni in the excerpt of the interview above, that too much anxiety is experienced or rather felt by most farmers when they do good things for others. The feeling of being part of the land so that you treat the land as wisely as possible is no longer visible in the behavior of farmers in general. Exploitative behavior in the form of boosting production without regard to soil and environmental conditions is a dry form of farmer morality in carrying out their farming business. Then the spirit of morality, such as the philosophy that farming is a way of life, farming is a calling, and so on, has been extinguished which has been the torch of farmers in navigating the agricultural world? If this is the case then there is a very deep meaning related to the fading of these moral values. As the value of morality fades, it can be a sign of the end of agricultural existence. Subsistence farmers, or more precisely, small farmers, have survived until this moment, none other than because of the moral spirit that farming is a calling, a way of life, and a moral responsibility. For subsistence farmers or small farmers, subsistence ethics really is an ideology that they like or dislike, they have to hold on firmly because that is the only reason they remain in the farmer's path. The subsistence ethic is like a faith that will guide small farmers back to the path of agriculture or to keep on walking the path of the world of agriculture.

It becomes a serious problem when the spirit of peasant morality is allowed to drift in narratives of modernism driven by the energy of desire, which are full of pseudo-images and objects, shrouded in shallow meanings and decorated by simulacrum. As Pak Joni said, the activities of farmers driven by the desire to get as much produce from agriculture as possible is nothing but the implementation of hedonic traits that are deliberately narrated by the world of modernity, namely a world that is considered to be able to give humans pleasure, fascination, ecstasy, despite everything instantaneous, temporary, and not lasting. The imagination of stability, success, and satisfaction which is created through the symbols of modernity is a very evil virus that will undermine the morality and spirituality of the peasants. This virus will slowly but surely remove from farmers the traits of obedience, submission, and discipline that have been the true soul and breath of farmers.

Returning the moral spirit of farmers is a necessity if one expects that the world of agriculture must exist. Pak Joni and Pak Purnomo and Pak Sardjito as role models in Sumberbrantas Village have the same opinion in responding to the recent decline in farmer morality. A moral movement is needed so that farmers can return to their moral character as farmers. That farming is not just about planting and harvesting as much as you can. Behind it all there is a responsibility to protect fellow human beings, to maintain the survival of fellow creatures of God, to maintain the balance, and sustainability of the ecosystem. The following are excerpts of an interview with Pak Joni regarding his opinion on carrying out a moral movement to return farmers to their natural path.

6. Conclusions

Actors or agents in the implementation of organic farming consist of four types, namely, type A actor, type B actor, type C actor, and type D actor. Type A actor is an actor who is oriented toward spiritual values so that organic farming praxis is implemented as form of worship and manifestation of gratitude for the gift of Allah SWT. Type B actors are actors who are oriented toward moral values, so that the praxis of organic farming is actualized as a form of wisdom, and moral responsibility for preserving noble values in managing land and other resources. Type C actors are actors who are oriented toward economic values, so that organic farming practices are implemented as an effort to accumulate rupiah profits based on the profit-loss calculation of the choices they have. Type D actors are actors who are oriented toward political values, so that the organic farming practice is carried out in order to maintain and gain power or legitimacy.

The initiation of the "Batu Go Organic" program which is loaded with various interests, places the implementation of organic farming development in Batu City at the intersection of projects and community movements. As a project, the values of organic farming are reduced to a meaning that deviates far from the true organic nature and philosophy. Organic farming is nothing more than a developmentalism approach that traps farmers in various forms of dependency, powerlessness and exploitation in other forms. However, the incident of reducing the attitudes and values of organic farming on the one hand has also created anxiety for some actors who are concerned about the decline in the morality of farmers in treating land, water and other resources.

The traces of the implementation of organic farming which are packaged in the "Batu Go Organic" program are not only reduced in meaning as a means for project accumulation and rhetoric, but also shackled in the grip of the market. The approaches and policy instruments issued by the government are in contrast to the spirit of independence which has become the spirit of the national program "Thousand Organic Villages" Laws and regulations such as regional regulations, regulations from the Minister of Agriculture, implementation guidelines or implementation guidelines do not reflect the spirit of building village independence at all, but rather the standards and qualifications of organic agricultural products. This is of course only market oriented which actually cannot be intervened directly by the government.

Author details

Hamyana Yana[1*], Kliwon Hidayat[2], Keppi Sukesi[2] and Yayuk Yuliati[2]

1 Agriculture Development Polytechnic of Malang, Malang, East Java, Indonesia

2 Brawijaya University, Malang, East Java, Indonesia

*Address all correspondence to: hams.lodaya@gmail.com

References

[1] Aji B, Ningrum V. Reorientasi Kebijakan Pertanian Organik Sesudah "Go Organik 2010" dan "Program Seribu Desa Pertanian Organik" di Indonesia. Malang, Indonesia: UB Press; 2020

[2] Giddens A. A Contemporary Critique of Historical Materialism. 2nd ed. Standford, CA: Stanford University Press; 1995

[3] Kinseng RA. Struktugensi: sebuah teori tindakan. Jurnal Sosiologi Pedesaan. 2017;**5**(2):1-11

[4] Layder D, Ashton D, Sung J. The empirical correlates of action and structure: The transition from school to work. Sociology. 1991;**25**(3):447-464

[5] Grim JA, editor. Indigenous Traditions and Ecology. Cambridge, MA: Harvard University Press; 2001. p. 33

[6] Marselle MR, Warber SL, Irvine KN. Growing resilience through interaction with nature: Can group walks in nature buffer the effects of stressful life events on mental health? International Journal of Environmental Research and Public Health. 2019;**16**(6):986

[7] Irvine KN, Hoesly D, Bell-Williams R, Warber SL. Biodiversity and spiritual well-being. In: Biodiversity and Health in the Face of Climate Change. Cham, Switzerland: Springer; 2019. pp. 213-247

[8] Wilson E. The Spiritual History of Ice: Romanticism, Science and the Imagination. Cham, Switzerland: Springer; 2003

[9] Alhamidi SK, Gustafsson M, Larsson H, Hillbur P. The cultural background of the sustainability of the traditional farming system in the Ghouta the oasis of Damascus, Syria. Agriculture and Human Values. 2003;**20**:231-240

[10] Yazdanpanah M, Tajeri Moghadam M, Javan F, Deghanpour M, Sieber S, Falsafi P. How rationality, morality, and fear shape willingness to carry out organic crop cultivation: A case study of farmers in southwestern Iran. Environment, Development and Sustainability. International Journal of Environmental Research and Public Health. 2021;**18**(10):5310. DOI: 10.1007/s10668-021-01523-9

[11] Coleman JS, Coleman JS, Farraro TJ. Rational Choice Theory: Advocacy and Critique. 1992

[12] Pham L, Shively G. Profitability of organic vegetable production in Northwest Vietnam: Evidence from Tan Lac District, Hoa Binh Province. Organic Agriculture. 2019;**9**:211-223

[13] Putra R, Suyatna H. Genealogi Kuasa dalam Kebijakan Pengembangan Pertanian Organik di "Wilayah Pardikan". Jawa. 2018;**5**(1):69-84

[14] Emirbayer M, Mische A. What is agency? American Journal of Sociology. 1998;**103**(4):962-1023

[15] Tscharntke T, Grass I, Wanger TC, Westphal C, Batáry P. Beyond organic farming–harnessing biodiversity-friendly landscapes. Trends in Ecology & Evolution. 2021;**36**(10):919-930

[16] Nascimbene J, Marini L, Paoletti MG. Organic farming benefits local plant diversity in vineyard farms located in intensive agricultural landscapes. Environmental Management. 2012;**49**:1054-1060

[17] Kalinina O, Cherkinsky A, Chertov O, Goryachkin S, Kurganova I, de Gerenyu VL, Giani L. Post-agricultural restoration: Implications for dynamics of soil organic matter pools. Catena. 2019;**181**:104096

[18] Setiawan I, Supyandi D, Rasiska S, Gunardi MJ. Pertanian Postmodern: Jalan Tengah Vertikal Generasi Era Bonus Demografi Membangkitkan Peradaban Nusantara. 1st ed. Penebar Swadaya; 2018

[19] Rigby D, Cáceres D. Organic farming and the sustainability of agricultural systems. Agricultural Systems. 2001;**68**(1):21-40

[20] Macilwain C. Organic: Is it the future of farming? Nature. 2004;**428**(6985):792-794

[21] Pretty JN. Participatory learning for sustainable agriculture. World Development. 1995;**23**(8):1247-1263

[22] Lockeretz W. What explains the rise of organic farming? In: Organic Farming: An International History. Wallingford, UK: CABI; 2007. pp. 1-8

[23] Mishra AR, Rani P, Pardasani KR, Mardani A. A novel hesitant fuzzy WASPAS method for assessment of green supplier problem based on exponential information measures. Journal of Cleaner Production. 2019;**238**:117901

[24] Hunt V, Layton D, Prince S. Diversity matters. McKinsey & Company. 2015;**1**(1):15-29

[25] Winnett Y. Go organik! Berangkat dari wacana revolusi hijau menuju pertanian berkelanjutan: siapa diuntungkan oleh pendekatan pertanian organik diarahkan ekonomi dan pemberdayaan sosial? Studi Physics Review E; 2011. Available from: http://www.ainfo.inia.uy/digital/bitstream/item/7130/1/luzardo-buiatria-2017.pdf

[26] Sumaryono. Hermeneutik: Sebuah Metode Filsafat. Kanisius; 1993

[27] Bauman Z. Liquid modernity. John Wiley & Sons, Polity Press; 2013

Chapter 15

Increasing the Value of Waste Hop Biomass by Composting: Closing the Nutrient Cycle on Hop Farms

Barbara Čeh, Lucija Luskar, Julija Polanšek, Ana Karničnik Klančnik and Žan Trošt

Abstract

After the harvest of each hectare of hop (*Humulus lupulus* L.), an average of 15 tons of fresh waste plant biomass, consisting of leaves and stems, is generated next to the harvest hall. On-site (on-farm) composting is an excellent way to manage excess biomass after the harvest, and it can help to establish a circular economy in the hop-growing sector. It is essential that the resulting compost is both nutrient-rich and safe for use, while also avoiding any potentially harmful leachate impacts. We have to establish the compost pile as soon as possible after the harvest, and to mix it several times during the thermophilic phase, related to temperatures measurements. After approximately 2 months, the pile will begin to cool down, at which point it is advisable to cover it with a semi-permeable membrane. Five months later, the compost is mature—ready to use as an organic fertilizer with app. 9.6 kg N, 1.6 kg P, and 5.9 kg K per ton, no phytotoxic properties, stable, and suitable for plant production (based on the radish germination index of GI 89%). The cress germination test (average of GI 196%) has demonstrated that it is a nutrient-rich and stimulating substrate.

Keywords: *Humulus lupulus* L., composting, compost fauna, nutrients content, circular economy, germination test, circular economy, resource efficiency

1. Introduction

The hop plant, *H. lupulus* L., is cultivated for its cones, which are an essential ingredient in the brewing industry. During the harvest, the plants are cut, and the aboveground biomass is removed from the fields to the harvesting machine, which harvests the hop cones; they are collected, dried, and packed for use in the brewing industry. The stems and leaves are left in this procedure as a by-product next to the harvest machine (see **Figure 1**), with the supporting twine intertwined within this waste material. In a single growing season, the hop plant can attain a height of up to 7 meters and as a climbing plant requires a guiding twine for support. Typically, plastic twines or iron strings in conjunction with plastic twine are utilized and, to a lesser

Figure 1.
Subsequent to the harvest of the hop cones, waste hop biomass in the form of hop stems and leaves is generated as a by-product and accumulates in close proximity to the harvest machine hall (photo: L. Luskar*).*

extent, also biodegradable twine made from renewable sources like jute, coconut husks, or polylactic acid.

Every harvest season, the hop industry generates on average 15 tons of waste hop biomass (fresh matter) from each harvested hectare of hop fields [1]. So 2600 hop farms with 26,500 hop fields in the European Union produce beside 50,000 tons of hop cones per year [2] also 400,000 tons of waste hop biomass, which is of good composition for on-site (on-farm) composting [1], although this use is not common. However, based on massive amounts of organic waste produced on hop farms, on-site composting of hop biomass needs to be considered because it offers a perfect way for nutrients and organic matter to be returned to the agricultural fields in the narrowest circle possible.

Hop biomass after harvest contains roughly 18% organic mass, 0.8% nitrogen, 0.3% potassium, and 0.1% phosphorus, with a carbon-to-nitrogen ratio of 13:1 [1]. The ratio between carbon and nitrogen in waste hop biomass is 13:1 when composting stems and leaves together and 23:1 when composting only stems [1].

Composting represents a conventional, low-investment technology for transforming biomass into a stabilized end product with low levels of readily degradable organic matter and without any phytotoxic effects on plants [3, 4]. This approach aligns with the need to develop new methods for reducing the use of chemical fertilizers [5, 6]. Hop biomass can be taken to industrial composting plants, where it is mixed with other organic materials and subjected to thermal and technical processing, after which it is sifted to produce high-quality and safe compost. Alternatively, on-site composting (i.e., on-farm composting) represents a much more cost-effective option for farmers, which, if properly managed, can be just as efficient as industrial composting.

2. Hop biomass on-site (on-farm) composting

Utilization of biodegradable twine as a plant support mechanism in hop fields during the growing season serves as a favorable predisposition for the appropriate composting of hop biomass. Before harvesting, the knife on the harvest machine has to be checked, sharpened, and set for cutting the stems into pieces smaller than 5 cm. During the harvest, the knife has to be cleaned several times per day. Before composting, the location of the placed compost pile has to be determined in line with national rules and regulations.

The compost pile has to be set right after harvest or no later than two days after harvest. Biomass could be collected on a tractor trailer and transported directly to the location of composting. Otherwise, the collecting biomass place could be right next to the harvesting hall, later loaded on the trailer, and taken to the location of composting. The optimal direction of compost pile placement is north-south.

The quantity of fresh waste hop biomass from a one-hectare hop field ranges between 13 to 20 tons, depending on variables such as the hop variety, agricultural practices, and weather conditions during the growing season, with an average of approximately 15 tons [7]. This quantity of biomass is sufficient to establish favorable conditions for successful composting.

The LIFE BioTHOP project has introduced a 100% biodegradable and 100% on-site compostable bio-plastic polylactic acid (PLA) twine, which offers improved solutions for managing hop biomass, including its use as a substrate for producing high-quality compost. When subjected to proper on-site composting, this twine is broken down into CO_2, water, and organic matter. To optimize composting characteristics, the stems of hop plants should be cut into pieces that are less than 5 cm in length (**Figure 2**). However, if plastic supporting twine is still being used by hop growers, the stems on the harvest machine should be cut into longer pieces, not shorter than 30 cm, so they can be fully sifted out by the sieving machine. This ensures that the final compost is completely free of plastic particles.

For optimal composting, it is recommended to shape the hop biomass pile into a trapezoidal form, with a height of approximately 3 meters and a maximum settled height of 1.5 meters. The slope of the pile should be even to prevent waterlogging and should be constructed in a manner that allows for proper drainage of precipitation, without being absorbed into the pile. The actual size of the pile may vary depending on the method of mixing, and hop growers may adjust the shape and size of the pile over time to accommodate their experience and mixing techniques.

Regular temperature measurements must be taken within the hop biomass pile. The initial mixing of the pile should be performed once the temperature has surpassed 60°C for three consecutive days or when a temperature of 65°C has been recorded for two consecutive days. The pile must be turned to ensure that previously exposed parts are mixed inside the pile. During the thermophilic phase, when the temperature exceeds 45°C, it is recommended to turn the pile at least once per week. If the temperature rises above 60°C, the pile should be turned twice per week. After each mixing, the pile should be re-shaped into a trapezoidal form [7].

Ensuring an appropriate temperature distribution within composting piles is a critical aspect of achieving effective composting and preventing pathogen contamination. Generally, the temperature in the core of the pile is higher than the outer layer (as shown in **Figure 3**). In our experiments, we observed a temperature difference of roughly 10°C between a depth of 30 cm and 1 m. The outer layer, which is approximately 30 cm thick, did not exceed 55°C in our trials. Neglecting to adequately sanitize this outer layer can lead to the proliferation of phytopathogens, which can spread to the rest

Figure 2.
It is recommended that hop stems be cut into smaller pieces during the harvesting process if biodegradable and compostable twine is present (photo: L. Luskar*).*

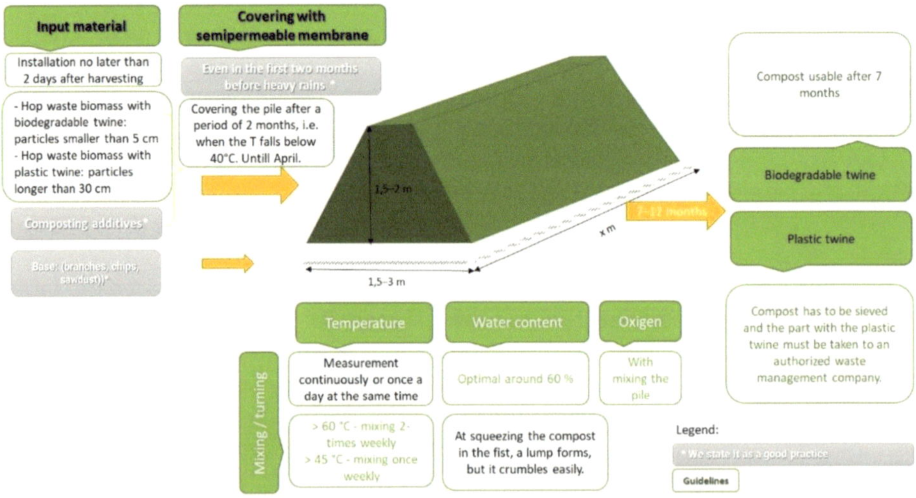

Figure 3.
*A diagram of on-farm hop biomass composting (*L. Luskar*).*

of the pile and the field soil if the compost is used there. Regularly turning the pile is essential for ensuring uniform composting and proper heat distribution throughout the biomass. Furthermore, closely monitoring temperature and moisture levels can help maintain optimal composting conditions and minimize the risk of pathogen contamination. Ultimately, effective management of composting piles is crucial for preventing the spread of plant diseases and promoting sustainable agriculture practices.

Composting leaves together with stems can increase the overall efficiency of the process. The addition of leaves can improve nutrient availability within the compost and prevent excessive drying caused by empty spaces within the pile. This extended thermophilic phase resulting from the inclusion of leaves is critical for both the degradation of PLA twine and the hygienization of the biomass. Therefore, composting the entire biomass after harvest is recommended to optimize the composting process. Two key factors for efficient composting are the combination of small biomass particles and frequent turning of the pile. The small particle size provides a larger surface area for microbial activity, leading to faster decomposition and composting. Frequent turning of the pile facilitates proper oxygen flow and moisture distribution, which are necessary for the growth of microorganisms and the prevention of odors and pathogen growth [8].

Overall, composting the entire biomass after harvest can provide greater benefits for soil fertility and sustainability. With the appropriate management practices in place, such as maintaining the appropriate moisture and oxygen levels and regularly turning the pile, the composting process can be optimized to achieve maximum efficiency and nutrient content. A diagram of on-farm hop biomass composting is presented in (**Figure 4**).

During the stabilization phase of composting, the temperature of the pile typically drops below 40°C. In hop biomass composting, this starts after around 2 months from the start, that is, in November. To optimize the composting process and prevent nutrients loss, it is advisable to keep the pile covered with a semi-permeable membrane during this phase (**Figure 5**). This helps to retain heat and moisture within the pile, prevent leaching,

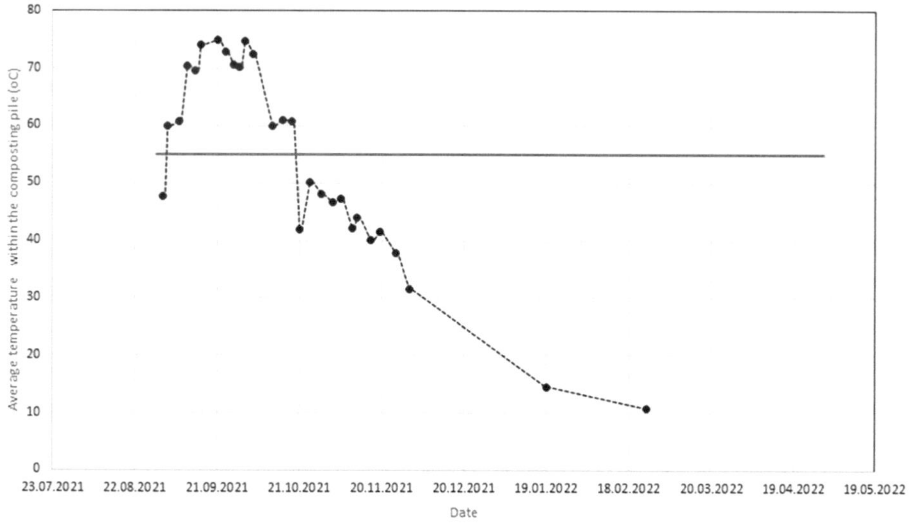

Figure 4.
Temperature in composting hop biomass pile with proper mixing procedure and time – An example.

Figure 5.
Composting hop biomass in winter, covered with a semipermeable membrane.

and facilitates optimal microbial activity. Once the compost has stabilized and the temperature has decreased to a level comparable to its surroundings, it can be considered for use. The core of the pile should have the same temperature as the surroundings and should exhibit a soil-like odor. Considering our experiments, this is in about seven months after the start of proper composting, in April of the following year.

3. Hop biomass compost characteristics

The characterization of hop biomass compost is typically determined using a range of chemical tests, including pH tests and assessments of ammoniacal nitrogen, organic C, total N, nitrate nitrogen, potassium, and phosphorus levels. In addition, water content is an important characteristic of compost that is commonly measured. Due to variations in input materials, such as differences in hop stem length and various additives that can be added at the start, including biochar and effective microorganisms, there may be differences in the chemical composition of hop biomass composts. Nevertheless, average values of key characteristics are presented in **Table 1**, along with a comparison to the original starting/fresh material. Final compost (dw) contains about 3–4% nitrogen, 0.3–0.4% phosphorus, 1.0–2.5% potassium, and 35–43% total organic carbon [8].

Legend: dry matter (DM), total nitrogen (TN), total phosphours (TP), total potassium (TK), total carbon (TC), nitrate nitrogen (NO_3-N), ammoniacal nitrogen (NH_4-N).

The pH range of mature compost is a crucial factor in determining its suitability for use in agricultural applications. According to Hachicha et al. [9] and Rynk et al. [10], the optimal pH range for mature compost is between 6.0 and 8.5. The pH of the hop biomass mature compost falls within this range at 7.8, while the pH of the hop biomass input material is around 6.5. Similar increase in pH has been observed in horticultural waste composting, as reported by Choy et al. [11]. Despite these general trends, it is important to note that the pH requirements of specific plants may vary [12].

	Start	End
pH	6.5	7.8** – 8.4*
DM (%)	27.8	31.2** – 31.3*
TP (%)[1]	0.28	0.38** – 0.46*
TK (%)[1]	1.67	1.08** – 2.07*
TC (%)[1]	48.1	22.8** – 34*
TN (%)[1]	2.6	2.7** – 3.5*
NO_3-N (mg/kg)[2]	0.8	376** – 1051*
NH_4-N (mg/kg)[2]	170	83* – 404**

Legend: dry matter (DM), total nitrogen (TN), total phosphorus (TP), total potassium (TK), total carbon (TC), nitrate nitrogen (NO_3-N), ammoniacal nitrogen (NH_4-N).
[1]*Measured in dry matter.*
[2]*Measured in fresh matter.*
Following the thermophilic phase lasting approximately two months, composting piles were covered with a semipermeable membrane for an additional five months (unpublished data of Institute of hop research and Brewing).
**Composting piles uncovered [8].*

Table 1.
The basic chemical characteristics of both the input material (fresh hop biomass after harvest) and the final/ mature compost produced after seven months of on-site composting.

The dry matter (DM) content of hop biomass compost typically undergoes a slight increase during the composting process, rising from an initial level of 27.8% to approximately 31.2% [8]. For mature compost to be effective in agricultural applications, it should have a DM content falling within the range of 30–50%, according to McFarland [13].

During the composting process, the average total nitrogen (TN) content of compost piles typically increases. Research has shown that biochar can be particularly effective in reducing nitrogen losses from composting materials, according to Steiner et al. [14]. In general, composts with a total nitrogen content greater than 2% are suitable for use as fertilizer, so compost from hop biomass covers this requirement [8].

The average phosphorus content (TP) of fresh hop biomass after harvest is typically around 0.28% in dry matter. During the composting process, it increases to around 0.38% in dry matter. Conversely, the average potassium content (TK) in hop biomass typically decreases during composting [8].

The expected range of potassium content in compost is typically reported to be between 0.6 and 1.7% [15]. The final hop compost pile analyzed in our study met this standard with a potassium content of 1.1% [8]. Adebayo et al. [16] reported a decrease in total potassium content during composting of food waste and yard trimmings, except in a closed system, where it initially increased before falling below the initial value in the substrate mixture. The total carbon (TC) content decreased significantly from an average of 48–23% of dry mass during composting, which is expected due to microbial immobilization of approximately 40% of available carbon, with the remaining 60% lost through respiration [17].

The average nitrate content (NO_3-N) in fresh hop biomass was found to be around 0.8 mg/kg and 375.9 mg/kg in the final compost [8]. This is consistent with nitrate being the final product of nitrogen mineralization [18] and an expected increase during composting. The starting material had an average ammoniacal nitrogen content (NH_4-N) of 169.6 mg/kg, while the final composts had 403.8 mg/kg [8]. Contrary to the expected decrease in ammonia levels during the maturation phase [19], our findings suggest an increase. Piles with added biochar had the highest nitrate and ammoniacal nitrogen content, indicating their potential to prevent nitrogen losses during composting [14, 20, 21]. The findings suggest that additives and small particle sizes contribute to nutrient retention in the pile; however, the most important factor is covering the pile with a semipermeable membrane throughout the maturation phase.

Compost stability and maturity are the main properties to characterize compost quality [22, 23]. However, chemical characterization alone is not enough to assess compost maturity. The phytotoxic effect on plants is related to immature compost, while low microbial respiration indicates compost stability [22, 23]. Germination and growth tests [24] are used to determine the effect of compost on plants. The method combines seed germination index and root elongation of cress seeds and garden radish (*Lepidium sativum* L. and *Raphanus sativus*) [25]. The number of germinated seeds is counted, and the overall length of seedlings (root) is evaluated. The number of germinated seeds and the length of the radicle are both affected by the compost extract, while the germination index (GI) describes both parameters compared to the control.

In 7 months, all the hop waste biomass composts had reached the mature phase. None of the composts from hop biomass in our research were phytotoxic (GI < 65) according to the Zucconi [25] criteria based on the germination index. The BioTHOP PLA twine inside has loosened its strength and was degraded. The degradation level was correlated with the level of shredding and differed due to the presence or absence of leaves. Taking the radish germination index (average of GI was 89%) into consideration, the composts in our study showed a substrate with no phytotoxic properties, stable and appropriate for plant production, whereas in the cress germination test, the composts showed nutrient-rich or stimulating substrates. The germination index was between 125 and 213 (average of GI 196%), and microbial respiration was between 0.1 and 0.2 C-CO_2/g compost/day [8].

When properly composted, hop waste biomass composts had an earthy smell with no phytotoxic effect on plant germination. The most promising treatment for on-site composting after the first year of trials is indicating to be good shredding (pieces smaller than 3 cm) and proper aeration/mixing in the thermophilic phase of the process. Previously considered agro-waste can now be used as an organic fertilizer, when properly on-site (on-farm) composted.

4. Hop biomass compost microbiological properties

Microorganisms are ubiquitous in the environment and play a pivotal role in the biochemical degradation of organic matter. These microbes are responsible for converting nutrients from organic to plant-available mineralized forms [26]. A single square centimeter of the plant leaf surface is typically colonized by approximately 10^6–10^7 bacteria [27], making plant material a significant source of microbial activity. Additionally, the soil serves as a reservoir for biological degraders, and when composting is carried out on soil, the microbes can enter the compost pile during the process.

The composting process is influenced by several factors, including the composition of the feedstock, moisture levels, oxygen content, pH, and temperature. As such, a comprehensive understanding of the microbiological processes at work is essential to supplement compost chemical composition analysis [28]. Soil compost amendments contribute to the general soil quality recovery and improvement of plant growing conditions by providing numerous ecosystem services, including replenishment of soil carbon stocks, increase of microbial activity and biodiversity, and restoration of plant nutriton [29]. Compost is normally populated by 3 general categories of

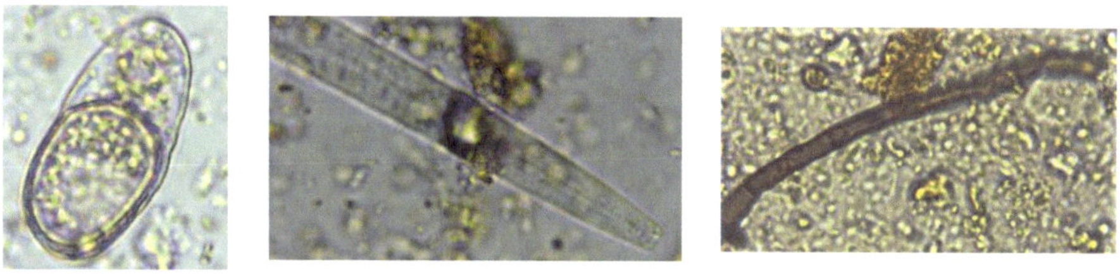

Figure 6.
Amoeba, nematode, and aerobic fungus, detected in mature hop biomass compost (photo: J. Kadunc*).*

microorganisms: bacteria, fungi, and actinomycetes. It is primarily the bacteria, and specifically the thermophilic bacteria, that create the heat of the compost pile [30].

During the composting process, the conditions in the pile, including temperature, aeration, moisture, pH, and substrate availability, are subject to constant change, resulting in stages of microbial consortia and their fluctuation [31]. The initial decomposers are mesophilic organisms such as bacteria and fungi. In the subsequent stage, thermophilic organisms, particularly actinomycetes, become dominant, and fungal populations decline. During the maturation phase of composting, a new mesophilic community develops, with actinomycetes remaining and fungi reappearing, along with cellulose-decomposing bacteria [32].

Only a sample study has been performed by now since the beginning of our research on the on-site hop biomass composting some years ago, to take the snapshot of hop biomass composts microbiological properties (**Figure 6**), in which we found out that the microbial world of composted hop biomass was dominated by bacteria [33]. Part of fungi mycelium was also found, but in general, diversity, which is main property of quality compost, was not high. The observed count of colonies forming units was approximately 10^6 CFU per gram of dry matter, falling within the anticipated range [34]. Fast-changing conditions in soil (heat, drought, moisture, and lightness) demand fast adaptation of microbes that can only be tackled by diversity. Due to their fast reproduction, the number does not play such an important role as their diversity. The work on the topic will continue within our research group on the composts from improved composting procedures [33].

5. Fauna in hop biomass compost

While processes of nutrient cycling are governed directly by microbes, such as bacteria and fungi, they are also affected by soil animals that live alongside them. Soil fauna affects decomposition processes both directly, through fragmentation and comminution of litter material, and indirectly, by altering microbial function through grazing of the soil microbial biomass and through excretion of nutrient-rich wastes. The movement of animals through soil influences the dispersal of microbial propagules attached to the animal body surfaces or transiting through their guts [35]. Invertebrates co-exist with the microbes and are essential to a healthy compost pile.

The mesofauna and macrofauna are a diverse group of organisms, varying in size from tiny mites to large insects, that inhabit the composting biomass under investigation. These organisms play a crucial role in decomposing the organic material and providing nutrients for the entire food web, which restores the natural balance in the compost. Obtaining rich, mature compost with a healthy population of fauna is essential because these organisms help to maintain the physical and chemical properties of the compost through their mechanical and chemical actions. By breaking down the organic material, they facilitate the nutrient cycling process, which is crucial for the growth of healthy plants. Therefore, it is crucial to ensure that the composting process allows for the growth and proliferation of these important organisms.

The fauna in the composting piles after five months from the start of the composting process (in winter) was dominated by springtails, mites, spiders, centipedes, soldier fly, and larva. The most numerous in all compost piles were springtails (at least ten in 2 g of compost) and mites (at least ten in 2 g of compost). A wide variety of arthropods were found in the compost pile, where effective microorganisms were mixed in at the start of composting. In the mature compost, in April, the most numerous were springtails, which were of various sizes and colors, and mites. Amoebas, earthworms, centipedes, spiders, larva, beetles, and insects were also detected (**Figure** 7).

Figure 7.
Fauna detected in hop biomass compost (photo: A. Karničnik Klančnik). A, B, C –springtails Collembola; D, E, F, G, H – Mites (Acarina); I – The initial stage in the development of an insect; J, K, L - beetles (Coleoptera); N, T – Spiders (Araneae); O – Pseudoscorpions; P – Earthworm; R – Larva; S – Ant; M, Š – Centipedes; T, U – Soldier Fly.

6. How to use hop biomass compost and why

To avoid the phytotoxic impact, which can delay seed germination or inhibit plant growth, compost should be mature and stable before being used as a fertilizer [36].

A ton of compost from hop biomass with an average moisture content of 70% contains 8.1 kg of total nitrogen (N), 1.14 kg of total phosphorus (P) or 2.6 kg of P_2O_5, and 3.24 kg of total potassium (K) or 3.8 kg of K_2O, according to current IHPS measurements [7]. If the compost pile is covered over the winter to be protected from leaching of the nutrients because of higher precipitation, the compost contains more nutrients. A ton of compost with an average moisture content of 70% contains 9.6 kg of total nitrogen (N), 1.56 kg of total phosphorus (P) or 3.57 kg of P_2O_5, and 5.93 kg of total potassium (K) or 7.15 kg of K_2O. A comparison with nutrients content in farmyard manure is presented in **Table 2**. However, it is advised to analyze each compost for the water content and main nutrients content in order to obtain accurate information before its use. By taking into account the soil analysis and nutrient removal from the soil with the certain crop, we can more accurately calculate how much compost can be used to fertilize a certain crop. We adhere to a five-year fertilization plan because in this way, we ensure the gradual achievement and maintenance of a good soil supply with nutrients and maintain soil fertility.

Compost can serve as an effective fertilizer, much like stable manure. The percentage of organic mass (over 35%) indicates that the compost is within the criteria of the first-class compost. Hop growers can load mature compost on the spreader and spread it across the field without problems. It can be utilized for basic fertilization in the spring or fall, by incorporating it into the soil. For example, it can be incorporated during basic tillage before sowing maize, which is an ideal timing as mature hop compost is typically available in April. Compost is also suitable for use in fertilizing hop fields, grasslands, at planting vegetables, and sowing cereals. However, in cases, where the hop biomass used in the composting process is infected with hop wilt, as a precautionary measure, we recommend that the compost is not returned to the hop fields but rather utilized in other fields or grasslands.

If we take into account the limitation of the annual nitrogen fertilization, which amounts to 250 kg/ha N with organic fertilizers, we can therefore apply a maximum of 30 t/ha of such compost in one year if the compost would be the only fertilizer. The limit of 250 kg/ha N applies to all types of organic fertilizers together, that is, for livestock fertilizers and all other types of organic fertilizers (digestate, compost, etc.) together [38]. So if we use any other organic fertilizer, then we must accordingly

Nutrient	Hop biomass compost with 70% moisture (kg/t)	Stable manure (kg/t) with 81.4% moisture*
N	8.1** - 9.6***	4.7
P_2O_5	2.6** - 3.6***	3.0
K_2O	3.9** - 7.2***	5.1

**The average moisture of hop biomass compost after 7 months was found around 70% according to IHPS measurements. Stable manure's average moisture is 81.4% [37].*
***Uncovered composting piles [8].*
****Following the thermophilic phase lasting approximately 2 months, composting piles were covered with a semipermeable membrane for an additional five months (unpublished data of the Institute of hop research and Brewing).*

Table 2.
Comparison of nutrients content in hop biomass mature compost and farmyard manure.

reduce the amount of compost. When deciding on the dose of fertilizer for an individual crop or vegetable, it is also necessary to take into account the limit values of the total intake of nitrogen with fertilizers (organic and mineral fertilizers) for individual types of agricultural plants per individual unit of agricultural land use in accordance with the plants' nitrogen needs and measures to reduce and prevent water pollution. In any case, it is recommended to have a fertilization plan, which must include a calculation of nitrogen fertilization needs, based on expected yield.

Regarding the use of compost, it is necessary to adhere to the decree on the protection of waters against pollution caused by nitrates from agricultural sources, guidelines for the implementation of water protection requirements against nitrate pollution from agricultural sources, keep an eye for proper application technique, timing, and cross-compliance.

To present a positive impact of mature hop biomass compost use for fertilization compared to fresh hop biomass incorporation, a pot experiment was set with Chinese cabbage (*Brassica rapa* L. subsp. *chinensis* (L.) Hanelt). Treatments were: K (control), 185 g substrate; SH (fresh hop biomass), 27 g fresh hop biomass +148 g substrate; and ZK (mature compost), 27 g mature compost +148 g substrate. In 4 days, there were significantly fewer plants emerged in SH compared to ZK and K, but later there were no significant differences between the treatments in the number of emerged plants. The above-ground biomass of the plants after 47 days, when the experiment was evaluated, was statistically significantly the most abundant in ZK, whose leaves were the most intensely green at the same time (**Figure 8**). There was no significant difference in biomass weight between K and SH, but the leaves of K were of paler green color. The content of nitrate in the leaves was significantly higher in ZK (130 mg/L) compared to K and SH (19 mg/L and 36 mg/L, respectively). The content of nitrate in the substrate was significantly higher in SH (<3.8 mg/L) compared to ZK and K (<3 mg/L). The content of ammonium nitrogen was significantly lower in K (0.3 mg/L) compared to SH and ZK (0.7 mg/L). The density and branching of the root system were the highest in ZK and the worst in SH (**Figure 9**).

Figure 8.
Fertilization pot trial with Chinese cabbage (Brassica rapa L. subsp. chinensis (L.) Hanelt) after 47 days from sowing; left: Fresh hop biomass (SK) mixed in substrate at sowing, middle: Mature hop biomass compost (ZK) mixed in substrate, right: Control (K) with pure substrate.

Figure 9.
Roots check in fertilization pot trial with Chinese cabbage (Brassica rapa L. subsp. chinensis (L.) Hanelt) after 47 days from sowing; left: Fresh hop biomass (SH) mixed in substrate at sowing, middle: Control (K) with pure substrate, right: Mature hop biomass compost (ZK) mixed in substrate.

7. Conclusions

Following the harvest of every hectare of hop (*Humulus lupulus* L.), an average of 15 tons of fresh waste plant biomass, consisting of leaves and stems, is generated next to the harvest hall [1]. This green waste is mostly seen as a waste and represents additional disposal cost to the farmers but can also be transformed into compost through on-site composting, which is a sustainable and effective means of repurposing this waste product [8]. To ensure a high-quality compost product with minimal environmental impact throughout the on-site composting process, advanced composting technology has been developed. This technology offers one of the most promising means for returning essential nutrients and organic matter to the agricultural land of the same farm. By recycling this waste product in this manner, farmers can enhance soil fertility and maintain the health and productivity of their crops while also minimizing the environmental footprint of their agricultural operations [39].

Adhering to professional composting guidelines is essential to ensure that the resulting compost is both nutrient-rich and safe for use while also avoiding any potentially harmful leachate impacts. One critical consideration in this regard is the amount of precipitation, as there is a strong linear correlation between this factor and the volume of leachate produced [40]. To minimize the risk of leachate-related issues, it is vital to implement appropriate measures. One of the most effective strategies is to cover the composting pile with a semipermeable membrane, particularly after the end of the thermophilic phase, which typically occurs approximately two months after the start of the process. This covering should remain in place throughout the maturation phase, which extends until April of the following year. By following these guidelines, farmers and composting operators can minimize leachate impacts, avoid the loss of nutrients, and therefore produce a high-quality, nutrient-rich compost product [7].

With on-site hop biomass composting, farmer gets his own organic fertilizer, which can be applied safely to his agricultural land, and in this way, the cycle of the nutrients and organic matter returned back to the agricultural land is the narrowest possible. The compost, prepared according to the professional guidelines is safe, contains nutrients, organic matter, many microorganisms, such as bacteria, actinomycetes, and fungi, as well as different fauna species, which enhance soil biodiversity when applied to the agricultural land.

With a bit of effort to guide a proper process of composting, farmer can save some money on the purchase of fertilizers and avoids landfill costs. And the last but not the least, a circular economy on farm is established.

Acknowledgements

This paper has been produced with the support of the EU LIFE program, as part of the BioTHOP project's After-LIFE program. The authors take sole responsibility for the contents of this paper, and it does not necessarily represent the views of the European Commission. The research was also financially supported by the municipalities of the Lower Savinja Valley, the Ministry of the Environment and Spatial planning of the Republic of Slovenia, and the Slovenian Hop Growers' Association. The work was also conducted under the auspices of the professional task, Technology of Hop Production and Processing, which was funded by the Ministry of Agriculture, Forestry, and Food of the Republic of Slovenia.

Author details

Barbara Čeh*, Lucija Luskar, Julija Polanšek, Ana Karničnik Klančnik and Žan Trošt
Slovenian Institute of Hop Research and Brewing, Žalec, Slovenia

*Address all correspondence to: barbara.ceh@ihps.si

References

[1] Čeh B, Luskar L, Čremožnik B. Hop biomass after harvest as input material for composting. Hop Bulletin. 2019;**26**:81-90

[2] International Hop Growers' Convention–IHGC. Economic Commission-Summary Reports. 2021. Available online: http://www.hmelj-giz.si/ihgc/doc/2021_NOV_IHGC_ECReport_public.pdf. (accessed on 14 December 2021)

[3] Hait S, Tare V. Transformation and availability of nutrients and heavy metals during integrated composting–vermicomposting of sewage sludges. Ecotoxicology and Environmental Safety. 2012;**79**:214-224

[4] Yu Z, Tang J, Liao H, Liu X, Zhou P, Chen Z, et al. The distinctive microbial community improves composting efficiency in a full-scale hyperthermophilic composting plant. Bioresource Technology. 2018;**265**:146-154

[5] Barrena R, Font X, Gabarrell X, Sánchez A. Home composting versus industrial composting: Influence of composting system on compost quality with focus on compost stability. Waste Management. 2014;**34**:1109-1116

[6] Martínez-Blanco J, Colón J, Gabarrell X, Font X, Sánchez A, Artola A, et al. The use of life cycle assessment for the comparison of biowaste composting at home and full scale. Waste Management. 2010;**30**:983-994

[7] Čeh B, Flis J, Luskar L, Polanšek J, Trošt Ž. Smernice za ravnanje s hmeljevino in njeno predelavo v kompost na kmetijskem gospodarstvu, ki se ukvarja s hmeljarstvom. 2022a. Available at: https://www.life-biothop.eu/wpcontent/uploads/2022/08/Smernice_hmeljevina-AVGUST-2022_FINAL-VERZIJA1.pdf. [Accessed: February 17, 2023]

[8] Luskar L, Polanšek J, Hladnik A, Ceh B. On-farm composting of hop plant green waste—Chemical and biological value of compost. Applied Sciences. 2022;**12**:4190

[9] Hachicha S, Sellami F, Cegarra J, Hachicha R, Drira N, Medhioub K, et al. Biological activity during co-composting of sludge issued from the OMW evaporation ponds with poultry manured, Physico-chemical characterization of the processed organic matter. Journal of Hazardous Materials. 2009;**162**:402-409

[10] Rynk R, Van de Kamp M, Willson GB, Singley ME, Richard TL, Kolega JJ, et al. On-farm composting handbook. Northeast Regional Agricultural Engineering Service. 1992;**54**:186

[11] Choy SY, Wang K, Qi W, Wang B, Chen CL, Wang JY. Co-composting of horticultural waste with fruit peels, food waste, and soybean residues. Environmental Technology. 2015;**36**(11):1448-1456

[12] Van der Wurff AWG, Fuchs JG, Raviv M, Termorshuizen AJ. Handbook for composting and compost use in organic horticulture. BioGreenhouse. 2016. 106 p

[13] McFarland MJ. Biosolids engineering. McGraw-Hill Education; 2001

[14] Steiner C, Das KC, Melear N, Lakly D. Reducing nitrogen loss during

poultry litter composting using biochar. Journal of Environmental Quality. 2010;**39**(4):1236-1242

[15] Bord na Móna. Analysis of greenwaste compost from Dublin city council and interpretation of results. Service–interpretation of results. Dublin; 2003. Available from: https://www.goctech.com/wpcontent/uploads/2016/11/Bord-na-Mona-Paper.pdf

[16] Adebayo OS, Kabbashi NA, Alam Z, Mirghani MS. Recycling of organic wastes using locally isolated lignocellulolytic strains and sustainable technology. Journal of Material Cycles and Waste Management. 2015;**17**:769-780

[17] Barrington S, Choinière D, Trigui M, Knight W. Effect of carbon source on compost nitrogen and carbon losses. Bioresource Technology. 2002;**83**(3):189-194

[18] Chefetz B, Chen Y, Hadar Y. Water-extractable components released during composting of municipal solid waste. In International Symposium on Composting & Use of Composted Material in Horticulture. 1997;**469**:111-118

[19] Riffaldi R, Levi-Minzi R, Pera A, De Bertoldi M. Evaluation of compost maturity by means of chemical and microbial analyses. Waste Management & Research. 1986;**4**(4):387-396

[20] Hua L, Wu W, Liu Y, McBride MB, Chen Y. Reduction of nitrogen loss and Cu and Zn mobility during sludge composting with bamboo charcoal amendment. Environmental Science and Pollution Research. 2009;**16**(1):1-9

[21] Theeba M, Husni MA, Samsuri AW, Robert TB, Illani ZH. Nutrient retention capacity of rice husk biocharcoal in co-composted poultry manure. Journal of Tropical Agriculture and Field Science. 2016;**44**(2):197-209

[22] Iannottil DA, Pang T, Tothl BL, Elwell DL, Keener HM, Hoitinkl HAJ. A quantitative respirometric method for monitoring compost stability. Compost Science and Utilization. 1993;**1**:52-65

[23] Tamás A, Vrânceanu N, Duşa M, Stan V. Steps in organic fraction of municipal solid waste composting and compost quality evaluation. Scientific Papers Series A Agronomy. 2019;**2**:144-153

[24] Luo Y, Liang J, Zeng G, Chen M, Mo D, Li G, et al. Seed germination test for toxicity evaluation of compost: Its roles, problems, and prospects. Waste Management. 2018;**71**:109-114

[25] Zucconi F. Evaluating toxicity of immature compost. Biocycle. 1981;**22**(2):54-57

[26] Coleman D, Crossley D, Hendrix PF. Fundamentals of Soil Ecology. Second Edition. 2004:1-386

[27] Lindow SE, Brandl MT. Microbiology of the phyllosphere. Applied and Environmental Microbiology. 2003;**69**(4):1875-1883

[28] Partanen P, Hultman J, Paulin L, Auvinen P, Romantschuk M. Bacterial diversity at different stages of the composting process. BMC Microbiology. 2010;**10**:94

[29] Pane C, Celano G, Villecco D, Zaccardelli M. Control of Botrytis cinerea, Alternaria alternata and Pyrenochaeta lycopersici on tomato with whey compost-tea applications. Crop Protection. 2012;**38**:80-86

[30] Sánchez A. A kinetic analysis of solid waste composting at optimal conditions. Waste Management. 2007;**27**(6):854-855

[31] Insam H, De Bertoldi M. Microbiology of the composting process. In: Waste management series. Vol. 8. Elsevier; 2007. pp. 25-48

[32] Bustamante MA, Suárez-Estrella F, Torrecillas C, Paredes C, Moral R, Moreno J. Use of chemometrics in the chemical and microbiological characterization of composts from agroindustrial wastes. Bioresource Technology. 2010;**101**(11):4068-4074

[33] Luskar L, Čeh B. Hop biomass composting approach impact on compost microbiological properties. Agricultural Research & Technology. 2021;**26**(2):1-6. ISSN 2471-6774

[34] Milinković M, Lalević B, Jovičić-Petrović J, Golubović-Ćurguz V, Kljujev I, Raičević V. Biopotential of compost and compost products derived from horticultural waste—Effect on plant growth and plant pathogens' suppression. Process Safety and Environmental Protection. 2019;**121**:299-306

[35] Cole L, Bradford MA, Shaw PJ, Bardgett RD. The abundance, richness and functional role of soil meso-and macrofauna in temperate grassland—A case study. Applied Soil Ecology. 2006;**33**(2):186-198

[36] Warman PR. Evaluation of seed germination and growth tests for assessing compost maturity. Compost Science and Utilization. 1999;**7**:33-37

[37] Verbič J, Babnik D, Sušin J. Koliko rastlinskih hranil vsebujejo živinska gnojila? 2017. Available from: https://www.govedo.si/files/jozev/Koliko%20rastlinskih%20hranil%20vsebujejo%20%C5%BEivinska%20gnojila.pdf. [Accessed on 17. February 2023]

[38] Uredba o varstvu voda pred onesnaževanjem z nitrati iz kmetijskih virov (Uradni list Republike Slovenije, št. 113/09, 5/13, 22/15 in 12/17). Available at: http://www.pisrs.si/Pis.web/pregledPredpisa?id=URED5124. [Accessed: February 17, 2023]

[39] LIFE BioTHOP. 2022. Available online: https://www.life-biothop.eu/

[40] Čeh B, Luskar L, Hladnik A, Trošt Ž, Polanšek J, Naglič B. The quantity and composition of leachate from hop plant biomass during composting process. Applied Sciences. 2022b;**12**(5):2375

Chapter 16

Potential of Anaerobic Digestates in Suppressing Soil-Borne Plant Disease

Mami Irie and Tomomi Sugiyama

Abstract

The use of anaerobically digested slurries (ADSs) is promising strategies in resource management and agricultural production. ADSs has a potential as alternatives to chemical fertilizer. ADSs from various source materials suppressed the growth of some plant pathogenic fungi *in vitro*. ADS filtrates did not suppress them, indicating the effect was caused not by water-soluble substances in ADSs. In pot experiment, the efficacy of ADS from dairy cow manure (AD) for Fusarium wilt of spinach was assessed. Applying 8% (w/w) of AD significantly reduced pathogen density in soil and promoted the growth of spinach. Inoculation of a bacterial isolate AD-3 from AD, which showed high suppressiveness against Fusarium spp. *in vitro*, effectively controlled the disease. Based on the results, AD-3 strain is related to the disease control ability of AD that belonged to *Bacillus velezensis*. ADSs can supply not only plant nutrients but also antagonistic microbes. For crop production, ADSs application would be effective to infected soil. It was effective for improving ADS handling that ADS was absorbed to foamed glass and dried at 60°C. To apply ADS to farmland for crop production, these findings are promising for sustainable agriculture.

Keywords: biological control, anaerobically digested dairy slurry, *Bacillus velezensis*, Fusarium, anaerobic digestion

1. Introduction

Fusarium wilt disease, caused by *Fusarium oxysporum* f. sp. *spinaciae* (Fos), is a serious soil-borne disease. *F. oxysporum* has high host specificity and is responsible for severe damage to economically important plants [1]. Chemical fungicides or soil fumigation is commonly used to control the disease. However, it is needed to develop alternatives to these conventional controls for sustainable agriculture. Organic amendments can be used to improve soil quality and manage soil-borne diseases [2]. Among organic amendments, compost has been studied the most commonly. Composting is a well-known method for recycling organic waste, and the final by-product can be used in agriculture [3]. There have been reported that compost has suppressive effects against soil-borne fungal diseases [4, 5]. Anaerobic digestion is another countermeasure for the treatment of agro-industrial waste and the organic parts of source-separated household waste. Anaerobic digestion produces biogas as a source of renewable energy and anaerobically digested slurry (ADS) that can be used as liquid fertilizer [6, 7]. Organic wastes from agriculture and industry, and the organic parts of source-separated household wastes are considered as biomass and the most dominant future renewable energy sources. Organic waste materials could be specific important resources because these sources do not compete with food crops in agricultural land usage. Anaerobic treatment is suitable for their utilization. The treatment is an attractive solution since producing biogas can be used as renewable energy and anaerobically digested slurry (ADS) can be applied to farmland as organic amendment. Scientific reports of anaerobic treatment, especially biogas, have increased rapidly in terms of renewable energy technology [8]. In general, ADS has a high concentration of plant nutrients such as nitrogen (N), phosphorus (P), and potassium (K), which are available in a suitable form for plants to absorb, meaning that ADS can be used as liquid fertilizer. Normally, it needs appropriate handling before discharging to river water because of high content of nutrients such as N, P, and K. This means, on the other hand, it has a potential as alternatives to chemical fertilizer. When considering land application of ADSs, risks and impacts for agricultural production should be concerned as the chemical composition including phytotoxic compounds and heavy metals is different among ADSs, depending on their primarily source materials [9, 10]. Therefore, the chemical composition of ADSs should be accurately assessed before being used as fertilizer. In Europe, there is the application rate of ADS recommending its use based on specific regulations and guidelines [11, 12]. There are some studies reported that ADS is a valuable alternative to fertilizer in agriculture [13, 14]. The self-sufficiency of fertilizer must be increased for sustainable agriculture and food security. The use of ADS has mainly focused on a function as nutrient availability, crop productivity, and reusing organic waste [15–17]. The role of ADS in plant disease control has been studied. For example, Tao reported that ADS sourced from pig manure suppressed the growth of plant pathogenic fungi *in vitro* [18], and Amari reported that the application of ADS to soil has been shown to suppress several plant diseases including *Ralstonia* spp. [19] and *Fusarium* spp. [19], and Cao founded the suppressive effect on *Phytophthora* spp. [20]. These reports focused on physicochemical properties as factors for plant disease control. Microorganisms in ADS may also have a suppressive effect. However, the disease control ability of organic amendments and ADS nutrient contents might be varied depending on the source materials. There are few findings indicating disease suppression of ADSs. Furthermore, it still has two difficulties for its handling and storage. For handling, ADS is a liquid, but it contains plant residues or sediments, and it is difficult to spray to farmland like liquid fertilizer; some attachment or special application agricultural machinery is needed. Hence, there are microbes in ADSs, which continuously digest remaining organic materials in ADSs, and methane gas emission has been continuing. Therefore, it is difficult to keep ADS in plastic bottles or plastic bags like chemical fertilizer. This made farmers difficult to use ADSs. In Japan, ADS is generally treated at attached wastewater treatment facility and released into river water. The first objective was to assess how the ADS generated from different source materials is suppressing *F. oxysporum* f. sp. *spinaciae* (Fos). We isolated bacteria from ADS and tested the effects of these against Fos *in vitro*. The second objective was to verify the suppressive effect of a selected bacterial isolate against Fusarium wilt of spinach *in vivo*. The third objective was to demonstrate the applicability of ADS not only antagonism against Fos but also against another plant pathogens. Furthermore, relatively new and old ADSs were used for the applicability experiment. The fourth objective was to improve the handling of ADS for agricultural use, ADS was absorbed to porous materials, and the antagonism of its material against seven kinds of plant pathogen was investigated. Furthermore, heat dry temperature effect of ADS absorbed material was investigated whether drying process reduces the suppression.

2. Materials and methods

2.1 Sampling and characteristics of ADSs

In this study, five ADSs were generated from different source materials: dairy cow manure (AD), sewage sludge (AS), food garbage (AF), *pigmanure* + *foodgarbage* (APF), and *sewagesludge* + *nightsoilsludge* + *foodgarbage* (ASNF). Mesophilic fermentation of anaerobic biological treatment in facilities occurred at 35°C. All digestates were taken from a methane fermentation tank running in continuous mode without dehydrating, and the slurry was used for experiments. After sampling, the slurry was stored in 20-L plastic tanks at 4°C refrigerator. The details of each ADS source materials rates are presented in **Table 1.**

2.2 Analysis of chemical properties of ADSs

The following parameters were determined in the fresh digestate samples: pH, analyzed by multi-function water quality meter (MM-60R TOADKK Corp.) followed the methods of analysis of feeds and feed additives [21]; dry matter content (DM) after drying the digestate sample at 105°C for 24 h followed the method of Japanese Industrial standards method (JIS) JIS K 0102 14.2 [22]; the volatile solids, which reflect the OM content, by loss on ignition at 500°C for 24 h followed the method of JIS K 0102 14.4 [22]. The total organic carbon (TOC) was analyzed as liquid samples using an automatic analyzer TOC-LCSH system (Shimadzu Corp.) followed the method of JIS K 0102 22 [22], total nitrogen (TN) followed the testing method for fertilizer 4.1.1.b [23], and total carbon (TC) followed the testing method for fertilizer 4.11.1.b [23] were measured in dried samples by NCH-22 (Sumika Chemical Analysis Service, Ltd.). C/N was calculated by the division of carbon by nitrogen. Ammonia and chlorine were analyzed by ion chromatograph LC-20 AD sp., detector CDD-10A, columns; IC-C4 for cation and IC-SA2 for anion (Shimadzu Corp.). After HNO3 digestion, the following elements were determined by inductively coupled plasma-atomic emission spectrometry (ICP-AES, iCAP 6200, Thermo Fisher Scientific) followed the testing method for fertilizer [23]: P, K, S, Na, Ca, Mg, Fe, Cu, Mn, Al, Zn, Pb, Cd, Cr, and Ni. Hg was analyzed by cold-vapor atomic absorption spectrometry followed the testing method for fertilizer 5.12.1 [23]. Lignin and cellulose were analyzed according to acid detergent method [21], and CEL/LIGN was calculated by the division of cellulose by lignin.

2.3 Antifungal activity of raw and filtrate ADSs

The plant pathogen *F. oxysporum* f. sp. spinaciae (Fos, MAFF: 103059) was used. Fos was cultured on potato-dextrose agar (PDA) (Nihon pharmaceutical Co., Ltd., Tokyo, Japan) in petri dishes (90×15 mm) at 25°C for 14 d and used as inocula. Raw and filtrate ADSs were used for the assay. Filtrate ADSs were prepared as follows: 10 mL of each raw ADS was centrifuged at 3000 rpm for 20 min, and the supernatant was filtered through a 0.2 μm disposable membrane filter (Toyo Roshi Kaisha, Ltd., Tokyo, Japan). The fungal colony was taken as a small colony disk using a 5-mm cork

Sample[1]	Rate of source materials (%)									
Name	Dairy cow manure	Sewage sludge	Food garbage	Pig manure	Night soil sludge	Treatment temperature	HRT (days)	Microbial immobilization method	Pre-Treatment	Facility location
AD	100					35–36	45	Anaerobic fluidized bed process	Solid liquid separation	Tochigi Pref.
AS		100				35–36	30	Anaerobic contact process	Gravitation enrichment	Yamagata Pref.
AF			100			35–36	20	Anaerobic fluidized bed process	Acid fermentation	Hokkaido
APF			10	90		35–36	30	Anaerobic fluidized bed process	Crush	Okayama Pref.
ASNF		74	9		17	35–36	30	Anaerobic fluidized bed process	Acid fermentation	Fukuoka Pref.

[1]*ADSs generated from different source materials. AD, dairy cow manure; AS, sewage sludge; AF, food garbage; APF, pig manure + food garbage; ASNF, sewage sludge + night soil sludge + food garbage*

Table 1.
Source material composition, processing conditions, and facility location for five anaerobically digested slurries (ADSs).

borer, and one side was placed on a PDA plate. Sterilized filter paper disks (5 mm) were placed in 10 mL of each raw and filtered ADS for 10 min to absorb it, and then placed on the other side of the PDA plates. A pathogen-only plate was used as a control. All treatments had five replicates. All plates were placed in incubator for 18 d at 25°C in the dark. After incubation, two plates were randomly selected from each treatment. Photos were taken of the plates, and the digital images were used to measure the diameter of any colonies developed from the mycelial disk, using ImageJ software (version 1.52, NIH, Bethesda, USA). The experiment was repeated twice.

2.4 Isolation of bacteria from five ADSs

The five ADSs were serially diluted with sterilized water, and the dilutions were spread onto nutrient agar (NA, Nissui Pharmaceutical Co., Ltd., Tokyo, Japan). After incubation for 2 to 4 d at 25°C in the dark, bacterial colonies that appeared on the plates were transferred on NA plates, and single colony isolates were obtained. These isolates were preserved in solution (10 g of skim milk and 1.5 g sodium L-glutamate monohydrate^{-1} 100 mL distilled water) at -20°C until use.

2.5 Antifungal activity of bacterial isolates from ADSs

Bacterial isolates from the five ADSs described in section above were used for confrontation assays. Small bacterial colony disks were removed from cultures on NA plates after 24–48 h by 5-mm cork borer and were placed on one side on PDA plates. Fos fungal colonies were also taken as a small colony disk in the same manner. A pathogen-only plate was used as a control. All treatments had five replicates. All plates were placed in incubator for 18 d at 25°C in the dark. After incubation, two plates were randomly selected from each treatment. Photographs were taken, and the digital images were used to measure the diameter of mycelial colonies developed from a mycelial disk, using ImageJ software (version 1.52, NIH, Bethesda, MD, USA).

2.6 Pot experiment

2.6.1 Preparation of Fos inoculum

Fos inoculum was prepared on potato-sucrose broth (200 g potato and 20 g sucrose L^{-1} distilled water) at 25°C by shaking the culture at 110 rpm for 5 d. The resulting spore suspension was filtered through double gauze to remove the hyphae and was then transferred into sterile 50-mL plastic tubes, centrifuged at 2000 rpm for 3 min, and then, the supernatants were discarded. The Fos inoculum was prepared with distilled water and adjusted to 1.0×10^6 to 1.0×10^7 bud-cells mL^{-1} using a hemocytometer (AS ONE Corp., Osaka, Japan).

2.6.2 Preparation of AD-3

As mentioned above, ADS had several effective bacteria against Fos among ADSs. Of the several bacteria in AD, AD-3, which suppressed Fos growth the most, was used for pot experiments. AD-3 cells were grown on NA plates for 24 h, and the culture was grown on nutrient broth media (5 g meat extract, 5 g NaCl, and 10 g peptone L^{-1} distilled water) at 25°C on a constant rotary shaker at 100 rpm for 48 h.

The suspension was transferred in sterile 50-mL plastic tubes, centrifuged at 3000 rpm for 10 min, and the supernatants were discarded (repeated twice). The AD-3 cell pellet was dissolved in sterilized distilled water. The AD-3 suspension was adjusted to an optical density of 1.0 at 600 nm using a spectrophotometer (Thermo Electron Corp., Waltham, MA, USA).

2.6.3 Pot assay of AD-3

The experiment was performed in pots (10.5×9 cm) containing the equivalent of 200 g of dried black loam soil. The soil was sieved through 2 mm mesh, and soil pH was adjusted to within the range 6.9 to 7.1 using hydrate lime. The amount of hydrate lime was calculated according to the Arrhenius equation. The final rates of N, P_2O_5, and K_2O were adjusted to 80 $kg\,ha^{-1}$ using ammonium sulfate, fused phosphate, and potassium chloride. Treatments of the pot experiment were as follows: F, only Fos-inoculated soil; $F + AD - 3$, soil amended with Fos and AD-3; and unamended, non-pathogen control. To prepare Fos-infected soil, the bud-cell suspension of Fos was inoculated into the soil to give a final concentration of 1.0×10^5 bud-cells g^{-1} dry soil, and pots were incubated for 5 d at 25°C in the dark. AD-3 suspension was added to the infected soil to give a final concentration of $1.0 \times 10^6\, CFUg^{-1}$ dry soil. All pots were incubated for 10 d at 25°C in the dark. During the incubation, soil moisture was maintained at 60% water-holding capacity (WHC) by spraying with distilled water. As for spinach cultivar, "Okame" (*Spinacia oleracea* L.; Takii Seed, Kyoto, Japan) with high susceptibility to Fos [24] was used. After incubation, spinach seeds were sown in each pot and the plants were grown in an incubator (day/night: 25/22°C, 12/12 h). All pots were irrigated daily to keep the soil moisture at 60% WHC. A total of 60 pots were prepared; 30 pots (15 pots of F and 15 pots of $F + AD - 3$) were used to estimate the density of Fos in the soil. Soil was sampled from three pots that were randomly selected per treatment at 0, 7, 14, 21, and 28 d after sowing, and the density of Fos was measured using Komada selective medium [25]. The other 30 pots (10 pots per treatment) were used to evaluate disease severity at 28 d after sowing. The disease symptoms were categorized using a disease index of 0, no infection; 1, 1–30% of leaves wilted; 2, 31–60% of leaves wilted; 3, 61–90% of leaves wilted; 4, more than 90% of leaves wilted or dead. Disease severity (DS) was calculated according to the following formula:

$$DS = \sum nd100/4T \tag{1}$$

where n = number of plants in each disease index, d = disease index, and T = total number of plants by treatments. The inhibitory effect (IE) of AD-3, as a percentage reduction in disease severity, was calculated according to the following formula:

$$IE(\%) = (100 - DSF + AD - 3)/DSFx100 \tag{2}$$

where $DSF + AD - 3$ = *disease severity of* $F + AD - 3$, DSF = *disease severity of* F. The pot experiments were repeated twice (Experiment 1, Exp 1; Experiment 2, Exp 2). In total, disease severity for 27 plants (T = 9) in Exp 1 and 21 plants (T = 7) in Exp 2 was investigated. R v. 3.5.3 software was used for statistical analyses. The antagonistic activity of ADSs and bacterial isolates against Fos *in vitro* were analyzed with Tukey's HSD test following one-way analysis of variance (ANOVA). The effect of inoculation of AD-3 on Fos density was analyzed using a t-test. The data for disease severity were arcsine transformed in advance, and Tukey's HSD test following one-way analysis of variance (ANOVA) was performed to reveal the efficacy of AD-3 against Fusarium wilt of spinach ($p<0.05$).

2.7 Genome sequencing

DNA extraction from cells was performed using the Promega Maxwell RSC PureFood GMO Kit (Promega Corporation, Madison, WI, USA). Genome sequencing was performed with the GridION X5 (Oxford Nanopore Technologies, Oxford, UK) followed by preparation of the genome library using a Rapid Barcoding Sequencing Kit (SQK-RBK004) (Oxford Nanopore Technologies, Oxford, UK). Read sequences were assembled using Unicycler (version0.4.8) [26].

2.8 Genome analysis

A genome sequence was annotated using DFAST (version 1.1.6, https://dfast.ddbj.nig.ac.jp/) [27]. AntiSMASH (version 5.0, https://antismash.secondarymetabolites.org/) [28] was used for the prediction of secondary metabolite gene clusters. A 16S rDNA sequence on the genome was analyzed to identify the strain using EzBioCloud (https://www.ezbiocloud.net/) [29]. Phylogenetic relationships were analyzed on the basis of 16S rDNA sequences using MEGA x (version 10.2.6) [30].

2.9 Pot experiment of AD

The pot experiment was conducted in pots (10.5 × 9 cm), each containing 200 g of dried black loam soil under incubator conditions at 25°C. The bud-cell suspension of Fos (1.0×10^5 bud-cells/g dry soil) was inoculated into black loam soil, and 8% (w/w) of AD was applied to non-infested and Fos-infested soil 10 days before sowing the spinach seeds. The pot experiment had four treatments, including the control: (1) Fos, pathogen-only control; (2) Fos + AD, soil amended with Fos and 8% (w/w) of AD; (3) AD, soil amended with 8% (w/w) of AD; (4) Ctrl, control with no amendments. Each treatment was replicated for 10 plants. All treatments of N, P_2O_5, and K_2O were adjusted as the equivalent of 80 kg ha^{-1}, and recommended fertilizer application rate. For AD-amended treatment, 8% (w/w) of AD was applied not to exceed the rate.

2.10 ADS applicable experiment against seven plant pathogens

The plant pathogens as shown in **Table 2** were used for testing the ADS applicability. Plant pathogen was cultured on potato dextrose agar (PDA) (Nihon pharmaceutical co., Ltd., Tokyo, Japan) in petri dishes (90×15 mm) at suitable temperature for each plant pathogen as shown in **Table 2** and used as inocula. Sterilized filter paper disks (7 mm) were placed in 10 mL of ADS for 10 min to absorb it and then placed on the other side of the PDA plates. A pathogen-only plate was used as a control. All treatments had five replicates. All plates were incubated for suitable days as shown in **Table 2** in the dark. After incubation, three plates were randomly selected from each plant pathogen. Photos were taken of the plates, and clear inhibition zone were judged by eyes. Two FW ADSs where its facility is in Kanagawa Prefecture were used to examine the applicability of fresh and old ADS, one is 3 month past after the 30 days HRT and sampling, and the other is 5 month past after the 30 days HRT and sampling. Mesophilic fermentation of anaerobic biological treatment in facility occurred at

	Plant pathogen name	MAFF No.	Incubation temperature (°C)	Incubation period (days)
1	*Pyrenochaeta lycopersici*	238,906	25	75
2	*Setophoma terrestris*	243,290	25	90
3	*Phialophora gregata* f. sp. Adzukicola	241,056	20	150
4	*Helicobasidium mompa* Nobuj. Tanaka	305,915	20	100
5	*Pyricularia oryzae*	101,418	25	120
6	*Fusarium graminearum*	240,353	20	30
7	*F. oxysporum Schlechtendahl* f. sp. *fragariae* Winks et Y.N. Williams	744,009	25	14

Table 2.
Tested plant pathogen list for applicability experiment.

35°C, and pre-treatment was crush method. After sampling, the slurry was stored in 20-L plastic tanks at 4°C.

2.11 Supporting material effect on the antifungal activity of ADS absorbed material

To improve ADS handling, ADS was absorbed to four industrial waste materials each; wood ash, steel-making slag, activated carbon, foamed glass diameter 3 to 5 mm, and soaked to ADS for 1 hour and the mixed ratio was as shown in **Table 3**. As a result of 4.1. newer ADS was used for the following experiments. The absorbed material was weighed as dried ADS equivalent to 23 mg each and put on one side of PDA media, and then, the fungal colony was removed as a small colony disk using a 7-mm cork borer and was placed on the other side of the PDA plate. A pathogen-only plate was used as a control. Tested plant pathogens was *F. oxysporum* Schlechtendahl f. sp. fragariae Winks et Y.N. Williams. All treatments had five replicates. Plant pathogen plates were incubated for 14 days 25°C in the dark. After the incubation period, three plates were randomly selected from each. Photos were taken of the plates and the inhibitory effect was judged by eyes.

2.12 Dry heat temperature effect on the antifungal activity of ADS adsorbed material

Adsorbed AF-ADS materials were weighed 2 g on aluminum foil and dried at 60, 80, 100, and 150°C in dry heat oven for 3 h each. After cooled to room temperature in a desiccator dried materials and non-heated material were weighed as dried AF-ADS equivalent to 23 mg each and put on one side of PDA media, and then, the fungal

Mixed ratio %(w/w)	Carrier material name			
	Wood ash	Steel-making slag	Activated carbon	Foamed glasses
Carrier material ratio	80	92.5	45	75
ADS ratio	20	7.5	55	25

Table 3.
Mixed percentage of carrier material and ADS.

colony was removed as a small colony disk using a 7-mm cork borer and was placed on the other side of the PDA plate. *F. oxysporum* Schlechtendahl f. sp. *fragariae* Winks et Y.N. Williams was selected as inocula since its mycelia growth rate is high. As a control a pathogen-only plate was used. All treatments had five replicates. All plates were incubated for 14 d at 25°C in the dark. After incubation, three plates were randomly selected from each treatment. Photos were taken of the plates, and the inhibitory effect was judged by eyes.

2.13 Bacteria and archaebacteria flora analysis for succession confirmation

In order to support the result of mentioned above, bacteria and archaebacteria flora were analyzed to determine whether ADS was adsorbed to the carrier material. The ADS absorbed foamed glass material was produced in the same manner as mixed ratio in **Table 3** and dried at 60°C for 1 h. AF-ADS and the AF-ADS absorbed foamed glass, and foamed glass were analyzed by a next-generation sequencer (Qiime) targeting 16S rRNA.

2.14 Applicability experiment of antifungal activity of AF-ADS absorbed material

AF-ADS was absorbed to two industrial waste materials: wood ash and foamed glasses diameter 3 to 5 mm for 1 h and mixed as shown in **Table 3**, and dried at 60°C for 1 h in dry heat oven. After cooled to room temperature in a desiccator, dried materials and non-heated material were weighed as dried ADS equivalent to 23 mg each and put on one side of PDA media, and then, the fungal colony was removed as a small colony disk using a 7-mm cork borer and was placed on the other side of the PDA plate. A pathogen-only plate was used as a control. Tested plant pathogens are shown in **Table 2**. All treatments had five replicates. Each plant pathogen plates were incubated for suitable days as shown in **Table 2** in the dark. After incubation, three plates were randomly selected from each plant pathogen. Photos were taken of the plates and the digital images were used to measure the diameter of any colonies developed from the mycelial disk, using Image J software.

3. Results

3.1 Chemical properties of ADSs

Application of organic amendments to the soil would improve their physical, chemical, and biological properties. The variability of these properties depends on characteristics of ADSs, reflecting the initial biomass inputs. The chemical properties of ADSs used in this study are showed in **Table 4**. Carbon and nitrogen are the important components (Michalzik et al. [51]; Jenkinson et al. [52]) and their relative ratio will affect agronomic use of amendments (Havlin et al. [53]). C/N ratio of slurries was found to be variable between 5 and 19. AD which feedstock was mainly dairy cow manure showed highest C/N ratio because dairy cows are ruminants, and its manure contains low ratio of easily degradable carbon and relatively higher ratio of persistent carbon. ASNF, APF, and AF showed greater mineral nitrogen fractions (43–52% total N to the total N fraction), suggesting that those could be used as fertilizer (Tambone et al. [12]; Paavola and Rintala [54]). In contrast, AD, AS, and AFPS had displayed lower mineral fraction (27, 35, and 36% each to the total N), and especially

Parameter	AD	AS	AF	APF	ASNF	Value range		
DM (%)	**2.6**	**2.1**	**1.8**	**1.2**	**1.8**	**1.5**	—	**45.7 [31–35]**
Organic matter (% DM)	37	62	33	7	51	36	—	75.4 [33, 34, 36, 37]
Total N (kg Mg^{-1} DM)	11	15	19	14	16	31	—	140 [36–38]
Total N (kg Mg^{-1} FM)	2.9	3.1	3.5	1.8	2.9	1.2	—	15 [39]
NH4-N (% FM)	0.08	0.11	0.15	0.09	0.13	1.5	—	6.8 [31, 32]
NH4-N (% DM)	0.03	0.05	0.08	0.07	0.07			—
NH4+ share on total N (%)	27	35	43	52	45	35	—	81 [35, 37]
TC (% DM)	21.2	21.4	8.8	17.2	15.9	36	—	45 [37]
C/N	19	15	5	12	10	2	—	24 [35, 36, 38]
TOC	1300	400	620	220	530			—
Soluble C/N	1.7	0.4	0.4	0.2	0.4	0.55	—	6 [40]
Total P (% DM)	1.5	3.7	1.4	1.2	2	0.6	—	1.7 [33, 34, 37]
Total P (kg Mg^{-1} FM)	0.4	0.8	0.2	0.2	0.4	0.4	—	2.6 [41]
Total K (% DM)	1.8	0.4	2.7	1.5	1.8	1.9	—	4.3 [39, 41]
Total K (kg Mg^{-1} FM)	0.5	0.1	0.5	0.2	0.3	0.4	—	11.5 [41]
Total Ca (kg Mg^{-1} FM)	0.8	0.4	0.7	0.1	0.4	1	—	2.3 [37, 41]
Total Mg (kg Mg^{-1} FM)	0.3	0.1	0.1	0.1	0.1	0.3	—	0.7 [37]
Total S (kg Mg^{-1} FM)	0.1	0.3	0.2	0	0.2	0.2	—	0.6 [42, 43]
Si (kg Mg^{-1} FM)	1.7	1.5	0.63	0.16	0.75			—
Mn (kg Mg^{-1} FM)	6.7	6.8	2.1	0.6	3.1	50	—	55 [44]
Fe (kg Mg^{-1} FM)	39	410	140	9	240			—
Na (kg Mg^{-1} FM)	280	85	1200	120	600			—
Cl (kg Mg^{-1} FM)	760	77	1900	220	930			—
Zn (kg Mg^{-1} DM)	54	71	11	16	56			<1800 [45, 46]
As (kg Mg^{-1} DM)	0.1	0.8	0.9	0.1	0.9			<50 [45, 46]
Cd (kg Mg^{-1} DM)	ND	ND	ND	ND	ND			<5 [45, 46]
Ni (kg Mg^{-1} DM)	0.4	1.4	0.6	ND	1.1			<300 [45, 46]
Cr (kg Mg^{-1} DM)	ND	ND	ND	ND	ND			<500 [45, 46]
Hg (kg Mg^{-1} DM)	0	0	0	ND	0			<2 [45, 46]
Pb (kg Mg^{-1} DM)	ND	1.4	ND	ND	0.6			<100 [45, 46]
Cu (kg Mg^{-1} DM)	3.8	20.5	2.2	3.2	11.1			<600 [45, 46]
Al (kg Mg^{-1} FM)	16	180	10	3	120			—
CEL/LIGN	0.44	0.51	0.36	0.82	0.38	0.22	—	1.71 [43, 44, 47–49]
pH	7.6	7.4	7.9	7.8	7.8	7.3	—	9 [36, 38, 50]

Source: *ADSs generated from different source materials. AD, dairy cow manure; AS, sewage sludge; AF, food garbage; APF, pig manure + food garbage; ASNF, sewage sludge + night soil sludge + food garbage.*

Table 4.
Chemical properties of ADSs.

AD showed the lowest caused by lower degradability of dairy manure digestate. This indicated these ADSs have a high potential of stability as organic amendment (Nkoa [55]). The P and K are also essential nutrients as fertilizer. AS contained a higher amount of total-P that could be used as P fertilizer. Total-K was highly contained in AD, while lower in AS, ASNF, and AFPS. For biochemical fractionation, cellulose/lignin ratio indicates the degree of humification of the organic materials (Nkoa [55]), and a value of 0.5 has been suggested as boundary value between fresh and mature wastes (Komlis and Ham [56]). According to the value, APF and AFPS were comparably fresh digestates. Na that is contained in food garbage compost may affect plant growth (Hargreaves et al., [57]). The value of salt contained in food waste could be up to 8% (DM) for composting and the compost can be applied to soil 10 t $ha^{-1}y^{-1}$ without phytotoxic effects (Jin et al. [49]). Applying this value to digestates, upper limit for soil application can be calculated 80 g m^{-2}, which means digestate can contain 6240 mg L^{-1} Na as upper limit concentration for 50 t ha^{-1} digestate application. Digestates ASNF, AF, APF, and AFPS containing food garbage as its substrates had sufficiently lower than the calculated value. Considering the land application of amendments derived from organic wastes, the risk of soil contamination by phytotoxic compounds (Boydston et al. [50]; Gough and Carlstorm [58]) and heavy metals (Alburquerque et al. [9]; Wong et al. [59]) should be concerned. A value of pH is an indicator to determine harmless of ADSs, since it controls the behavior of heavy metals (Kapanen and Itavaara [60]). All the ADSs showed a safety value from 7.4 to 7.9. Composition of heavy metals in ADSs showed quite low heavy metal contents comparing with upper limit based on Japanese Fertilizers Regulation Act and recommended limit of Japanese Central Union of Agricultural Co-operatives. Although Mn is an essential component for plant growth, excess Mn in the soil would interfere with activities of the other mineral components such as Ca, Mg, Fe, and P (Clark [61]). The concentration of Mn in digestates can as high as 50–55 ppm (Sahm [44]), and all the ADSs contained sufficiently lower value of Mn. Besides heavy metals, micronutrients such as Cu and Zn also have risks for agricultural soil once in excess in the soil. Both nutrients are generally used as additives in livestock feeds to promote livestock growth and prevent the livestock disease (Nicholson et al. [62]; Alburquerque et al. [9]). The concentration of Cu and Zn in AD, APF, and AFPS was lower, and thereby, these ADSs could be used for land application.

3.2 Antagonistic activities of raw and filtrate ADSs *in vitro*

The five raw ADSs (AD, AF, AS, APF, and ASNF) significantly suppressed mycelial growth when compared with the control (**Table 5**). Of these, AD, AS, AF, and ASNF produced a clear inhibition zone. In contrast to the raw ADSs, the filtrate ADSs did not suppress mycelial growth (**Table 5**).

3.3 Antagonistic activities of raw and filtrate ADSs *in vitro*

Overall, 32 strains were isolated from the five ADSs. In descending order, nine isolates were obtained from AD, and named AD-3, AD-6, and AD-8 significantly suppressed the growth of Fos (**Figure 1**). Seven isolates were obtained from AF, and named AF-1, AF-3, and AF-5 significantly suppressed the growth of Fos (**Figure 1**). Six isolates each were obtained from APF and ASNF, and named APF-1 and ASNF-4 significantly suppressed the growth of Fos (**Figure 1**). Although four isolates were obtained from AS, no isolates suppressed the growth of Fos (**Figure 1**).

Sample*	Mycelial growth (mm)**			
	Raw		Filtrate	
AD	56.1	b	85.6	NS
AS	55.9	b	85.2	
AF	60.8	b	85.5	
APF	57.6	b	86.3	
ASNF	53.5	b	86.4	
Control	82.9	a	83.3	

**ADSs generated from different source materials. AD, dairy cow manure; AS, sewage sludge; AF, food garbage; APF, pig manure + food garbage; ASNF, sewage sludge + night soil sludge + food garbage.*
***The same letter indicates no significant difference based on Tukey's HSD ($p < 0.05$; $n = 4$). NS indicates not significant.*

Table 5.
Antifungal activity of raw and filtrate anaerobically digested slurries (ADSs) against Fos, indicated as mycelial growth by co-culture test.

3.4 Effects of AD-3 on Fusarium wilt of spinach

The pot experiments were repeated twice (Exp 1 and Exp 2) and the results were similar. Disease severity with F + AD-3 was significantly *($P < 0.05$)* lower than that with F (**Table 6**). The inhibitory effect of F + AD-3, as a relative reduction percentage in disease severity compared with that of F, was 64.3% (Exp 1) and 44.1% (Exp 2). As for Fos density in soil, there were no significant differences between F and F + AD-3 in either experiment (**Figure 2**). Fos density during cultivation was in the range of 4.3–4.8 $CFUg^{-1}$ dry soil (Exp 1) and 3.9–4.4 $\text{CFU}g^{-1}$ dry soil (Exp 2).

3.5 Identification of AD-3

The morphological analysis showed that the strain AD-3 was a Gram-positive, rod-shaped bacterium. Based on the 16S rRNA gene sequence analysis, the closest species to strain AD-3 was *B. velezensis,* showing 99.8% similarity. Additionally, phylogenetic analysis of the 16S rDNA indicated that strain AD-3 was positioned in the same group as other *B. velezensis* strains (**Figure 3**).

3.6 Efficacy of AD on Fusarium wilt of spinach

Application of organic amendments contributes to reduce the addition of chemical fertilizer and improve soil productivity and crop performance. Additionally, that has been found to cause controlling soil-borne diseases [63]. In this study, the effects of AD on Fusarium wilt disease incidence and plant growth were assessed. At the end of the bioassay, the plant dry weight of Fos + AD was 1.3 times higher than that of Fos treatment. Furthermore, AD-treated non-infested soil also showed 1.2 times better plant growth compared to the control (**Table 6**). Thus, application of AD showed the effectiveness of plant growth in both the Fos-infested soil and the non-infested soil. AD application showed significantly positive effect on spinach growth (**Table 7**). The factors of suppressive effect of AD and the existence of the strain AD-3 in AD might be related. Even N, P_2O_5, K_2O levels were same to the other treatment. However, the

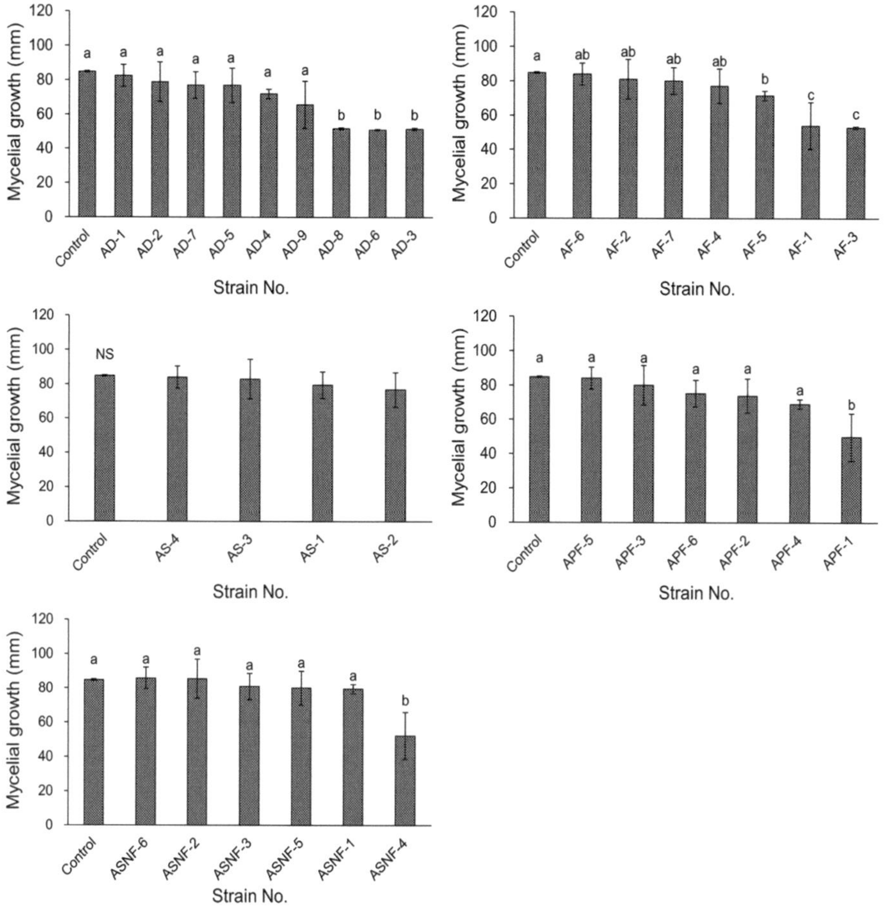

Figure 1.
Antifung*al activity of bacterial isolates from five anaerobically digested slurries (ADSs) against* F. oxysporum *f. sp. spinaciae (Fos) indicated as mycelial growth (bars ± SD) by co-culture test. AD, dairy cow manure; AS, sewage sludge; AF, food garbage; APF, pig manure+food garbage; ASNF, sewage sludge + night soil sludge + food garbage. SD, standard deviation of the mean. Bars with the same letter are not significantly different based on Tukey's HSD test (P<0.05; n=3).*

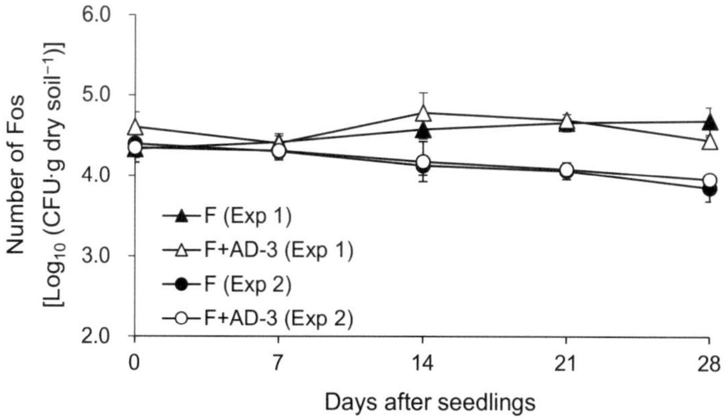

Figure 2.
Density of F. oxysporum *f. sp. spinaciae (Fos) (plots ± SD) in soil after seedling emergence. SD, standard deviation of the means. Each plot represents the average of three replicates (pots). The experiment was repeated twice (Exp 1 and Exp 2).*

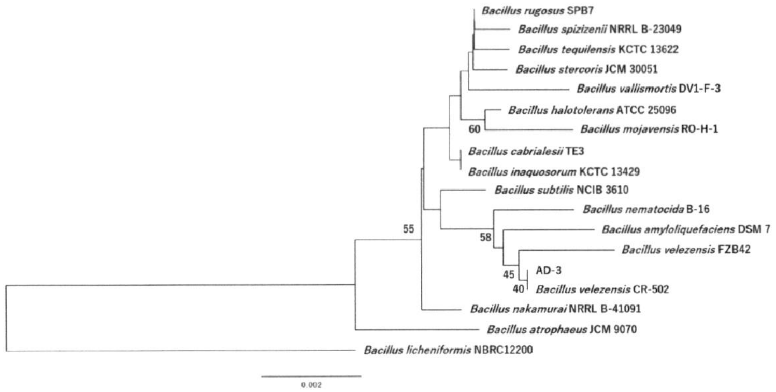

Figure 3.
Phylogenetic tree based on 16S rDNA gene sequences of AD-3 strain and related species using neighbor-joining analysis. Bacillus licheniformis *NBRC12200 served as an outgroup. Scale bar refers to a phylogenetic distance of 0.002 nucleotide substitutions per site. Bootstrap values were obtained based on 1000 replications.*

Treatment[a]	Relative plant growth[b]
Fos	0.76 ± 0.24
Fos + AD	0.99 ± 0.40
AD	1.23 ± 0.36
Ctrl	1.00 ± 0.25

[a]*Fos, inoculation of Fos alone; Fos + AD, soil amended with Fos and 8% (w/w) of AD; AD, soil amended with 8% (w/w) of AD; Ctrl: control.*[b]*Relative plant growth was calculated based on plant dry weight of control treated as 1.*

Table 6.
Effects of application of AD on spinach growth in plant dry weight. Values are expressed as means ± SD (n = 10).

number of AD-3 derived from 8% (w/w) of AD was estimated to be less than 1.0×10^4 CFU/g dry soil. In addition, AD contained some bacterial candidates suppressing Fos, as shown in **Figure 1**. The application of antagonistic microbes with organic amendments enhanced the disease control ability more efficiently than did the use of a single strain alone [64, 65]. Combining the results, we concluded the efficacy of application of AD for Fusarium wilt disease control was not only by a single microbe but also by the microbial community in AD.

3.7 ADS applicable experiment against seven plant pathogens

The relatively new and old AF ADSs significantly suppressed mycelial growth when compared with the control (**Figure 4**). Both AF-ADSs produced a clear inhibition zone against all tested plant pathogens. ADS suppressive effect had kept for at least 7 months.

3.8 Effect of carrier material to antifungal activity of ADS adsorbed materials

Figure 5 shows that carrier material wood ash and foamed glass produced clear zone showing antifungal activity. Therefore, these two materials were used as carrier material.

Factor	Df	Sum Sq	Mean Sq	F value	Pr(>F)	
Fos	1	0.559	0.5593	4.914	0.033	*
AD	1	0.522	0.5221	4.587	0.039	*
Fos x AD	1	0	0	0	0.996	
Residuals	36	4.097	0.1138			

**Signif. code: 0.05.*

Table 7.
*Application effects of AD and Fos on spinach growth in plant dry weight * significantly different two-way analysis of variance ($P<0.05$).*

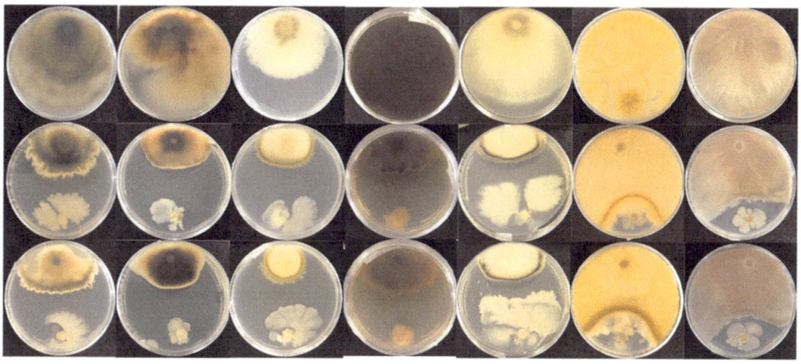

Figure 4.
Antifungal activity of relatively new and old ADSs against seven plant pathogens as mycelial growth by co-culture test.

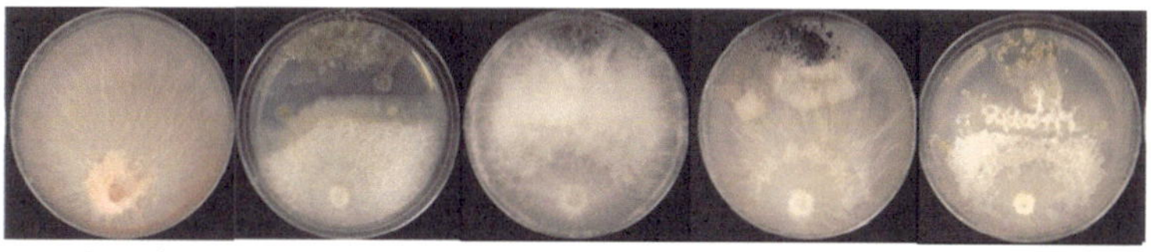

Figure 5.
Effect of carrier material to antifungal activity of ADS adsorbed materials against F.oxysporum *Schlechtendahl f. sp. fragariae winks et Y.N. Williams.*

3.9 Dry heat temperature effect on the plant pathogen suppression of ADS adsorbed material

Comparing to control non-heated, 60, 80, and 100°C produced a clear inhibition zone. In contrast, dry heat temperature 150°C did not suppress mycelial growth (**Figure 6**). Dry heat temperature until 100°C did not reduce absorbed ADS material

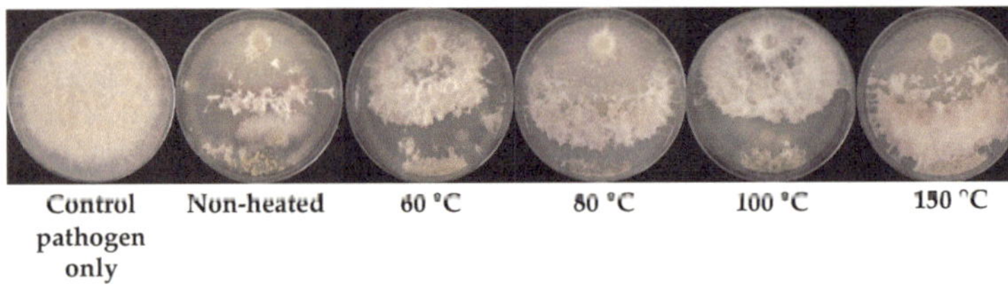

Figure 6.
Dry heat temperature effect on the plant pathogen suppression of ADS adsorbed material.

suppressive effect. Effective microbes could survive until 100°C. From the view of energy cost and plant pathogen suppression, 60°C was reasonable and effective.

3.10 Bacteria and archaebacteria flora analysis for succession confirmation

Figure 7 shows the results of bacterial flora analysis using a next-generation sequencer (Qiime) targeting 16S rRNA. As shown in **Figure 7**, when the bacterial flora of AF-ADS and the ADS absorbed foamed glass were compared, the AF-ADS absorbed foamed glass was found to inherit most of the flora of the AF-ADS. On the other hand, foamed glass bacterial flora was different from that of AF-ADS and AF-ADS absorbed foamed glass. It was clarified that the bacteria contained in the AF-ADS were adsorbed to the carrier material almost as they were.

3.11 Antifungal activity of ADS adsorbed materials against seven kinds of plant pathogen, indicated as relative mycelial

The two ADS absorbed materials significantly suppressed mycelial growth when compared with the control (**Table 8**). Both absorbed ADSs produced a clear inhibition zone against all tested plant pathogens. However, there was the physical condition difference. Wood ash could not keep the foam and be muddy after mixed with ADS. On the other hand, foamed glasses could keep the structure and easy to handle.

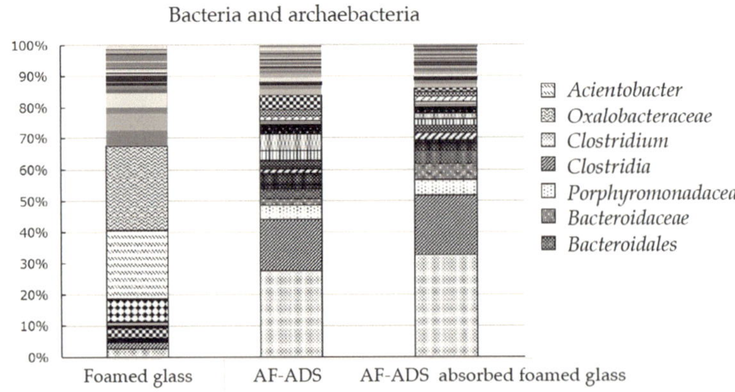

Figure 7.
Bacteria and archaebacteria flora analysis for succession confirmation of AF-ADS absorbed foamed glass.

	Treatment						IE[2]
	F		F + AD-3		Untreated		
Exp 1[1]	77.8	a	27.8	b	0	c	64.3
Exp 2	95.8	a	53.6	b	0	c	44.1

[1]The pot experiment was repeated twice (Exp 1 and Exp 2).[2]Values followed by the same letter with in a row are not significantly different according to Tukey's HSD test P < 0.05; n = 9 (Exp 1) and n = 7 (Exp 2).

Table 8.
Disease severity (DS) of Fusarium wilt of spinach and inhibitory effect (IE) 28 d after sowling.

Plant pathogen name	Wood ash		Foamed glasses	
P. lycopersici	**41**	**b**	**49**	**b**
S.terrestris	56	b	65	b
P. gregata f. sp. adzukicola	18	a	11	a
H. mompa Nobuj. Tanaka	59	b	74	c
P. oryzae	23	a	16	a
F. graminearum	82	c	84	c
F. oxysporum Schlechtendahl f. sp. fragariae Winks et Y.N. Williams	57	b	76	c

Table 9.
Antifungal activity of anaerobically digested slurries (ADSs) against plant pathogens, indicated as relative mycelial growth to pathogen growth only by co-culture test.

4. Discussion

The effect of the strain AD-3, isolated from AD, on Fusarium wilt disease incidence was assessed. In this study, approximately 1.0x 10^4 CFU g^{-1} dry soil was made and inoculated with AD-3 to achieve a concentration of 1.0x 10^6 CFUg^{-1} dry soil. The pathogen density of the infected soil was mostly consistent with previous studies examining the effect of antagonistic bacteria against Fusarium diseases [66, 67]. AD-3 inoculation into Fos-infected soil effectively reduced disease severity (by 64.3 and 44.1% in the two experiments). Thus, strain AD-3 effectively suppressed Fusarium wilt of spinach. Strain AD-3 belongs to the *B. velezensis* group. Recently, *B. velezensis* was reclassified as a synonym of several species including *Bacillus amyloliquefaciens subsp. plantarum, Bacillus methylotrophicus,* and *Bacillus oryzicola* [68, 69]. *B. velezensis* has been frequently isolated from soil [70, 71], rivers [72], and fermented food [73], and is accepted as a safe biological resource [73]. In this study, AD-3 was isolated from ADS sourced from dairy manure. Many strains of *B. velezensis* showed biocontrol effects against plant pathogens and have been applied for controlling common diseases of tomato, cucumber, lettuce, and wheat [74–77]. Our results revealed that AD-3 had the ability to control spinach Fusarium wilt. *B. velezensis* FZB42, which is close to the strain AD-3, which has gene clusters associated with the synthesis of secondary metabolites according to antimicrobial activity [69, 78, 79]. The strain AD-3 had a gene cluster related to surfactin synthesis. This result corresponds with those of Palazzini et al. [80]. Surfactin is an important lipopeptide in the suppression of plant disease [81]. Yokota et al. reported that although lipopeptides produced from *B. subtilis* suppressed Fusarium yellows, slight pathogen density reduction was shown [82]. We obtained similar results; inoculation with AD-3 significantly suppressed Fusarium wilt of spinach though a reduction in pathogen density was not observed. Five different source ADSs suppressed the growth of Fos *in vitro*. There are several reports about suppressive factors that affect plant pathogens. For example, Amari et al. reported that confrontation assay *in vitro* showed that ammonia and acetic acid in the slurry were the main factors influencing disease suppression [19]. Tao et al. reported that the supernatant of centrifuged ADS had a lower suppressive effect than raw ADS [18] and our results support this finding: Filtrate ADSs did not suppress Fos growth *in vitro* (**Tables 2** and **9**). Several antagonistic bacteria were isolated from ADSs (AD, AF, APF, and ASNF), except AS. Based on the results, the bacteria presence is important for the inhibitory effect of ADSs. The application of antagonistic microbes with organic amendment could provide more effective plant disease control than the use of a single strain alone [64, 65]. ADS showed its applicability of antagonism effects not only Fos but also the other tested seven plant pathogens and showed positive effect on plant growth. To utilize ADSs in crop production, further study is needed to assess the effects and dynamics of AD-3 when AD is applied to infected soil. To improve the ADS handling, ADS absorbed to foamed glass as porous material was the most effective among tested support materials: wood ash, steel-making slag, activated carbon, and foamed glass. ADS absorbed foamed glass was found to inherit most of the flora of the ADS. After foamed glass soaked into ADS for 1 h, and dried at 60°C for 1 h, the material maintained its antifungal activity. To apply ADS to farmland for crop production, these findings are promising for sustainable agriculture.

5. Conclusions

The use of ADS has mainly focused on a function as nutrient availability, crop productivity, and reusing organic waste. Thus, far reports focused on physicochemical properties as factors for plant disease control. We investigated that microorganisms in ADS also have a suppressive effect. The use of ADS for plant disease control has not been well studied compared with other types of organic amendments such as compost. Filtrate ADSs did not suppress Fos growth *in vitro*. Several antagonistic bacteria were isolated from ADSs (AD, AF, APF, and ASNF), except AS. Based on the results, the presence of bacteria is important for the inhibitory effect of ADSs. ADSs can supply not only plant nutrients but also antagonistic microbes. For crop production, ADSs application would be effective to infected soil. It was effective for improving ADS handling that ADS was absorbed to foamed glass and dried at 60°C. To apply ADS to farmland for crop production, these findings are promising for sustainable agriculture.

Acknowledgements

I am grateful to Tomomi Sugiyama for collaboration of this work. **Funding:** This research was supported by grants from Japan Association for Livestock New Technology.

The authorship criteria are listed in our Authorship Policy: https://www.intechopen.com/page/authorship-policy.

Conflict of interest

The authors declare no conflict of interest. The funders had no role in the design of the study; in the collection, analyses, or interpretation of data; in the writing of the manuscript, or in the decision to publish the results.

Author details

Mami Irie[1]* and Tomomi Sugiyama[2]

1 Tokyo University of Agriculture, Tokyo, Setagaya, Japan

2 NARO Institute of Vegetable and Floriculture Science, Tsukuba, Japan

*Address all correspondence to: mami-o@nodai.ac.jp

References

[1] Fravel D, Olivain C, Alabouvette C. *Fusarium oxysporum* and its biocontrol. New Phytologist. 2003;**157**:493-502

[2] Vida C, de Vicente A, Cazorla FM. The role of organic amendments to soil for crop protection: Induction of suppression of soilborne pathogens. The Annals of Applied Biology. 2020;**176**:1-15

[3] Banegas V, Moreno JL, Moreno JI, García C, León G, Hernández T. Composting anaerobic and aerobic sewage sludges using two proportions of sawdust. Waste Management. 2007;**27**: 1317-1327

[4] Bonanomi G, Antignani V, Pane C, Scala F. Suppression of soil-borne fungal diseases with organic amendments. Journal of Plant Pathology. 2007;**89**: 311-324

[5] Bonanomi G, Ippolito F, Scala F. A "black" future for plant pathology? Biochar as a new soil amendment for control-ling plant diseases. Journal of Plant Pathology. 2015;**97**:223-234

[6] Neshat SA, Mohammadi M, Najafpour GD, Lahijani P. Anaerobic co-digestion of animal manures and lignocellulosic residues as a potent approach for sustainable biogas production. Renewable and Sustainable Energy Reviews. 2017;**79**:308-322

[7] Mao C, Feng Y, Wang X, Ren G. Review on research achievements of biogas from anaerobic digestion. Renewable and Sustainable Energy Reviews. 2015;**45**:540-555

[8] Baştabak B, Koçar G. A review of the biogas digestate in agricultural framework. Journal of Material Cycles and Waste Management. 2020;**22**: 1318-1327

[9] Alburquerque JA, de la Fuente C, FerreCosta A, Carrasco L, Cegarra J, Abad M, et al. Assessment of the fertiliser potential of digestates from farm and agro-industrial residues. Biomass and Bioenergy. 2012;**40**: 181-189

[10] Provenzano MR, Iannuzi G, Fabbri C, Senesi N. Qualitative characterization and differentiation of digestates from different biowastes using FTIR and fluorescence spectroscopies. Journal of Environmental Protection. 2011;**2**:83-89

[11] Monlau F, Sambusiti C, Ficara E, Aboulkas A, Barakat A, Carrere H. New opportunities for agricultural digestate valorization, current situation and perspectives. Energy & Environmental Science. 2015;**8**:2600-2621

[12] Tambone F, Scaglia B, D'Imporzano G, Schievano A, Salati V, Adani F. Assessing amendment and fertilizing properties of digestates from anaerobic digestion through a comparative study with digested sludge and compost. Chemosphere. 2010;**81**:577-583

[13] Weiland P. Biogas production: Current state and perspectives. Applied Microbiology and Biotechnology. 2010; **85**:849-860

[14] Xu M, Xian Y, Wu J, Gu Y, Yang G, Zhang X, et al. Effect of biogas slurry addition on soil properties, yields, and bacterial composition in the rice-rape rotation ecosystem over 3 years. Journal of Soils and Sediments. 2019;**19**: 2534-2542

[15] Yu FB, Luo XP, Song CF, Shan SD. Concentrated biogas slurry enhanced soil fertility and tomato quality. Acta Agriculturae Scandinavica Section B Soil and Plant Science. 2010;**60**:262-268

[16] Abubaker J, Risberg K, Pell M. Biogas residues as fertilisers-effects on wheat growth and soil microbial activities. Applied Energy. 2012;**99**: 126-134

[17] Theuerl S, Herrmann C, Heiermann M, Grundmann P, Landwehr N, Kreidenweis U, et al. The future agricultural biogas Plant in Germany: A vision. Energies. 2019;**12**:396

[18] Tao X, Shang B, Dong H, Chen Y, Huang H. Effects of anaerobically digested pig slurry on in vitro growth crop pathogenic fungi. In: Animal Production Technology. International Conference of Agricultural Engineering CIGR-AgEng, The International Commission of Agricultural Engineering (CIGR), 2012: Agriculture and 18 Engineering for a Healthier Life. Valencia, Spain; July, 2012. p. C-0927

[19] Amari M, Toyota K, Islam MS, Masuda K, Kuroda T, Watanabe A. Effect of the addition of anaerobically digested slurry to soil and hydroponics on soil-borne plant disease (in Japanese). Soil Microorganisms. 2008;**62**:106-113

[20] Cao Y, Chang Z, Wang J, Ma Y, Yang H, Fu G. Potential use of anaerobically digested manure slurry to suppress Phytophthora root rot of chilli pepper. Scientia Horticulturae. 2014;**168**:124-131

[21] Japan scientific feed association. Methods of analysis in feeds and feed additives. Food and Agricultural Materials Inspection Center. 2009;**2010**: 745. Available from: http://www.famic.go.jp/ffis/feed/bunseki/bunsekikijun.html Accessed 2023-06-12

[22] Japan industrial standard committee 2013. Available from: https://www.standardsuniversity.org/wp-content/uploads/Japans_Standardization_Policy_2013.pdf (Accessed 2023-06-12)

[23] Food and Agricultural Materials Inspection Center Available from: http://www.famic.go.jp/ffis/fert/bunseki/sub9_shiken2021.html (Accessed 2023-06-12)

[24] Katsube K. Studies on Fusarium Wilt of Spinach [Spinacia oleracea], caused by Fusarium oxysporum f. sp. spinaciae. Bulletin of the Iwate Agricultural Research Center. 2001. pp. 1-61, Available online: https://agriknowledge.affrc.go.jp/RN/2030651657.pdf. (accessed on June 2023)

[25] Komada H. Development of a selective medium for quantitative isolation of Fusarium oxysporum from natural soil. Review of Plant Protection Research. 1975;**8**:114-124

[26] Wick RR, Louise JM, Gorrie CL, Holt KE. Unicycler: Resolving bacterial genome assemblies from short and long sequencing reads. PLoS Comput. Biol. 2017;**13**:e1005595

[27] Tanizawa Y, Fujisawa T, Nakamura Y. DFAST: a flexible prokaryotic genome annotation pipeline for faster genome publication. Bioinformatics. 2018;**34**(6): 1037-1039. DOI: 10.1093/bioinformatics/btx713

[28] Medema MH, Blin K, Cimermancic P, de Jager V, Zakrzewski P, Fischbach MA, et al. AntiSMASH: Rapid identification, annotation and analysis of secondary metabolite biosynthesis gene clusters in bacterial and fungal genome sequences. Nucleic Acids Research. 2011; **39**:W339-W346

[29] Yoon SH, Ha SM, Kwon S, Lim J, Kim Y, Seo H, et al. Introducing EzBioCloud: A taxonomically united data-base of 16S rRNA and whole genome assemblies. International Journal of Systematic and Evolutionary Microbiology. 2017;**67**:1613-1617

[30] Kumar S, Stecher G, Li M, Knyaz C, Tamura K. MEGA X: Molecular evolutionary genetics analysis across computing platforms. Molecular Biology and Evolution. 2018;**35**:1547-1549

[31] Svoboda N, Taube F, Wienforth B, Klub C, Kage H, Herrmann A. Nitrogen leaching losses after biogas residue application to maize. Soil and Tillage Research. 2013a;**130**:69-80

[32] Svoboda N, Taube F, Wienforth B, Klub C, Wienforth B, Kage H, et al. Crop production for biomass and water protection. A trade-off? Agriculture, Ecosystems and Environment. 2013b; **177**:36-47

[33] Teglia C, Tremier A, Martel JL. Characterization of solid digestates: Part 1, review of existing indicators to assess solid digestates agricultural use. Waste and Biomass Valorization. 2011a;**2**:43-58

[34] Teglia C, Tremier A, Martel JL. Characterization of solid digestates: Part 2, assessment of the quality and suitability for composting of six digested products. Waste and Biomass Valorization. 2011b;**2**:113-126

[35] Gutser R, Ebertseder T, Weber A, Schraml M, Schmidhalter U. Short-term and residual availability of nitrogen after long-term application of organic fertilizers on arable land. Journal of Plant Nutrition and Soil Science. 2005;**168**: 439-446. DOI: 10.1002/jpln.200520510

[36] Möller K, Stinner W, Deuker A, Leithold G. Effects of different manuring systems with and without biogas digestion on nitrogen cycle and crop yield in mixed organic farming systems. Nutrient Cycling in Agroecosystems. 2008;**82**:209-232

[37] Voća N, Kricka T, Cosic T, Rupic V, Jukic Z, Kalambura S. Digested residues as a fertilizer after the mesophilic process of anaerobic digestion. Plant, Soil and Environment. 2005;**51**:262-266

[38] Fouda S. Nitrogen availability of biogas residues. [Ph.D. thesis] Technische Universitat Munchen. Available from: https://mediatum.ub.tum.de/doc/1078371/document.pdf [accessed on 08.04.2023]

[39] Poetsch EM, Pfundtner E, Much P. Nutrient content and hygienic properties of fermentation residues from agricultural biogas plants. In: Lúscher A, Kessler W, Huguenin O, Lobsiger M, Millar N, Suter D, editors. Land Use Systems in Grassland Dominated Regions. Proceedings of the 20th General Meeting of the European Grassland Federation. Luzern, Switzerland: Organizing Committee of the 20th General Meeting of the European Grassland Federation Swiss Grassland Society; 2004;**68**:1055-1075. Available online https://www.europeangrassland.org/fileadmin/documents/Infos/Printed_Matter/Proceedings/EGF2004_GSE_vol9.pdf

[40] Bernal MP, Paredes C, Sanchez-Monedero MA, Cegarra J. Maturity and stability parameters of composts prepared with a wide range of organic wastes. Bioresource Technology. 1998; **63**:91-99

[41] Möller K, Schulz R, Müller T. Substrate inputs, nutrient flows and nitrogen loss of two centralized biogas plants in southern Germany. Nutrient Cycling in Agroecosystems. 2010;**87**: 307-325

[42] Szendrey LM. Start-up and operation of the Bacardi cooperation anaerobic filter. In: Wentworth RL, Stafford DA, Wheatley BI, Edelmann WE, Lettinga G, Minoda Y, et al., editors. Proceedings of the 3rd

International Symposium on Anaerobic Digestion. Watertown, Mass: Evans and Faulkner, Inc.; 1983. pp. 365-377

[43] Khan AW, Trottier TM. Effect of sulfur-containing compounds on anaerobic degradation of cellulose to methane by mixed cultures obtained from sewage sludge. Applied and Environmental Microbiology. 1978;**35**: 1027-1034

[44] Sahm H. Biologie der Methan-Bildung (biology of methane formation). Chemie Ingenieur Technik. 1981;**53**: 854-863

[45] Environmental Quality Standards for Soil Pollution. Available from: https://www.env.go.jp/en/water/soil/sp.html [accessed on 08.04.2023]

[46] Kumazawa K. Sewage sludge and wastewater for use in agriculture. In: Proceedings of Consultants Meetings. Vienna, Austria, 1997. Available from: https://inis.iaea.org/collection/NCLCollectionStore/Public/29/009/29009781.pdf?r=1. [accessed on 08.04.2023]

[47] Tambone F, Genevini P, D'Imporzano G, Adani F. Assessing amendment properties of digestate by studying the organic matter composition and the degree of biological stability during the anaerobic digestion of the organic fraction of MSW. Bioresour Technol. 2009;**100**:3140-3142. DOI: 10.1016/j.biortech.2009.02.012

[48] Chantigny MH, Angers DA, Bélanger G, Rochette P, Eriksen-Hamel N, Bittman S, et al. Yield and Nutrient Export of Grain Corn Fertilized with Raw and Treated Liquid Swine Manure. 2008; DOI: 10.2134/agronj2007.0361

[49] Jin Y, Goto M, Irie M, Yamaguchi T, Ushikubo A. Evaluation of oil and salt effects on food waste composting and plant growth (in Japanese). Journal of the Japanese Agricultural Systems Society. 2008;**24**:175-182

[50] Boydston RA, Collins HP, Vaughn SF. Response of weeds and ornamental plants to potting soil amended with dried distillers grains. Hortscience. 2008;**43**: 191-195

[51] Michalzik B, Kalbitz K, Park JH, Solinger S, Matzner E. Fluxes and concentrations of dissolved organic carbon and nitrogen – A synthesis for temperate forests. Biogeochemistry. 2001;**52**:173-205

[52] Jenkinson DS, Andrew SPS, Lynch JM, Goss MJ, Tinker P. The turnover of organic carbon and nitrogen in soil. Philosophical Transactions of the Royal Society of London. Series B. 1990;**329**: 361-368

[53] Havlin JL, Kissel DE, Maddux LD, Claassen MM, Long JH. Crop rotation and tillage effects on soil organic carbon and nitrogen. Soil Science Society of America Journal. 1990;**54**:448-452

[54] Paavola T, Rintala J. Effects of storage on characteristics and hygienic quality of digestates from four co-digestion concepts of manure and biowaste. Bioresource Technology. 2008; **99**:7041-7050

[55] Nkoa R. Agricultural benefits and environmental risks of soil fertilization with anaerobic digestates: A review. Agronomy for Sustainable Development. 2014;**34**:473-492

[56] Komlis DP, Ham RK. The effects of lignin and sugars to the aerobic decomposition of solid wastes. Waste Management. 2003;**23**:419-423

[57] Hargreaves JC, Adl MS, Warman PR. A review of the use of composted

municipal solid waste in agriculture. Agriculture, Ecosystems and Environment. 2008;**123**:1-14

[58] Gough RE, Carlstrom R. Wheat gluten meal inhibits germination and growth of broadleaf and grassy weeds. Hortscience. 1999;**34**:269-270

[59] Wong JWC, Li GX, Wong MH. The growth of Brassica chinensis in heavy metal-contaminated sewage sludge compost from Hong Kong. Bioresource Technology. 1996;**58**:309-313

[60] Kapanen A, Itavaara M. Ecotoxicity tests for compost applications. Ecotoxicology and Environmental Safety. 2001;**49**:1-16

[61] Clark RB. Plant response to mineral element toxicity and deficiency. In: Christiansen MN, Lewis CF, editors. Breeding Plants for less Favorable Environments. New York: Wiley; 1982. pp. 71-142

[62] Nicholson FA, Chambers BJ, Williams JR, Unwin RJ. Heavy metal contents of livestock feeds and animal manures in England and Wales. Bioresource Technology. 1999;**70**:23-31

[63] Bonanomi G, Antignani V, Capodilupo M, Scala F. Identifying the characteristics of organic soil amendments that suppress soilborne plant diseases. Soil Biol. Biochem. 2010; **42**:136-144. DOI: 10.1016/j.soilbio.2009.10.012

[64] Ling N, Xue C, Huang Q, Yang X, Xu Y, Shen Q. Development of a mode of application of bioorganic fertilizer for improving the biocontrol efficacy to fusarium wilt. BioControl. 2010;**55**: 673-683

[65] El-Hassan S, Gowen S. Formulation and delivery of the bacterial antagonist *Bacillus subtilis* for management of lentil vascular wilt caused by *fusarium oxysporum* f.sp. lentis. Journal of Phytopathology. 2006;**154**:148-155

[66] Cao Y, Wang J, Wu H, Yan S, Guo D, Wang G, et al. Soil chemical and microbial responses to biogas slurry amendment and its effect on fusarium wilt suppression. Applied Soil Ecology. 2016;**107**:116-123

[67] Moreno-Velandia CA, Izquierdo-García LF, Ongena M, Kloepper JW, Cotes AM. Soil sterilization, pathogen and antagonist concentration affect biological control of fusarium wilt of cape gooseberry by *Bacillus velezensis* Bs006. Plant and Soil. 2019;**435**:39-55

[68] Dunlap CA, Kim SJ, Kwon SW, Rooney AP. *Bacillus velezensis* is not a later heterotypic synonym of *Bacillus amyloliquefaciens*; *Bacillus methylotrophicus*, *Bacillus amyloliquefaciens* subsp. *plantarum* and *Bacillus oryzicola* are later heterotypic synonyms of *bacillus velezensis* based on phylogenomics. International Journal of Systematic and Evolutionary Microbiology. 2016;**66**:1212-1217

[69] Fan B, Wang C, Song X, Ding X, Wu L, Wu H, et al. *Bacillus velezensis* FZB42 in 2018: The gram-positive model strain for plant growth promotion and biocontrol. Frontiers in Microbiology. 2018;**9**:2491

[70] Jin Y, Zhu H, Luo S, Yang W, Zhang L, Li S, et al. Role of maize root exudates in promotion of colonization of *bacillus velezensis* strain S3-1 in rhizosphere soil and root tissue. Current Microbiology. 2019;**76**:855-862

[71] Wu X, Shan Y, Li Y, Li Q, Wu C. The soil nutrient environment determines the strategy by which *bacillus velezensis* HN03 suppresses fusarium wilt in

banana plants. Frontiers in Plant Science. 1801;**2020**:11

[72] Ruiz-García C, Béjar V, Martínez Checa F, Llamas I, Quesada E. *Bacillus velezensis* sp. nov., a surfactantproducing bacterium isolated from the river Vélez in Málaga, southern Spain. International Journal of Systematic and Evolutionary Microbiology. 2005;**55**:191-195

[73] Min SC, Yong JJ, Bo KK, Yu KP, Kim CK, Dong SP. Understanding the ontogeny and succession of *bacillus velezensis* and *B. subtilis* subsp. subtilis by focusing on kimchi fermentation. Scientific Reports. 2018;**8**:7045

[74] Cao Y, Pi H, Chandrangsu P, Li Y, Wang Y, Zhou H, et al. Antagonism of two plant-growth promoting *Bacillus velezensis* isolates against *Ralstonia solanacearum* and *fusarium oxysporum*. Scientific Reports. 2018;**8**:1-14

[75] Xu Z, Shao J, Li B, Yan X, Shen Q, Zhang R. Contribution of bacillomycin D in *Bacillus amyloliquefaciens* SQR9 to antifungal activity and biofilm formation. Applied and Environmental Microbiology. 2013;**79**:808-815

[76] Chowdhury SP, Dietel K, Rändler M, Schmid M, Junge H, Borriss R, et al. Effects of *Bacillus amyloliquefaciens* FZB42 on lettuce growth and health under pathogen pressure and its impact on the rhizosphere bacterial community. PLoS One. 2013;**8**:e68818

[77] Cai XC, Liu CH, Wang BT, Xue YR. Genomic and metabolic traits endow *bacillus velezensis* CC09 with a potential biocontrol agent in control of wheat powdery mildew disease. Microbiological Research. 2017;**196**:89-94

[78] Koumoutsi A, Chen XH, Henne A, Liesegang H, Hitzeroth G, Franke P, et al. Structural and functional characterization of gene clusters directing nonribosomal synthesis of bioactive cyclic lipopeptides in *Bacillus amyloliquefaciens* strain FZB42. Journal of Bacteriology. 2004;**186**:1084-1096

[79] Chen XH, Koumoutsi A, Scholz R, Schneider K, Vater J, Süssmuth R, et al. Genome analysis of *Bacillus amyloliquefaciens* FZB42 reveals its potential for biocontrol of plant pathogens. Journal of Biotechnology. 2009;**140**:27-37

[80] Palazzini JM, Dunlap CA, Bowman MJ, Chulze SN. *Bacillus velezensis* RC 218 as a biocontrol agent to reduce Fusarium head blight and deoxynivalenol accumulation: Genome sequencing and secondary metabolite cluster profiles. Microbiological Research. 2016;**192**:30-36. DOI: 10.1016/j.micres.2016.06.002

[81] Ongena M, Jacques P, *Bacillus lipopeptides*: Versatile weapons for plant disease biocontrol. Trends in Microbiology. 2008;**16**:115-125. DOI: 10.1016/j.tim.2007.12.009

[82] Yokota K, Hayakawa H. Impact of antimicrobial lipopeptides from bacillus sp. on suppression of fusarium yellows of tatsoi. Microbes and Environments. 2015;**30**(3):281-283. Available online DOI: 10.1264/jsme2.ME15062

Chapter 17

Composting in Our Primary School

Ornela Maria Munoz Millet

Abstract

One of the most innovative features that we have put into practice in our school is the creation and use of composting with organic waste from the school canteen. Thanks to this school composting program, we have been able to implement this practice in our school. First of all, we talk about the program: the objectives, the contents, the methodology, and the evaluation of it. Secondly, we talk about the implementation of the program at school, the composting process, and the tasks and organizational aspects of the program. Finally, we talk about the conclusions and contributions to the teaching and learning process of the students. It is a very didactic practice for the construction of environmental attitudes and capacities in our students and for the sensitization with the environment and the acquisition of sustainable habits.

Keywords: primary school, composting, organic waste, circular economy, implementation

1. Introduction

Today, we perceive an increase in the production of urban solid waste that tends to exceed the capacity of urbanization and will constitute a more challenging financial and environmental conflict for humanity [1]. The production of waste and the lack of regulation in its final degradation have a negative effect for several reasons: (a) the environment and the resources that are produced from it; (b) public health, as the increasing possibility of diseases in individuals who are regularly exposed to pollution; (c) finances, households, and the country [1].

Keeping future generations in mind, people explore different means to harmonize progress and environmental safeguard. The school encourages the introduction of sustainable environmental practices, especially when students are exposed to sensory experiences that reinforce education, being more susceptible to environmental influences [2]. In addition, the creation and use of school gardens as a pedagogical resource, in all courses, would be useful from the point of view of education for sustainability [3]. UNESCO [4] has already advanced us some of the most interesting contributions that are obtained in the use of this educational resource such as: (1) addressing three of the four learning contents considered most urgent: climate change, biodiversity, production, and consumption sustainable [4]; (2) put into practice the three types of learning considered most appropriate: participatory and collaborative learning, problem-based learning, and learning that adopts a critical approach [5]; (3) develop general skills in SE: critical analysis, systemic reflection, collaborative decision-making, and sense of responsibility toward present and future generations [4].

Proper waste management is considered essential [6]. A fundamental strategy is its use through biological treatment options such as composting, which allows closing the material cycle, reducing the impact caused by its management, and contributing to the sustainability of agroecosystem production [7]. It is important to know how to propose a distribution in the accumulated composting, which consists of arranging areas in which an air-free zone or free air space can be created so that there is good ventilation irrigation during the duration of the waste later that guarantees an efficient movement for the development of aerobic bacteria for a relevant and effective evolution.

Gardens in schools are educational meeting places that can help to complete training, as well as increase academic results and nutrition of students [8, 9]. In addition, they learn to conserve the environment, enjoy better nutrition, and optimize relationships between people [10]. Other studies talk about the potential that these have in academic results, mainly in science subjects [11]. A more modern investigation shows how school gardens can help group unification and collaborative work [12].

2. Contextualization of the program

This program is being carried out at an early childhood and primary education school, which is located in a northern suburban area of the town of Águilas, in Murcia, Spain. It is a preferential educational care center with 13 classes: 3 for early childhood education and 10 for primary education. The school is equipped with a canteen and school transport and has 182 students, most of whom are of Moroccan origin and of gypsy ethnicity. Also counting on some immigrant students from South America and Romania, mostly from the neighborhood called "El Labradorcico."

2.1 Family situation at school

In general, families have a high degree of marginalization and have difficulties at an economic, social, health, and cultural level.

Currently, 77 students use the school canteen, all of them on scholarships. This generates a large amount of organic waste, which is collected daily by the students and used to make compost for our garden.

In general, families present a high degree of marginalization and have difficulties such as:

- Economic: Most of the families do not have a permanent job, are unemployed, or eventually work in the fields.

- Social: Families tend to be in some cases very large, are unstructured, or are single parent.

- Sanitary: They present a lack of hygiene due mainly to the lack of infrastructure in the house.

- Cultural-emotional: Families have a very low cultural level, so they cannot help their children with their school tasks.

This primary school remains faithful to its ideals of social assistance and integration of the families that take their children to study at this center and transcending the educational aspect, to go to the root of the great economic, social, and health problems that affect a large part of the families of our students. We are aware that our students live a harsh reality that implies students leave their academic formation very soon, so, at the school, we try to work with different projects and measures to give emotional stability as something essential. With all this, we seek their integration into society and increase the number of possibilities of success for future teachings and professions.

We find irregular situations of absenteeism. On many occasions, the parents work and do not bring the students to the school, other times, it is due to the great insouciance of the parents. In certain periods, immigrant families return for a time to their country of origin and do not report their absence.

2.2 Programs at school

There are many different programs, which are being carried out at school. We have a library and encouragement of reading program, an information and technology program, a maths program, a healthy life program, and the one we are describing in this chapter: "Composting in our Primary School."

3. Description of the program

Food waste basically consists of organic components such as lipids, carbohydrates, and proteins, which can be digested into different carbonic forms. This is very important taking into consideration that nearly 1.3 billion tons of food wastes are generated every year all around the world [13].

3.1 Introduction

Ecological practices in society are increasingly on the rise, both at the level of consumption and of habits and care for the environment. And it is that without these practices our planet is adrift, punished by the millions of tons of waste that are generated each year. These residues, together with chemicals and polluting gases, endanger our health and are directly responsible for an increase in illnesses. It is enough to look at the situation of many natural sites around the world, which are increasingly affected by pollution, both fauna and flora. All these problems are interconnected with each other, and their study is perfectly integrable in the work with our ecological garden and with all areas of Early Childhood and Primary Education.

Therefore, it is essential to educate in the ecological, in the use of natural and homemade fertilizers, in the creation of compost, in the prohibition of chemical agents, in the recycling, reduction, and reuse of containers, in the respect of biodiversity, in the practice of a circular economy that makes both students and the entire educational community aware of the use of resources. All this can be worked through the school garden, a space for entertainment and enjoyment, relaxation, and at the same time of responsibility and effort for our students.

3.2 Objectives of the program

When we considered starting an organic garden in our school, we must have been very clear about what objectives we want to achieve based on content that can be worked on globally through the different areas of the curriculum. These objectives are: (a) To know the techniques of organic farming, knowing how to choose the crops and the rotations that best adapt to the type of soil and the climate. (b) To appreciate the quality and nutritional properties of the products from the organic garden. (c) To learn to value biodiversity. (d) To raise awareness of the importance of putting the "3 Rs" into practice: reduce, reuse, and recycle; highlighting above all the 1st (reduce), as it is the one that has the greatest positive effect on our garden and planet. (e) To learn the operation and implementation of the composting process. (f) To acquire the theoretical and practical knowledge of how to plant in seedbeds. (g) To acquire basic notions about the operation of a greenhouse. (h) To build and decorate elements and spaces of a childish nature that brings a touch of color and joy to our garden. (i) To make use of new technologies to disseminate the work carried out. (j) To work on values and attitudes such as coexistence, autonomy, solidarity, and cooperative work. (k) To promote initiative, responsibility, and critical thinking. (l) To conduct a study and follow-up, where students collect information on different plants and insects. (m) To promote a healthy diet. (n) To learn about the world of work through the different jobs and tasks that can be carried out in the school garden.

3.3 Program's contents

To achieve these objectives, it is necessary to plan some content, which has been timed in an appropriate and progressive way, and based on the needs of the educational center. The contents to work on are: (a) Organic farming: basic notions and importance. (b) The mediterranean diet: characteristics and properties of food from the garden. (c) The biodiversity of a school garden. (d) The "3 Rs" (reduce, reuse, and recycle). (e) Composting: fundamental principles. (f) Difference between planting in the ground and in seedbeds. (g) Operation of a greenhouse. (h) Construction and organization of plots of different shapes, depending on the type of crop we want to plant. (i) Manufacture of decorative elements for our orchard. (j) Use of new technologies for the search for information and the dissemination of the work carried out. (k) Respect for the environment: space, animals, plants, and people. (l) Realization of a weekly orchard diary, where we can keep track of: animals, plant development, and rainfall. (m) Agriculture: importance and types of tasks.

3.4 Methodology

On the other hand, at a methodological level, we have based our work on carrying out experiences that allow students to connect with reality and with their previous knowledge.

It is very interesting to turn the students into researchers, who are in charge of searching for information, exploring, and finding solutions to the problems that arise. This type of work makes children ask questions, take risks, contribute ideas, and make decisions, which will make their learning much more reflective and motivating.

As for the organization of the students, the most advisable thing is to make small groups, each one in charge of carrying out a task, so that from time to time they change roles and at the end of the session they have carried out all the proposed tasks equally. Working in small groups allows us to: safely use the material resources available to us, make the most of time that students do not get overly tired of doing tasks, develop collaborative learning when carrying out research or data collection activities, and so on.

In conclusion, the work in a space, such as the school garden, must be based on the meaningful learning, close to the student, researcher, and collaborative, and at the same time autonomous, where students learn through manipulative, experiential, and fieldwork activities. All the activities that are carried out are included in the classroom schedules, and therefore are evaluable.

3.5 Evaluation of the program

Regarding aspects related to evaluation, we will divide it into three types: initial, continuous, and final. Regarding the initial evaluation, we must bear in mind that to start our work we must always start from the previous knowledge of the students. To get to know them, we can prepare questionnaires, brainstorm ideas, or assemblies, among others. However, it is interesting to use the "rotating folio." For its implementation, we can group students into groups of three or four components, favoring coeducation at all times. A member of the group begins to write her part or her contribution on a "rotating" sheet of paper, about the previous ideas that she has about the garden or any type of question or topic that we are interested in addressing. Meanwhile, the others notice how the classmate does it, they can help him, correct him, encourage him, and then he passes it to the classmate next to him in a clockwise direction so that he writes his part of the task on the sheet of paper, thus one by one until all the members of the team have participated in the development of the task. Each student can write the part of it with a different color, the name at the top of the work will be written in the same color. In this way, we can easily see the contribution of each one.

Regarding the continuous evaluation, we will evaluate the different activities that are proposed: searching for information, monitoring the growth of plants, studying insects, writing recipes, etc. Most of these activities are included in the garden diary, but in the same way, we can select the best worked to publish them on our blog.

For the final evaluation, a self-assessment and a co-assessment will be carried out, and we will assess the work done by the students in their diary and their contributions to the blog. In addition, each teacher will evaluate a series of basic criteria such as those detailed below: (a) Knows the techniques of organic farming. (b) Appreciate the quality and nutritional properties of the products of an organic garden. (c) Know the essential foods of the mediterranean diet. (d) Values biodiversity as an essential aspect for life itself and as an advantage for the garden. (1) Be aware of the importance of putting the "3 Rs" into practice. (2) Know how the composting process works. (3) Acquire basic knowledge of how to plant in seedbeds. (4) Know how a greenhouse works. (5) Collaborate in the construction of elements and spaces for children. (6) It makes use of new technologies to disseminate the work carried out. (7) Acquire values and attitudes such as coexistence, autonomy, solidarity, and cooperative work. (8) Track and collect data in your garden diary. (9) Recognizes the importance of the work of farmers for our day to day.

4. Implementation of the program at school

Bearing in mind that one of our general objectives for the school garden is "to learn how the composting process works and how to put it into practice," we are going to focus on this aim and see how it is carried out in our school.

Composting in schools is economical, instructive, enjoyable, and useful. By creating compost that is nutrient-rich and helps to nourish the soil to encourage the growth of new life, you can teach people about the eco life cycle and sustainable living.

All of your food scraps can be recycled on-site as soon as they are thrown away. Fruits, vegetables, meat, eggs, fish, and dairy products can all be properly composted together with any other raw or cooked food waste that your school generates (**Figure 1**).

Figure 1.
Composting at canteen school.

Composting is a natural process, but for it to happen, certain conditions must be met. To the delight of kids and some adults, energy is released during this natural process, heating the mixture, and generating steam.

Compost is a natural fertilizer created from the action of bacteria, fungi, and worms on the organic or biological waste that we generate at home or in the school canteen (food scraps, dry plants, etc.). It has a triple function: (1) Serve as fertilizer to improve the properties of the land. (2) Serve as food for plants and, at the same time, we recycle the waste generated by using it for compost. (3) It helps preserve and improve the fertility of the soil and is a stable product with a pleasant smell (as long as the process is adequate) (**Figure 2**).

We must take into account that about 40% of the waste in our home is organic matter, most of it of plant origin, and susceptible to biodegradation. It is important to teach children to manage and take advantage of them through simple techniques. Dumping is more expensive every day, so composting on a small and large scale represents a more sustainable management of waste, and therefore savings.

Composting is a basic pillar when considering that our garden is ecological. Other types of fertilizers such as the different types of manure (horse, goat,

Figure 2.
Compost.

chicken...) are equally valid, but they do not offer us the possibility of "creating" our own compost, building the compost bin, preparing organic waste (wet and dry material), keep track of the process, detect incidents, propose solutions... In short, the creation of composting allows us endless learning possibilities that we must take advantage of and teach our students so that from school they can extrapolate it to small-scale homes.

Through composting, we are committed to the cycle of organic matter, we involve and take responsibility for children, developing and assimilating the concept of circular economy, a more sustainable model than the traditional "linear economy" (obtain, produce, consume and produce). This concept aims to increase the useful life of the products, taking advantage of the waste generated, and reintroducing it into the system in the form of new raw materials.

4.1 Composting process

In order to carry out the composting process, the first thing we must have is an adequate space to put the compost bin. This should be placed inside the school garden, in contact with the ground, so that insects, bacteria, and fungi have easy access to the waste. In addition, we must protect it from the wind, and from the sun's rays in summer, it is advisable to place it under deciduous trees that will give us the necessary sun (in winter) and shade (in summer).

Next, we need to build the compost bin. In our case, we have installed a wooden one, approximately 1 m^3. It is made with wooden slats separated by about 4 cm, to favor oxygenation. There is nothing on the ground part to facilitate the entry of insects, and on the top part, there is a wooden cover to prevent water from falling in case of rain. However, on the front side and next to the ground, the wooden strip is mobile sliding upwards when we want to remove the compost that is generated. In addition, there is an annex with an explanatory poster of the waste that can be thrown away and those that cannot (**Figure 3**).

Once the compost bin is finished, it is time to start dumping the organic waste that is generated in the school canteen. Before commenting on how we organize ourselves for this task, we must know that, in the compost bin, we are going to work with two types of waste: wet, rich in water and nitrogen (remains of fruits and vegetables...); and dry, composed mainly of carbon (dry branches and leaves, straw, cardboard...). The ratio of both should be two parts of wet material to one of dry material in order to maintain adequate humidity. To this task of adding dry and wet material, we must add the task of mixing and aerating, thus oxygenating our future compost and controlling its humidity. We will carry it out with a long stick hoe every time we make a contribution to the compost.

It is also important that in the first filling of the composter, we will prepare a bed at the bottom with thick woody material to facilitate air circulation. Whenever possible, the compost bin should be kept at least half full.

To fully understand the composting process, we must know that it is mainly divided into two phases: (a) Decomposition and degradation, where bacteria and fungi generate heat in their activity. The temperature can reach up to 70°C, which decreases as the activity of the microorganisms slows down. With this rise in temperature, we manage to kill any pathogenic organism present, and we make the seeds present in the remains sterilized. (b) Cooling and maturation: here the bacteria work at a temperature below 30°C. This favors the appearance of small animals such as

Figure 3.
Compost bin.

earthworms, mealybugs, and insects that feed on microorganisms, plant debris, and various invertebrates, thus contributing to the formation of mature compost.

4.2 Tasks and organizational aspects of composting

Composting brings a series of tasks and organizational aspects that we must consider. It is important to be clear about who is going to carry out each of them. In this sense, at our primary school, we follow a series of actions every day that will finally put an end to the dumping of waste in the compost bin: (1) Our chef pours all the vegetable waste of the day into a bucket. (2) In the same way, each class must have its "composting bin" where to throw the waste generated from their breakfasts. Each student is responsible for chopping and pouring that residue into the bucket. (3) Upon returning from recess, "the compost patrol" (the class in charge of compost that week) goes through the classes with various buckets to collect the waste from the entire school. These residues, together with those from the canteen, will be thrown in the compost bin at the end of the day. (4) The caretaker takes the waste from the canteen to the class in charge of managing it. (5) The students review the waste (on tables with tablecloths prepared for it) to check that there is none that negatively affects the compost and chop them up as much as possible. (6) Ten minutes before the end of the day, the students go to the garden and dump the wet waste generated in the dining room and in the classrooms. (7) We pour dry material if necessary. (8) With the help of a long stick hoe, we remove a little and oxygenate.

Finally, 6–8 months after the start of the process, we will use a sieve to separate the materials that have not yet been completely composted (fresh compost) from the mature compost, which we will leave to rest for a few days, with a pleasant smell of forest soil (**Figure 4**).

Figure 4.
Recycling.

5. Conclusions and contributions to the teaching and learning process of the students

The composting program has benefited the entire educational community of our school for raising awareness about the environment, resource management, and cooperative work.

It is well known that having an organic garden in a school is all beneficial for the teaching-learning process. Even more so, when within this program, we include the compost that directly favors the reduction of the amount of organic rubbish from the school canteen that goes directly to landfills, reduces the use of inorganic fertilizers and we save irrigation water; due to the water retention capacity of the compost, it provides the necessary nutrients for the development of plants in a natural way and reduces the costs of transporting waste with the consequent benefits for citizens.

All this has been transmitted to the entire educational community so that awareness of the impact that our daily acts have on the environment has been systemically created, and thus have alternatives to recycle waste and use it to generate raw material.

The process of implementing this composting program has been very pleasantly received by the entire educational community. The school canteen staff feels very satisfied to be able to reuse all the organic material in the garden on a daily basis. Students and teachers are responsible with their functions of collecting the organic elements and seeing how, little by little, the compost is formed in order to fertilize each of the plots, without the need to buy or use external fertilizers. The students, and even adults, are interested in learning about the process of making compost since even if organic remains are piled up it never smells bad.

In short, the composting program has been an ideal activity for the entire educational center, a perfect way to bring the entire educational community closer to nature, to transmit values of responsible consumption, to recycle, and to create respect for the environment. And the most important thing of all is that, with the completion of this activity, a delicious reward is obtained: growing our fruits and vegetables with that flavor of before that now only maintains quality organic products.

Author details

Ornela Maria Munoz Millet
Education Faculty, University of Murcia, Murcia, Spain

*Address all correspondence to: ornelamaria.munoz@murciaeduca.es

References

[1] Cahe EJ, Dante J. Multi-criteria analysis and selection of urban solid waste management proposals. Ibero-American Journal of Ecological Economy (REVIBEC). 2019;**2**:53-66

[2] Eloy GR, Santos ACM, Caetano GL, Perdigão M, Gontijo HM. Horta ecológica e compostagem como educação ambiental desenvolvida na Fundação Crê-Ser em João Monlevade/MG. Research, Society and Development. 2019;**8**(2). DOI: 10.33448/rsd-v8i2.763

[3] Zuazagoitia Rey-Baltar D, Aragón L, Ruiz GA, Eugenio GM. ¿Podemos cultivar este suelo? Una secuencia didáctica para futuros maestros contextualizada en el huerto. Investigación En La Escuela. 2021;**103**:32-47. DOI: 10.12795/IE.2021.i103.03

[4] United Nations Educational, Scientific and Cultural Organization (UNESCO). Road Map for Implementing the Global Action Program on Education for Sustainable Development. Paris: UNESCO; 2014

[5] United Nations Educational, Scientific and Cultural Organization (UNESCO). Shaping the Future we Want: A Decade of Education for Sustainable Development. Paris: UNESCO; 2014

[6] Williams ID. The importance of education to waste (resource) management. Waste Management. 2014;**11**(34):1909-1910

[7] Wezel A, Casagrande M, Celette F, Vian JF, Ferrer A, Peigne J. Agroecological practices for sustainable agriculture: A review. Agronomy for Sustainable Development. 2014;**34**:1-20

[8] Desmond D, Subramaniam A. Revisiting Garden-Based Learning in Basic Education. Rome, Italy: FAO; 2004

[9] Özer Ö, Wei W. Strategic commitments for an optimal capacity decision under asymmetric forecast information. Management Science. 2006;**52**(8):1238-1257. DOI: 10.1287/mnsc.1060.0521

[10] Haros B, Garcia T, Californias S. School garden: Educational strategy for life. Ra Ximhai. Journal of Society, Culture, Development. 2013;**9**:543-558

[11] Williams D, Dixon S. Impact of garden-based learning on academic outcomes in schools: Synthesis of research between 1990 and 2010. Review of Educational Research. 2013;**83**(2):211-235

[12] Santiz G. The School Garden, an Opportunity to Strengthen Collaborative Work and Integration among Primary School Students. San Cristóbal de Las Casas, Mexico: The Colegio de la Frontera Sur; 2018

[13] Awasthi SK, Sarsaiya S, Awasthi MK, Liu T, Zhao J, Kumar S, et al. Changes in global trends in food waste composting: Research challenges and opportunities. Bioresource Technology. 2020;**299**:122555